Zoophysiology   Volume 27

Zoophysiology

U.M. Norberg

Vertebrate Flight

Mechanics, Physiology, Morphology,
Ecology and Evolution

With 103 Figures

Springer-Verlag Berlin Heidelberg New York
London Paris Tokyo Hong Kong

Dr. Ulla M. Norberg
Department of Zoology
Division of Zoomorphology
University of Göteborg
Box 25059
S-400 31 Göteborg

QP
310
.F5
N67X
1990

ISBN 3-540-51370-1 Springer-Verlag Berlin Heidelberg New York
ISBN 0-387-51370-1 Springer-Verlag New York Berlin Heidelberg

Typesetting: International Typesetters Inc., Makati, Philippines
Printing: Druckhaus Beltz, Hemsbach
Binding: J. Schäffer, Grünstadt
2131/3145-543210 – Printed on acid-free paper

Dedicated to Åke, Peter, and Björn

Preface

It has been great fun to write this book, even though it has taken longer than planned, and occasionally been exasperating. The most difficult problem was deciding what to exclude among so many interesting things, because the available material usually exceeded the space. Because a book like this covers so many aspects, each component must be limited.

This book is intended for graduate and undergraduate students as well as professional scientists who want to work with animal flight or to gain some insight into flight mechanics, aerodynamics, energetics, physiology, morphology, ecology and evolution. My aim has not been to give the whole mathematical explanation of flight, but to provide an outline and summary of the main theories for the understanding of how aerofoils respond to an airflow. I also hope to give the reader some insight into how flight morphology and the various wing shapes have evolved and are adapted to different ecological niches and habitats.

I would never have written or completed this book without the encouragement and help of my family and my friends. I owe tremendous thanks to my husband Åke Norberg for endless discussions on various problems within flight mechanics, morphology, ecology and evolution, and for reading and commenting on the manuscript, and to my sons Peter and Björn (born 1974 and 1977) for patience and encouragement rather than complaints when my time with them was limited. I also thank Peter for checking the references. I am most grateful to M. B. Fenton, in particular, for reading and commenting on the whole manuscript, and to C. J. Pennycuick, J. M. V. Rayner, H. D. J. N. Aldridge, P. J. Butler, G. Goldspink, and F. R. Hainsworth for reviewing various parts of it. Special thanks go to G. R. Spedding for allowing me to use his exellent stereo-photographs of the vortex wake of a pigeon. I am indebted to the series editor, G. Neuweiler, for asking me to write the book and for his patience, and to various colleagues who have encouraged the writing by asking questions like "When will the book be ready?". I thank Academic Press, Harvard University Press, The Company of Biologists Limited, The Royal Society, The University of Chicago Press, The Zoological Society of London and Springer-Verlag for permission to reproduce figures from their journals. The Swedish Natural Science Research Council has generously supported my research, much of which has been included in various sections of the book.

It is impossible to write a book without any errors. Although the manuscript has been checked by specialists, the book may still contain several mistakes which are my responsibility.

Askim, Summer 1989   ULLA M. NORBERG

VII

Contents

Chapter 1

Introduction

Flight in animals includes *gliding*, or passive, flight without wingstrokes, which costs a minimum of energy, and *flapping*, or active, flight. Active flight is a very efficient way to transport a unit of mass over a unit of distance, even though it requires extremely high power output (work per unit time). Flying vertebrates can move more quickly than running ones. While the cheetah, which is the fastest of animals, can achieve 18 body lengths per second, a swift can achieve 67, a chaffinch 72, and a starling 80 (Alerstam 1982; Kuethe 1975). This is not far from a jet airplane, which at Mach 3 reaches about 100 lengths per second.

It is interesting that for a given body size, flying is a far cheaper way to move a unit distance than is running, but more expensive than swimming (Schmidt-Nielsen 1972). Consider a 15-g migrating passerine bird which easily can fly 1000 km non-stop in about 24 h, a feat far beyond the abilities of comparable sized animals (e.g. a mouse) either in terms of endurance or in speed. Apart from such enormous advantages, flight permits an animal to cross water, deserts and other inhospitable environments, and to do so at heights where the temperature is suitable for cooling purposes. Flight also allows an animal to reach otherwise inaccessible foraging sites. There are several bird species that can fly, swim and dive, such as auks and ducks, but there are also swimming and diving birds that have lost their flight ability, such as penguins and the flightless cormorant in the Galapagos, as have also several terrestrial island birds.

The obvious advantages of flight make it easy to understand why flying animals have undergone such dramatic adaptive radiations. Insects comprise the most diverse and numerous class of animals (about 750,000 species), while flying reptiles showed considerable diversity of forms and included the largest flying animal known. Birds include more than 8000 species, and the 850 + species of bats make up the second largest group of mammals.

The size range of flying birds spans four orders of magnitude, from about 1.5 g to 15 kg, while that of bats goes from about 1.5 g to 1.5 kg and that of insects from about 1 μg to 20 g. The largest flying birds thus weigh ten times as much as the largest bat and 750 times the largest flying insect. Why are there no larger bats or larger insects? Morphological and metabolic constraints to maximum size may be more strict in bats than in birds, but the size variation may also be linked to differences in foraging strategies. Pterosaurs may have ranged from about 4 g to as much as 75 kg.

Flying insects had existed for perhaps 200 million years before pterosaurs and birds began to fly. Insects show far greater variation than flying vertebrates and insect flight spans a wider range of *Reynolds numbers*, which indicate the ratio of inertial forces to viscous forces in a flow. The very small insects operate in the lower range of Reynolds numbers, where viscous forces are dominant.

The flight of larger insects and vertebrates is in the high range of Reynolds numbers, and their flight characteristics can be understood by aircraft flight aerodynamics.

Wings of pterosaurs, birds and bats evolved from the forelimbs of certain land animals. Scales evolved into feathers, skin flaps into membranes. The forelimb bones became thin or thin-walled with weblike inner structure for stiffness. The feathers became strong and rigid, and several mechanical arrangements contributed to the rigidity of the wing. Without a considerable size and high degree of articulation the wing was useless as a flight structure.

The first flying vertebrates probably used gliding as a cheap way of moving between foraging sites, and as escape from predators. Birds evolved from small warm-blooded coelurosaurian dinosaurs and the first known bird *Archaeopteryx lithographica* (about 180–225 million years old) has been called a good glider that could not actively fly. Today most biologists disagree with the view and cite several morphological features indicating that *Archaeopteryx* could actively fly. The features include a complex structure and asymmetry of the wing feathers indicating an aerodynamic function (Feduccia and Tordoff 1979; Norberg 1985), and a large cerebellum — the centre of neuromuscular control. *Archaeopteryx* may have made short flights among trees and used its long tail for control and stability.

The oldest known fossil bat *Icaronycteris index* lived about 50 million years ago and resembled modern bats and may be closely related to the ancestors of all bats (Jepsen 1970). There is no direct evidence that the two suborders of bats, the Megachiroptera and the Microchiroptera, have a common bat ancestor (Smith 1977). Several morphological and behavioural traits suggest that bats are diphyletic (e.g. Smith 1977; Pettigrew 1986; Scholey 1986).

The pterosaurs (order Archosauria) of the Jurassic and Cretaceous survived from about 180 until 65 million years ago. Some species were small, maybe tree-living, and some were enormous, with a largest estimated span of 11 to 12 m for *Quetzalcoatlus northropi* (Langston 1981). Pterosaurs were adapted for active flight but the largest ones were probably mostly gliders and soarers (Bramwell 1971; Padian 1983).

1.1 Gliders

Very few recent animals are pure parachuters (descendent angle $> 45°$), but many vertebrates are good gliders (descendent angle $< 45°$; Fig. 1.1) travelling from tree to tree and voluntarily steepening their descent angle. Escape from predation may have furthered the evolution of parachuting and gliding abilities. But optimal foraging over some vertical zone in trees also may have been an important factor. Even for animals with good ability of powered flight it takes less energy to climb and hop upward in a tree and fly downward to the next one than to do the reverse (R.Å. Norberg 1981a). Parachuting or gliding animals are represented in all five vertebrate classes and have thus evolved independently.

2

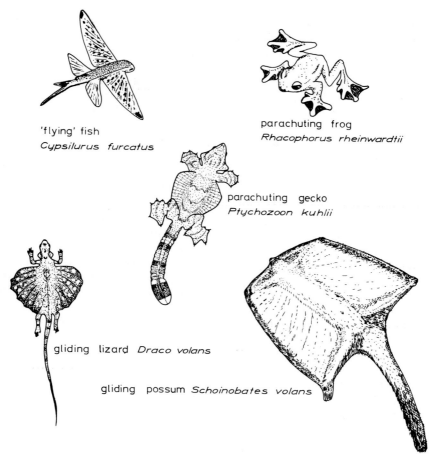

'flying' fish
Cypsilurus furcatus

parachuting frog
Rhacophorus rheinwardtii

parachuting gecko
Ptychozoon kuhlii

gliding lizard *Draco volans*

gliding possum *Schoinobates volans*

Fig. 1.1. Gliding and parachuting vertebrates

Marine *"flying fishes"* of the families Exocoetidae and Hemirhamphidae, such as the two-winged *Exocoetus* and four-winged *Cypsilurus* (Fig. 1.1), occur in the tropical oceans. Flying fish have large pectoral fins, thin bodies and enlarged lower lobes of the caudal fins and apparently "fly" to escape predators.

"Flying frogs" of South-East Asia, Australasia, South and Central America belong to the families Rhacophoridae (Fig. 1.1) and Hylidae. Many of these species can only parachute (e.g. *Phrynohyas venulosa*). Some flying frogs rely on spreading their webbed feet to achieve a flight surface, but some can flatten their bodies so that the entire animal contributes to lift generation.

Among reptiles, many of the pterosaurs were undoubtedly good gliders and some other archosaurs could glide, for example *Icarosaurus* and *Daedalosaurus*, which had flight surfaces formed from elongated ribs, as in the recent genus *Draco* (Agamidae with about 20 species; Fig. 1.1). The lizard-like Triassic reptile

3

Sharovipteryx (initially *Podopteryx*) *mirabilis* had an elongated head and hind-limbs, and a patagium that reached from the hindlimbs to the base of the tail and possibly a smaller flap between the forelimbs and body (Gans et al. 1987). *Ptychozoon*, the "flying" geckos, are parachuters with broad flaps of skin on either sides of their bodies. The flaps are spread out by the limbs during flight, along with smaller flaps bordering the neck and tail, webbed feet, and flattened body and tail. A colubrid snake (*Chrysopelea*, from Borneo) can draw in the ventral surface of its body to form a deep concavity, so the body can be used as an aerofoil.

A gliding wing formed by a fold of skin stretched between the fore and hind legs has been evolved independently in three orders of mammals: Marsupialia (three genera of flying phalangers), Dermoptera (two species of *Cynocephalus*, flying lemurs) and Rodentia with Sciuridae (with about 12 genera) and Anomaluridae (with three genera), all being arboreal. Most gliding mammals are tropical forest species, but the flying squirrels occur throughout the Holarctic region, and also in Nearctis. *Cynocephalus* is the only species with no part of the body extending beyond the membrane, for even the digits are included. Gliding rodents have a cartilaginous spine attached to the forelimbs, stretching out the membrane and somewhat increasing the span. This spine extends from the wrist in Sciuridae and from the elbow in Anomaluridae. The size range of gliding mammals (10 g to 1.5 kg) is similar to that of bats.

Several large birds are good gliders. Those with low wing loadings (large relative wing area) can glide slowly with low sinking speed. Some bats can glide, but they seldom do. They are less able than birds to flex their wings during flight while maintaining good aerodynamic performance, and have therefore not the same ability to vary the wing area to control gliding.

1.2 Active Flyers

Most recent birds and all bats fly actively and, like other animals, they occupy species-specific niches, but the time separation (day and night foraging, respectively) makes their competition for food rather low, and wing design has converged between many birds and bats. The type of habitat a flying or gliding animal chooses to live in, as well as its way of exploiting the habitat, are closely related to its body size, wing form, flight style, flight speed, and flight energetics. Many active fliers are aerial insectivores foraging in open areas, at high altitudes, closer to the ground, or near to vegetation; others hunt among vegetation, hovering, gleaning or otherwise manoeuvering. Some species commute and forage on ground, in trees, in water or other substrates and have less need of manoeuvrable and agile flight. Some species perform long or short migrations. Natural selection is likely to act towards a wing structure that minimizes the power required to fly at the speed and style optimal for the animal, and is assumed to result in some near-optimal combination among these variables. The optimal flight speed varies with the flight goal and with the type and abundance

4

of food. Some birds, such as albatrosses and vultures, can soar for hours with a minimum energy cost, while hummingbirds hover when foraging, which is the most energy-demanding type of flight. Such different flight activities require very different wing morphology (Fig. 1.2). To understand how flying animals work, their physiology, morphology, ecology and wing function must be known.

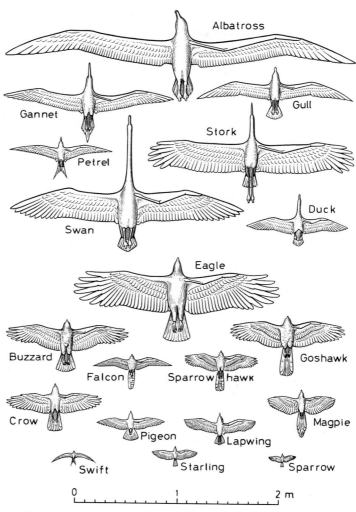

Fig. 1.2. Wing shapes of birds. (After Herzog 1968)

1.3 Outline of the Book

Vertebrate flight includes a large number of categories, all interconnected in a complicated network. In this book, my aim is to present the basic principles of flight including the main requirements on body functions and morphology and the main correlations between structure, form, and function in ecological and evolutionary perspectives. The book does not give a detailed account of every category, for each by itself could be expanded into a book.

The following chapter (2) gives the basic aerodynamics and power requirements for gliding and for flapping flight. The literature on animal flight is often difficult to understand, so here I introduce the basics of flight mechanics and give references to useful and more advanced studies. Flapping animal flight can be understood by applying quasi-stationary aerodynamic theories developed for aeroplane wings (*blade element theory*) and airscrews or propellers (*momentum theory*). But part of flapping flight is better explained by the *vortex theory*, which is based on an idealized pattern of the induced air flow behind a flapping wing and therefore is not bound by the constraints of steady-state aerodynamics. These theories are described.

The metabolic cost of flight depends on the mechanical power requirements and on the mechanical efficiency of the flight muscles in converting metabolic energy into mechanical work. Flight physiology (Chap. 3) is complex and includes several functions, such as respiratory mechanics and gas exchange, circulation, temperature regulation and water loss. Many estimates of flight metabolism have been based on physiological measurements, and I have reviewed and compared different methods.

Chapter 4 describes a number of gross morphological characters appropriate for flight analyses, most of which are used in later chapters. I recommend how various measurements should be taken, and in what units, because variation in these practices has complicated the literature on flight morphology.

Chapters 5–7 treat the kinematics and aerodynamics of gliding, soaring and migration. The gliding chapter concludes with a section on stability and control of movements, essential for all types of flight. Different soaring and migration methods are described and related to wing designs.

Hovering is the most expensive form of animal flight, but it is used by most insects and several smaller birds and bats. Hummingbirds hover differently from other birds and the bats, and have different wing morphology. The kinematics, aerodynamics, energetics and morphology of hovering animals are given in Chapter 8, concluded by recipes for power calculations for ecologists and other biologists trying to estimate hovering costs.

Similar treatment has been given in the following chapter (9) on forward flight, including manoeuvring, energy-saving modes of flight (e.g. intermittent flight and ground effect), take-off and landing. Recipes for power calculations are also given here.

What are the optimal flight speeds, kinematics, and wing sizes and shapes are questions treated in Chapter 10, which also addresses scaling and flight constraints.

Since flight is expensive, one might expect that the flight system has been subjected to strong selection to enhance flight performance. The muscle fibres, the musculo-skeletal systems, and wing structures are designed for different flight modes, and these main morphological characters are treated in Chapter 11, where I discuss the controversial ideas about pterosaur locomotion.

In Chapter 12 the flight types and wing forms are related to the ecological lifestyle of the animal. Birds and bats are adapted to occupy different lifestyles which mean flying in different ways and which are associated with different wing morphology.

In Chapter 13 I discuss the two main theories for explaining and under-standing the evolution of flight, the "ground-up" theory and the "trees-down" theory. The first starts from a bipedal, running animal that gradually began to fly actively without any intermediary gliding stage. The second begins with a climbing animal that started to glide between trees and then gradually to fly with wingstrokes. An aerodynamic model that I have developed shows how a tran-sition from gliding to active flight is mechanically and aerodynamically feasible, and that gliding was a probable intermediary stage between non-flight and active flight. The model applies to the transition from gliding to flapping flight regardless of whether gliding began in trees or followed upon a leap after a fast horizontal run. This chapter also considers various potential selection forces acting throughout the entire evolutionary process towards fully powered flight. The last chapter (14) gives some reflections on the problems of flight.

Chapter 2

Basic Aerodynamics

2.1 Introduction

The connection between wing and body shape and the elicited aerodynamic forces can be treated in detail by theory, but aerodynamic theory is not an exact science. It contains approximations, simplifications, and empirical coefficients, and the picture is complicated for a flapping wing, in which angles, velocities, and shape change instantaneously. Therefore, animal flight is complicated, but can be understood by the application of theories of airplane (fixed) wings and airscrews or propellers (blade-element, momentum-jet and vortex theories).

The blade-element theory of propellers describes the forces experienced by a wing in steady motion. In quasi-steady analyses it is generally assumed that the instantaneous forces of a flapping wing are those corresponding to steady motion at the same instantaneous velocity and attitude. Using this theory one can obtain the mean aerodynamic forces generated by the wings during a wingbeat cycle. The vortex theory of flight is applicable when steady, quasi-steady and unsteady effects prevail, making it the most reliable theory.

Aerodynamic theory is covered in several handbooks, for instance, von Kármán and Burgers (1935), Shapiro (1955), Prandtl and Tietjens (1957a,b), Milne-Thomson (1958), von Mises (1959), and Durand (1963) and the aerodynamic principles governing animal flight summarized in Pennycuick (1968a, 1969, 1972a, 1975), Kuethe (1975), Lighthill (1975, 1977), Alexander (1977, 1983), Rayner (1979a,b,c), Childress (1981), Ellington (1978, 1980, 1984a-f), and Norberg (1985a), as a sampling. Earlier papers on animal flight were concerned with flight kinematics (e.g. Marey 1887; Guidi 1938; Brown 1948), while many recent ones deal with the cost of flight (Pennycuick 1968a, 1969, 1975, 1989; Tucker 1968a,b, 1969, 1970, 1972, 1973; Thomas and Suthers 1972; Weis-Fogh 1972, 1973; Greenewalt 1975; Norberg 1976a; Oehme et al. 1977; Torre-Bueno and Larochelle 1978; Ellington 1978, 1984a-g; Rayner 1979a,b,c; Carpenter 1985, 1986). Relations between flight energetics and the morphology and ecology of animals are treated in many papers, such as Feinsinger et al. (1979), U.M. Norberg (1979, 1981a,b, 1987), Rayner (1981), Norberg and Rayner (1987). A bibliography by Rayner (1985a) encompasses biomechanics, aerodynamics, physiology, ecology and many other aspects of biology in relation to flight. I recommend it to all those who study any aspect of animal flight.

In this and following chapters many symbols are needed for the various aerodynamic principles and analyses. These symbols are listed in Table 2.1.

Table 2.1 List of symbols

A	Length of an ellipse; muscle cross-sectional area	D'_{ind}	Effective induced drag
A_b	Parasite power factor	$D_{ind.f}$	Induced drag in formation flight
A_d	Downstroke planar area	$D_{ind.s}$	Induced drag in solo flight
A_e	Equivalent flat-plate area	D_{par}	Parasite drag
A_o	Vortex sheet area	D_{pro}	Profile drag
A_u	Upstroke planar area	D_w	Wing drag
A_w	Profile power factor	$D_{w.h}$	Horizontal component of wing drag
AR	Aspect ratio	DLW	Doubly labelled water method
a	Wake spacing; fraction of time	d	Constant
a_d	Wake spacing in downstroke	dh	Height of a section
a_u	Wake spacing in upstroke	dl	Length of a section
a_{opt}	Optimal fraction of time	dS	Area of a section
B	Height of an ellipse; induced power factor	E	Energy
BMR	Basal metabolic rate	E_s	Energy saving
b	Wingspan	e	Energy equivalent; induced drag efficiency
b_2	Effective wingspan	e_h	Aerodynamic efficiency of hovering
C	Thermal conductance; cost of transport; constant	F	Force; resultant aerodynamic force
Ca_{O_2}	Oxygen content of arterial blood	F'	Local aerodynamic force
Cv_{O_2}	Oxygen content of venous blood	F''	Resultant aerodynamic force of wing feather
C_D	Drag coefficient	F_d	Resultant aerodynamic force in downstroke
C_{Df}	Friction drag coefficient	F_u	Resultant aerodynamic force in upstroke
$C_{D.ind}$	Induced drag coefficient		
$C_{D.par}$	Parasite drag coefficient	F_w	Resultant aerodynamic force of wing
$C_{D.pro}$	Profile drag coefficient		
C_F	Force coefficient	F_m	Muscle force
C_L, C_L'	Lift coefficient	F_o	Maximum muscle force in isometric contraction
C_{L1}	Lift coefficient at V_{mp}		
$C_{L.max}$	Maximum lift coefficient	F_{pa}	Force of parallel-fibred muscle
C_{mr}	Minimum cost of transport	F_{pa}'	Force component of parallel-fibred muscle
C.P.	Centre of pressure		
c	Wing chord; climbing period; constant	F_{pe}	Force of pennate muscle
$\bar{c}$	Mean wing chord	F_{pe}'	Force component of pennate muscle
$\hat{c}$	Non-dimensional (normalized) chord		
		$F_{I_{O_2}}$	Concentration of inspired air
D	Drag	$F_{E_{O_2}}$	Concentration of expired air
D'	Average drag	FG	Fast-glycolytic fibres
D_b	Body drag	FMR	Field metabolic rate
$D_{b.h}$	Horizontal component of body drag	FOG	Fast-twitch oxidative-glycolytic fibres
$D_{b.v}$	Vertical component of body drag	f	Pumping frequency; Initial fat content; feathering parameter; fulcrum
D_f	Friction drag	f_h	Heart rate
D_{f1}	Friction drag for a flat plate	F_i	Lift-impulse frequency
D_{ft}	Friction drag for a turbulent boundary layer	f_m	Mass flow
D_h	Horizontal drag	f_r	Respiration rate
D_{ind}	Induced drag	f_w	Wingbeat rate
		$f_{w.max}$	Maximum wingbeat rate

9

Table 2.1. Cont.

$f_{w,min}$	Minimum wingbeat rate	M_1	Body mass at start of flight
g	Acceleration of gravity	M_2	Body mass at end of flight
H	Vertical force	Mg	Body weight
$\dot{H}_{cc}$	Cutaneous convection	M_r	Rolling moment
$\dot{H}_{ce}$	Cutaneous evaporation	MR	Metabolic rate
$\dot{H}_e$	Evaporative heat loss	m	Muscle mass; limb mass
$\dot{H}_{ne}$	Heat loss/gain by non-evaporative means	m^*	Specific muscle mass
		$\dot{m}_b$	Rate of body-mass loss
$\dot{H}_p$	Heat production	$\dot{m}_d$	Mass loss due to defecation
$\dot{H}_{rc}$	Respiratory convection	$\dot{m}_e$	Evaporative water loss
$\dot{H}_{re}$	Respiratory evaporation	$\dot{m}_f$	Mass loss by fuel consumption
$\dot{H}_{rl}$	Long-wave radiation		
$\dot{H}_{rs}$	Short-wave radiation	m_m	Muscle mass
H_s	Heat storage	m_p	Pectoralis mass
h	Height	m_s	Supracoracoideus mass
I	Moment of inertia; force impulse; intermediate fibre type	m_v	Wing virtual mass
		m'_v	Virtual mass of a wing element
I_b	Moment of inertia of body	m_w	Mass of one wing
I_o	Ideal lift impulse	m'_w	Mass of a wing element
I_w	Moment of inertia of wings	$\hat{m}'_w$	Non-dimensional (normalized) mass of a wing element
J_m	Muscle moment		
K	Spacing ratio function		
K_O	Fractional turnover rate of O^{18}	$\dot{m}_{wl}$	Mass loss of water
		$\dot{m}_{wp}$	Mass gain by metabolic water production
K_D	Fractional turnover rate of D		
k	Induced drag factor; conversion factor; constant; inverse power density of mitochondrium	$\dot{m}_{CO_2}$	Mass loss by CO_2 release
		$\dot{m}_{O_2}$	Mass gain by O_2 consumption
		N	Mean body water content
L	Lift force; length; circuit	n	Wingbeat frequency; number of animals
L'	Average lift; lift force per unit wingspan	n_c	Number of chord lengths
L'_q	Quasi-steady lift per unit wingspan	P	Power
		P^*	Specific power
L_{pa}	Fibre length of parallel-fibred muscle	P_a	Power available
		P_{am}	Absolute minimum power (ideal animal)
L_{pe}	Fibre length of pennate muscle	P_b	Power for internal body functions
L_v	Vertical lift		
$L_{v.d}$	Vertical lift in downstroke	P_{bound}	Power for bounding flight
$L_{v.u}$	Vertical lift in upstroke	P_{climb}	Climbing power
$L_{v.w}$	Vertical wing lift	P_{desc}	Descending power
L/D	Lift:drag ratio	P_{flap}	Power required for flapping in a bound
(L'/D')	Effective lift/drag ratio		
$(L'/D')_{max}$	Maximum effective lift/drag ratio	P_{fold}	Power in passive phase of a bound
l	Length	P_{hov}	Power for hovering
l_{aw}	Armwing length	P_{hov}^*	Specific power for hovering
l_{hw}	Handwing length	P_i	Power input
l_w	Wing length	P_{ind}	Induced power
l_x	Distance behind leading edge	$P_{ind.g}$	Induced power with ground effect
M	Body mass; force component in stroke plane	$P_{ind.RF}$	Induced power in Rankine-Froude model

Table 2.1. Cont.

P_{iner}	Inertial power	r_b	Radius of gyration of body
$P_{iner.\,u}$	Inertial power in unsteady motion	r_w	Radius of gyration of wing
		r_{min}	Minimum circling radius
$P_m{}^*$	Mass-specific power output of muscle	r'_{min}	Absolute minimum circling radius
P_{max}	Maximum power output	S	Aerofoil area; wing area;
$P_{max}{}^*$	Maximum specific power output	S′	Wingstrip area
		S_{aw}	Armwing area
$P_{min}{}^*$	Minimum specific power output	S_b	Body frontal area
		S_{bs}	Body surface area
P_{mp}	Power required at minimum power speed	S_d	Wingdisk area; vortex ring area
P_{mr}	Power required at maximum range speed	$S_{d.\,proj}$	Horizontal projection of wingdisk area
P_{par}	Parasite power	S_{hw}	Handwing area
P_{pro}	Profile power	S_r	Wingstrip area
$P_{pro.\,hov}$	Hovering profile power	S_w	Area of one wing
P_r	Power required to fly	SMR	Standard metabolic rate
P_{sum}	Sum of aerodynamic power components	SO	Slow-twitch (oxidative) fibres
P_{tot}	Total power	ST	Slow-tonic fibres
P_{und}	Power for undulating flight	s	Stagnation point; wingtip spacing; spacing parameter
$P_v{}^*$	Volume-specific power output of muscle	T	Thrust; time; wingstroke period
p	Pressure	T′	Horizontal thrust
p_0	Pressure far above and far below wing	T_a	Ambient temperature
		T_b	Mean body temperature
p_a	Pressure immediately above wing	T_c	Thrust in a climb
		T_i	Wingtip-shape index
p_b	Pressure immediately below wing	T_l	Wingtip-length ratio
		T_s	Wingtip-area ratio
p_s	Pressure at stagnation point	T_w	Wake period
Q	Heat flow; muscle work	t	Time
Q_a	Aerodynamic torque	t_s	Wingstroke time
Q_d	Downstroke wake momentum	u_t	Wingtip velocity
Q_I	Inertial torque	V	Forward speed; muscle volume
$Q_m{}^*$	Mass-specific work of muscle	V′	Circulation speed
Q_u	Upstroke wake momentum	$\hat{V}$	Ventillation rate
$Q_v{}^*$	Volume-specific work of muscle	V_a	An animal's airspeed
		V_b	Cardiac output
q	Dynamic pressure; (mitochondrial volume)/ (myofibril volume)	V_c	Climbing speed
		V_f	Flapping velocity
		V_g	Gliding speed; an animal's groundspeed
R	Core radius; red fibre type	V'_g	Gliding speed
R′	Vortex-ring radius	$V_{g.\,min}$	Minimum gliding speed
RE	Respiratory exchange ratio	V_h	Horizontal speed; head-wind velocity
Re	Reynolds number		
Re_{crit}	Critical Reynolds number		
RMR	Resting metabolic rate	V_{mp}	Minimum power speed
RQ	Respiratory quotient	$V_{mp.\,g}$	Minimum power speed with ground effect
r,r′	Circling (turning) radius; wing radius	V_{mr}	Maximum range speed
$\hat{r}$	Non-dimensional (normalized) radius		

Table 2.1. Cont.

V_{mr}	Maximum range speed	α_{max}	Maximum angular
$V_{mr.g}$	Maximum range speed with		acceleration; maximum
	ground effect		roll acceleration
V_{ms}	Minimum sink speed	α_{stall}	Stalling angle
V_{O_2}	Oxygen consumption	α_{roll}	Angular roll acceleration
$V_{O_2.min}$	Oxygen consumption at V_{mp}	β	Stroke plane angle;
V_{opt}	Optimal speed		regression coefficient
V_p	Velocity potential	β_r	Angle between vortex-sheet
V_r	Resultant air speed		plane and horizontal
$V_{r.d}$	Resultant air speed in	Γ	Circulation
	downstroke	Γ'	Local circulation
$V_{r.u}$	Resultant air speed in	$\hat{\ }$	Normalized circulation
	upstroke	Γ_o	Circulation at medium
V_s	Velocity at stagnation		plane of wingspan
	point; heart-stroke volume;	Γ_i	Circulation induced by
	sinking speed		wake vorticity
$V_{s.min}$	Minimum heart-stroke volume	Γ_{max}	Maximum circulation
V_{st}	Sinking speed in a turn	Γ_q	Quasi-steady circulation
V_{stall}	Stalling speed	Γ_q'	Quasi-steady circulation
V_T	Tidal volume		per unit wingspan
V_t	Tailwind velocity; forward	Γ_r	Circulation due to
	speed in a turn		profile rotation
V_{th}	Rate of climb in a thermal	Γ_t	Translational circulation
V_w	Wind speed	γ	Wing's positional angle
$V_{w.h}$	Speed of headwing	$\bar{\gamma}$	Mean positional angle
$V_{w.t}$	Speed of tailwind		of wing
v	Axial ring velocity; muscle	γ_{max}	Maximum positional angle
	fibre volume		of wing
v_c	Myofibril volume	γ_{min}	Minimum positional angle
v_t	Mitochondrial volume		of wing
W	Aerodynamic work; white	δ	Boundary layer thickness
	fibre type	ε	Angle decrease due to
W_t	Total aerodynamic work		downwash; muscle strain;
w	Induced velocity;		core radius
	downwash velocity	$\dot{\varepsilon}$	Muscle strain rate
w'	Air velocity far below a	η	Muscle mechanical
	hovering animal		efficiency
w_d	Downwash velocity	$\varnothing$	Stroke amplitude (total wing
w_t	Tangential velocity at		excursion)
	vortex-core edge	θ	Glide angle; descent angle;
w_u	Upwash velocity		and between L and D
w_{RF}	Induced velocity in Rankine-	λ	Total roll moment of inertia
	Froude model	λ_1	Angle between D_{pro}
X_1, X_2	Constants		and horizontal
x	Weight support fraction	λ_2	Angle between D_{par}
Y	Flight range		and horizontal
y	Wingspan location; ordinate	μ	Air viscosity; muscle ratio;
α	Angle of attack; angle of fibre		muscle factor; size constant
	insertion; ordinate intercept	ν	Kinematic viscosity
α'	Angle of incidence	ρ	Air density
α_o	Effective angle of incidence	ρ_m	Muscle density
α_1	Angle between profile	σ	Muscle stress; interference
	drag and horizontal		coefficient; spatial correc-
α_2	Angle between parasite		tion factor
	drag and horizontal		

Table 2.1. Cont.

τ	Downstroke ratio; rolling torque	ψ_d	Downstroke wake inclination
		ψ_u	Upstroke wake inclination
τ_a	Aerodynamic torque	ω	Angular velocity
Φ	Angle of bank	$\dot{\omega}$	Angular acceleration
φ	Climb angle	ω_{max}	Maximum angular velocity
ψ	Resultant airspeed angle; climb angle	ω_{min}	Minimum angular velocity

2.2 The Flow Around an Aerofoil

The forces acting on an aerofoil moving through the air depend, among other things, upon the flow pattern around the object. Animal wings act as aerofoils and a flying animal must generate a momentum flow. The reaction of this flow is equal but opposite to the forces and accelerations acting on it. At high Reynolds numbers (Sect. 2.2.2), as in most vertebrates, this momentum of the air can only be transported as a vortex. Animal wings generate weight support and thrust due to the presence of a bound vortex across the wing. To understand how forces are elicited to lift and propel the animal through the air, we have to see how an aerofoil affects the airflow.

I will begin by discussing the steady flow of a fixed wing, a flow where pressure, air density, magnitude and direction of velocity etc. are functions of the positions in space but independent of time. The motion of a given air particle need not be uniform, but the velocity at any given spot must be constant. Thus, this does not mean that there is an unchanging state of motion at any fixed point. The forces on a body are the same whether it is at rest and the air streams towards it or whether the air is at rest and the body moves through it with an equal, constant, and opposite velocity.

2.2.1 Bernoulli's Equation

The air particles in a steady flow move in the same manner along the same path hence making up permanent curves called *streamlines* (Fig. 2.1a). At each point of a streamline the tangent has the direction of the velocity. The mass of air moving through a cross section of the streamtube per unit time is constant along the tube. This constant mass equals ρVdS, where ρ is air density, V the velocity of the air at a normal section of the tube, and dS the area of this section. Therefore, we obtain the *condition of continuity*

$$\rho_1 V_1 dS_1 = \rho_2 V_2 dS_2.$$

Let us now look at a small cylinder of air whose axis is formed by a short streamline segment (Fig. 2.1b). If dl is the length and dS the cross-sectional area of the cylinder, then the mass of the air in the cylinder is ρ dl dS and the weight is $g \rho$ dl dS, where g is acceleration of gravity. According to Newton's second law

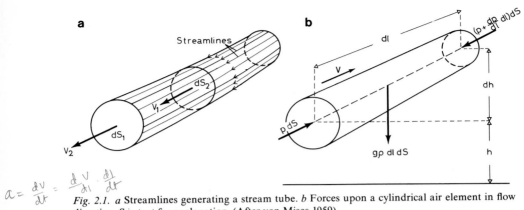

$a = \dfrac{dV}{dt} = \dfrac{dV}{dl} \cdot \dfrac{dl}{dt}$

Fig. 2.1. a Streamlines generating a stream tube. b Forces upon a cylindrical air element in flow direction. See text for explanation. (After von Mises 1959)

of motion, the forces acting on the air element in any direction equal the mass of the air times its acceleration in this direction. Hence,

$$\rho \, dl \, dS \times V \, dV/dl = p \, dS - [p + (dp/dl) \, dl]dS - g \, \rho \, dS \, (dh/dl) \, dl,$$

where $V \, dV/dl = d/dl(V^2/2)$ is acceleration of the air, p dS the pressure at one end of the element and $[p + (dp/dl) \, dl]dS$ the pressure at the other end of the element. The equation can be written as

$$d/dl(V^2/2g + h + p/g\rho) = 0,$$

where h is the elevation of the stream tube above some horizontal plane of reference and $g\rho$ is the constant specific weight of the air; therefore, along each streamline $(V^2/2g + h + p/g\rho)$ has a constant value, or

$$V^2/2 + gh + p/\rho = \text{const.} \tag{2.1}$$

This is *Bernoulli's equation* for an incompressible steady flow. In the theory of flight it is permissible to ignore not only the influence of the compressibility of the air but also the accompanying effects on air density. The term h represents the effect of gravity in Eq. (2.1) and must be omitted, so along each streamline we have

$$V^2/2 + p/\rho = (1/\rho) \, (\rho V^2/2 + p) = \text{const.} \tag{2.2}$$

Consider an aerofoil with asymmetric profile as in Fig. 2.2a. Some streamlines will pass above the aerofoil and some below it. They are separated from one another by one streamline that meets the front end (leading edge of a wing) at some point s (stagnation point), and they meet again at the rear end (trailing edge). At the stagnation point the velocity V_s is zero. If the pressure is p_s at the stagnation point and the pressure and velocity of the undisturbed flow (at a distance from the aerofoil) are p and V, respectively, the Bernoulli's equation (2.1) says that

$$(1/\rho) \, (\rho V_s^2/2 + p_s) = (1/\rho) \, (\rho V^2/2 + p),$$
$$(1/2)\rho V^2 = p_s - p. \tag{2.3}$$

14

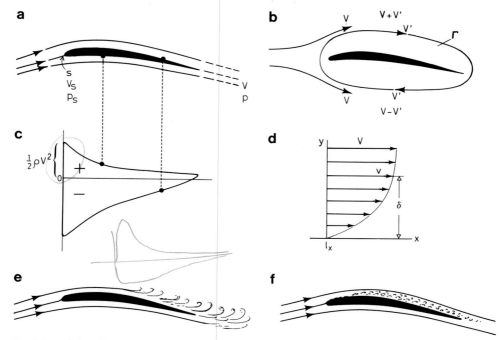

Fig. 2.2. *a* Airflow distribution around a typical wing profile during flight at a small angle of attack. *s* is the stagnation point where air velocity is V_s and pressure is p_s. V is velocity and p is pressure of the undisturbed air at a great distance from the aerofoil. *b* Airspeeds at a wing profile during flight; V is velocity of the free airstream meeting the wing and V' velocity of bound circulation *r*. *c* Pressure distribution around a typical wing profile at a small angle of attack. The ambient pressure is taken as zero pressure. (After Betz 1963). *d* Velocity gradients in the boundary layer. δ is the boundary layer thickness (along y-axis) measured from the wing surface to a point where the velocity $v = 0.94V$ (along x-axis). (After Betz 1963). *e* An increase in pressure at the rear causes the flow to separate from the wing. *f* Turbulent boundary layer, caused by surface roughness, keeping the flow above the boundary layer laminar

The expression $q = (1/2)\rho V^2$ (Fig. 2.2c) is called the stagnation pressure or *dynamic pressure;* it determines the magnitude of the force which the air exerts upon the aerofoil.

The motion of air relative to an asymmetric aerofoil can be represented as a flow of velocity V with a circulation of velocity V' superimposed on it. The velocity over the wing thus is V + V', and under the wing V − V' (Fig. 2.2b). If the pressures immediately above and below the wing are p_a and p_b, respectively, the pressure difference becomes

$$p_a - p_b = (1/2) \rho (V + V')^2 - (1/2) \rho (V - V')^2 = 2\rho VV'.$$

This pressure difference, acting across an aerofoil of area S, gives an upward directed *lift* force, which is

$$L = (p_a - p_b) S = 2\rho VV'S. \tag{2.4}$$

15

It takes some time (distance) to build up the circulation around the wing, so if the wing starts to move abruptly from rest, the lift does not reach its full value until it has travelled about one transverse length of the aerofoil (one wing-chord length).

2.2.2 Reynolds Number

The forces acting on an aerofoil moving through the air include, but are not limited to, the flow pattern around the aerofoil. The flow pattern depends on the viscosity μ and density ρ of the air, the velocity V of the undisturbed flow (the speed of the body relative to distant parts of the air), and on the size l (a length) of the body in the direction of flow, but also on the ratio of the inertial force and frictional force of the air (see, for example, Alexander 1983). We will now see what conditions make the streamline patterns similar for two different-sized but geometrically similar bodies.

To find which combination of μ, ρ, V and l has the same dimension as a force, we consider the dimensions of these parameters. From Newton's second law it follows that force is mass times acceleration, so the dimension of a force is [F] = mass × length/time^2 = M LT^{-2} (Table 2.2). The dimensional formulae for the other parameters are μ = M L^{-1}T^{-1}, ρ = M L^{-3}, V = L T^{-1}, and l = L, so the combination of the force components can be written as

$$[F] = \mu^a\rho^bV^cl^d \text{ or}$$
$$M\,L\,T^{-2} = M^{a+b}L^{-a-3b+c+d}T^{-a-c},$$

where a-d are the exponents to be solved. Equality of the corresponding exponents in the left and right sides of the equation gives a + b = 1, –a – 3b + c + d = 1, and –a – c = –2. Insertion of b = 1 –a and c = d = 2 – a gives

$$[F] = \mu^a\, \rho^{1-a}\, V^{2-a}l^{2-a} = \rho\, V^2\, l^2\, (\rho Vl/\mu)^{-a}.$$

Table 2.2. Dimensional formulae for some variables in the mass-length-time (MLT) system and the equivalent force-length-time (FLT) system, and their units

Quantity	MLT	FLT	SI units
Length	$[L]$	$[L]$	m
Area	$[L^2]$	$[L^2]$	m^2
Mass	$[M]$	$[FL^{-1}T^2]$	kg
Density	$[ML^{-3}]$	$[FL^{-4}T^2]$	kg m^{-3}
Moment of inertia	$[ML^2]$	$[FLT^2]$	kg m^2
Force, weight	$[MLT^{-2}]$	$[F]$	N = kg m s^{-2}
Pressure, stress	$[ML^{-1}T^{-2}]$	$[FL^{-2}]$	Pa = N m^{-2}
Work, energy, heat, work	$[ML^2T^{-2}]$	$[FL]$	J = N m
Power, energy consumption	$[ML^2T^{-3}]$	$[FLT^{-1}]$	W = J s^{-1}
Metabolic rate, specific power	$[L^2T^{-3}]$		W kg^{-1} = m^2 s^{-3}
Dynamic viscosity	$[ML^{-1}T^{-1}]$	$[FL^{-2}T]$	Pa s = N s m^{-2}
Kinematic viscosity	$[L^2T^{-1}]$	$[L^2T^{-1}]$	m^2 s^{-1}
Time	$[T]$	$[T]$	s
Frequency	$[T^{-1}]$	$[T^{-1}]$	s^{-1}
Velocity	$[LT^{-1}]$	$[LT^{-1}]$	m s^{-1}
Acceleration	$[LT^{-2}]$	$[LT^{-2}]$	m s^{-2}

Thus, any force on a body moving through the air must consist of the factor $\rho V^2 l^2$ and a second factor that is a function of $\rho V l/\mu$. The latter is the ratio (inertial force/frictional force), which, according to the *law of mechanical similarity*, must be the same at any instant for two geometrically similar bodies. This ratio is also known as the *Reynolds number* (Re),

$$\text{Re} = \rho V l/\mu = V l/\nu, \tag{2.5}$$

where $\nu = \mu/\rho$ is the kinematic viscosity of the fluid. The flow pattern around a wing depends on the ratio of the inertial force and the frictional force of the fluid, and thus on the dimensionless Reynolds number.

Greater kinematic viscosity means higher losses due to friction and reduced importance of inertia. The motions of the fluid around two geometrically similar bodies moving with the same orientation and without acceleration are kinematically similar only if their Reynolds numbers (Re) are the same. The kinematic viscosity of the air is influenced by altitude and temperature; it decreases at higher pressures and increases with temperature.

Aircraft operate at Re's up to the hundreds of millions (10^8), whereas in most birds and bats Re ranges from 10^4 to 10^5. The size and shape of body, wings and tail are adapted to the conditions characteristic of each range. Tiny insects operate at a lower range, far less than 100, where viscosity forces are overwhelming.

The length l in Eq. (2.5) is generally taken to be the length of the aerofoil in the direction of flow, the wing chord, while V is the velocity of the aerofoil relative to distant parts of the air. Instead of the factor l^2 in the expression $\rho V^2 l^2$ any area can be used to characterize the size of the aerofoil, for instance wing area S. Let $2 \times C_F$ be the function of (Vl/ν), then

$$F = (1/2)\rho V^2 S C_F. \tag{2.6}$$

$(1/2)\rho V^2$ is known as the dynamic pressure of (cf. Sect. 2.2.1). C_F is called the force coefficient; it is dimensionless and depends on the Reynolds number. Hence, *any force which the air exerts on an object can be expressed as the product of the dynamic pressure, an area, and a coefficient that depends on the Reynolds number but also on the shape and orientation of the object.* The forces for a whole wing section can be represented by a single force acting at the chordwise centre of lift, which according to thin aerofoil theory is generally about a quarter of the chord behind the leading edge of the aerofoil.

2.2.3 Boundary Layer

Very near the wing surface the air is retarded due to friction. In the adhesion layer the velocity relative to the surface is zero in all directions, even if the free flow velocity is quite high. The velocity increases with increasing distance from the surface until the velocity of the free flow is reached (Fig. 2.2d). This transition is continuous rather than stepwise because the viscosity of the air is responsible for frictional forces within the air which slow down the flow, where there are large differences in velocity. The thin layer of air within which this rapid velocity increase takes place is the *boundary layer,* which can be laminar, turbulent, or laminar at the front and turbulent at the rear of the aerofoil. The boundary layer decreases in thickness with increasing Reynolds number, becoming very thin at high Reynolds numbers as long as the flow is laminar.

At the lower range of Reynolds numbers (such as for the smallest birds and bats and insects about the size of dragonflies) the boundary layer is laminar, or smooth, and the skin friction drag will be comparatively small. A laminar boundary layer can also be achieved at high Reynolds numbers, but only if the surface is very smooth or if suction is applied through the surface. Otherwise the boundary layer will become turbulent, which dramatically increases skin friction. This is undesirable in flight except at high angles of incidence (e.g. during take-off, landing and manoeuvres) when a turbulent boundary layer can be advantageous, because it tends to prevent separation of the flow from the wing (stall). In spite of higher drag, the increase in lift will permit slower speeds.

The pressure distribution on a typical aerofoil (Fig. 2.2c) shows how the airflow and hence pressure distribution are affected by the boundary layer. The effect is slight if the boundary layer is kept laminar (= thin), but a turbulent boundary layer is much thicker, consumes more energy, and is associated with greater drag.

The thickness δ of the boundary layer (Fig. 2.2d) increases from zero at the stagnation point and reaches its maximum at some distance from the leading edge, where the pressure peak occurs. There are several different definitions of the boundary layer thickness (see, for example, Prandtl and Tietjens 1957b). It may be defined arbitrarily as the distance from the surface where the velocity differs by 1% from the velocity of the outer flow, but 6% is also used. Another definition is obtained by taking the intersection of the asymptote and a straight line through the origin of the velocity-distribution diagram in Fig. 2.2d such that the areas on both sides of the curve are equal. Taking the latter as a definition of the thickness of the boundary layer, the thickness δ at a distance l_x from the front edge of a flat plate moving with Reynolds number Re is $\delta = 3.4 l_x (Re)^{-1/2}$ when flow is laminar and $\delta = 0.37 l_x (Re)^{-1/5}$ when flow is turbulent (Prandtl and Tietjens 1957b).

For large Reynolds numbers, i.e. large velocities or dimensions and small kinematic viscosity (as for air), the inertia forces are more important than the viscous forces. Conversely, for small Reynolds numbers viscous forces are important and inertia forces small. When the boundary layer reaches a critical (minimum) thickness (at some critical Reynolds number) or some disturbance occurs, the laminarity gives way to turbulence so that, besides a longitudinal motion, the particles in the boundary layer also move back and forth transversely or whirl. The turbulent boundary layer has a much higher velocity gradient, giving a larger frictional drag. For subcritical thickness and Reynolds number the boundary layer is laminar, while for supercritical values it is turbulent; both forms of streams have positive and negative effects.

The boundary layer remains laminar over the aerofoil as long as the critical Reynolds number (somewhere around 6×10^4 for an arched profile) is not exceeded. The more the Reynolds number is below the critical value, the more stable the laminarity becomes. For laminar flow the wing should be profiled so

18

that the subpressure above the wing decreases slowly from a maximum at the stagnation point to a minimum as far back as possible, rising again only at the rear. Every pressure increase in the direction of flow is unfavourable for keeping the boundary layer laminar, particularly at high Reynolds numbers. With increasing velocity of the free airstream (i.e. with increasing Reynolds number) the location of the transition from a laminar to a turbulent flow moves towards the leading edge. Local disturbances, such as areas of roughness, can make the flow turbulent ahead of the transition point. In the case of a very stable laminar boundary layer the drag is almost independent of the relative roughness, but with a turbulent boundary layer the drag increases with the roughness (Schmitz 1960).

Although the boundary layer (either laminar or turbulent) moves with the airflow, it does so more slowly towards the rear, where it is continuously renewed from the incident flow. At the back of the wing the boundary layer flow is impeded by the increase in pressure, which causes the flow there to separate and form an eddying wake (Fig. 2.2e), which produces a high pressure drag determined primarily by the kinetic energy in the eddies. But if the boundary layer is turbulent due to surface roughness, there is a constant interchange of momentum between the rapid outer and slower inner layers, so the layers close to the surface receive kinetic energy from the free external flow. Therefore, compared to a laminar one, a turbulent boundary layer is better at maintaining a backward flow against the pressure increase towards the rear part of the aerofoil. This makes the total eddying dead-water region smaller with less pressure drag as a result (Fig. 2.2f).

2.3 Blade-Element and Momentum Jet Theories

The fundamental unit in the *blade-element theory* is the propeller blade, or wing, element, which is a chordwise strip of a wing between the radial distances r and r + dr from the wing base (Fig. 2.3a). This theory has frequently been used to describe animal flight when steady-state or quasi-steady-state conditions have been assumed to prevail. To calculate the aerodynamic forces during flapping flight, the wing has to be divided into a number of elements which are treated separately. The velocities and forces at each strip are different and vary during the wingbeat cycle, so the total forces elicited by a wing during a wingbeat cycle are the sum of the forces produced by each wing strip in the different phases of the wingstroke.

The essence of hovering flight is the production of a vertical force which must balance the animal's weight. As in forward flight, the wings must create low pressure above and high pressure below the wings, and a jet is produced corresponding to the momentum imparted to the air by the reaction of the wing forces. This momentum jet is identical with the vortex wake of the wings (see Sect. 2.4), and the air velocities therein can be determined by the momentum theory (Rankine-Froude theory) of propellers, initiated by Rankine (1865) and further developed by Froude (1889).

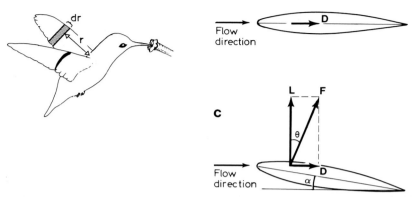

a **b**

c

Fig. 2.3. *a* The *stippled area* shows a chordwise wing element of width *dr* at distance *r* from the wing base. *b* The drag force *D* acting on a symmetrical body when its longitudinal axis is aligned along the direction of flow. *c* The forces acting on an aerofoil oriented at an angle *α* to the direction of motion. *F* is the resultant aerodynamic force, conventionally resolved into a lift force, *L*, perpendicular to the resultant direction of movement, and a drag force, *D*, backward along the direction of flow. The lift:drag ratio (*L/D*) is determined by tan *θ*

2.3.1 Lift and Drag

When a symmetrical body is exposed to a uniform, incompressible airflow, the forces exerted lie in the direction of motion (Fig. 2.3b) and are called drag D,

$$D = (1/2)\rho V^2 S C_D, \tag{2.7}$$

where S is a characteristic area of the body and C_D is called the drag coefficient. But when a symmetrical body moves with its long axis at an angle α to the air or if the body is asymmetrical, the forces act at an angle to the path of the object. The *angle of attack* α is the angle between the wing chord and the resultant airstream. Consider a wing with an assymetrical cross section and the upper side more convex than the lower one. The airflow around such an aerofoil becomes faster over it than under it, resulting in lower pressure on the upper surface than on the lower one. Because of the pressure difference of the whole aerofoil (from Bernoulli effects) the force F (acting at the quarter chord point that is, at a point 25% of the chord length from leading edge) becomes directed upwards and slightly backwards (Fig. 2.3c). This force is conventionally resolved into a drag component D along the airflow path and a component normal to it. Since this latter force normally is directed upwards, it is called the lift force L, which is

$$L = (1/2)\rho V^2 S C_L, \tag{2.8}$$

where C_L is the lift coefficient. From Fig. 2.3c and Eqs. (2.6), (2.7) and (2.8), it follows that L = Fcosθ and D = Fsinθ, and hence C_L = $C_F$cosθ and C_D = $C_F$cosθ. The lift is a result of the downward deflection (imparted downward

20

momentum) of the air passing over the wing. The downward air velocity influences the relative airflow over the wings and hence their lift.

The lift and drag coefficients, C_L and C_D, are dimensionless numbers that indicate the capacity of an aerofoil to generate lift and drag at a given angle of attack. They are dependent on the shape of the aerofoil, the Reynolds number, and the downwash angle, the angle through which air is deflected at the rear of the wing. The downwash angle, in turn, depends on the angle of attack of the wing (see Sect. 2.4.2). Figure 2.4a shows the relationships among the coefficients, Reynolds number and angle of attack for fixed-wing aircrafts. As α increases, C_L increases to a certain value but at the cost of a higher C_D. When a critical α (stalling angle α_{stall}) is exceeded, the airstream separates from the wing's upper surface with a sudden fall of C_L and increase of C_D. The drag coefficient for streamlined bodies falls slowly as the Reynolds number increases over a wide range of Reynolds numbers, and has a minimum value at a critical Reynolds number (usually between 6×10^4 and 2×10^5 for arched profiles) (Fig. 2.4b).

A good aerofoil maximizes the pressure difference between its upper and lower side and minimizes the drag (often maximizing the lift:drag ratio, L/D). The animal must either do mechanical work with its muscles to overcome drag (a retarding force) or must descend (glide or use partially powered flight) through the air at an angle where a component of its weight balances the drag.

The drag experienced by an animal is the sum of the *induced drag* D_{ind} and the *pressure (form) and friction drag* of the wings (conventionally termed profile drag D_{pro}) and of the body (termed parasite drag D_{par}). The induced drag is the drag component incurred because lift is being produced and arises from the downwash airflow (see Sect. 2.4.3).

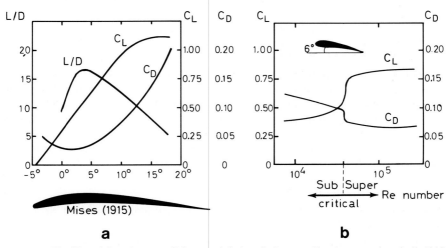

Fig. 2.4. *a* The lift and drag force coefficients and their ratio for a profile of an aeroplane built 1915 by R. von Mises. (von Mises 1959) *b* Diagram showing the relationships between lift and drag coefficients and Reynolds number for a particular wing profile. Lift coefficient increases and drag coefficient falls rapidly when Reynolds number increases above its critical value. (After Schmitz 1960)

Skin friction is due to viscous shearing stresses in the boundary layer, which loses energy because of these stresses and the flow separates from the wing before reaching the trailing edge. The wake produced by this separation has low pressure, giving rise to the pressure drag, a component that usually is very small for a thin curved aerofoil. For a flat plate moving in its own plane nearly all the drag is due to the shearing stresses the air exerts on the plate's surface. However, for a plate moving perpendicularly to its plane and leaving wide wakes of disturbed air behind, most of the drag is pressure drag.

Induced drag can be derived according to the Rankine-Froude axial *momentum-jet theory* (see also Sect. 8.3). In this theory the wake is considered as a steady jet with a uniform axial velocity across any cross-sectional area, and represents the *ideal* wake for a propeller with *minimum* induced velocity. For hovering and slow flight this theory far underestimates the induced drag, but for fast flights it may give approximate values. Here I will briefly describe the momentum-jet theory.

Consider an aerofoil of span b moving horizontally with speed V. It is assumed that the aerofoil affects the air in a vertical circle of diameter b around the aerofoil and that all the air passing this circle is given a downward *induced velocity* w. The mass of air passing through this circle of area $\pi b^2/4$ (called wing disk area S_d) is, in unit time, $\rho V \pi b^2/4$. The air is given momentum at a rate $\rho V w \pi b^2/4$, which must equal the lift L, so that $w = 4L/\rho V \pi b^2 = L/\rho V S_d$. The rate at which the air is given kinetic energy is $(1/2) \times$ (mass) $\times$ (velocity^2) and is $\rho V w^2 \pi b^2/8 = 2L^2/\rho V \pi b^2$. Some of the work done against drag in propelling the aerofoil must be used to give this kinetic energy to the air. The associated drag is the induced drag, D_{ind}, and because the rate of working (power) = (drag) $\times$ (velocity) we obtain

$$D_{ind} = 2L^2/\rho V^2 \pi b^2 = L^2/2\rho V^2 S_d. \tag{2.9}$$

The average, effective, induced drag, D'_{ind}, is usually written as

$$D'_{ind} = 2k\,(Mg)^2/\rho V^2 \pi b^2, \tag{2.10}$$

where k is the induced drag factor, which would be equal to 1 in the ideal case of an elliptical spanwise distribution [as in Eq. (2.9); see further Sects. 8.3 and 8.5]. In airplane wings k is 1.1 to 1.2. Mg is the weight of the animal (where M is body mass and g acceleration owing to gravity), which equals the vertical lift in horizontal and hovering flight (see below).

The induced drag coefficient C_{Dind} can be expressed as a function of the lift coefficient C_L:

$$C_{Dind} = k\,C_L^2 S/\pi b^2 = k\,C_L^2/\pi AR, \tag{2.11}$$

where S is wing area and $AR = b^2/S$ is the *aspect ratio* of the wing (cf. Sect. 4.3). Thus, for a given lift coefficient, an increase of aspect ratio (or wingspan) decreases the induced drag coefficient by spreading the wake.

For a streamlined body and a "good" wing profile the pressure drag is negligibly small, and the residual drag (profile drag D_{pro} + parasite drag D_{par}) is mainly friction drag which tends to fall as the Reynolds number increases. According to Eq. (2.7) the friction drag is

$$D_f = (1/2)\,\rho V^2 S C_{Df}, \tag{2.12}$$

where C_{Df} is the friction drag coefficient.

The amount of skin friction developed on the two surfaces of the aerofoil can be determined by boundary-layer theory. Skin friction drag is the rate of transfer of momentum from the surface of the animal to the mass of air in the boundary layer. According to laminar boundary layer equations by Blasius, the friction drag developed on the two sides of a flat plate of length c (equivalent to the chord of an aerofoil) and width 1 (the halfspan of an aerofoil, $l = b/2$) is $D_f = 2 \times 0.664l(\mu\rho V^3 c)^{1/2}$ (e.g. Prandtl and Tietjens 1957b). Since Reynolds number of this plate is $Re = \rho Vc/\mu$ [Eq. (2.5)], insertion of this expression of D_f in Eq. (2.12) gives

$$C_{Df} = 1.33(Re)^{-1/2},$$

since $S = bc$. The friction drag for a flat plate then becomes

$$D_{f1} = 0.66(Re)^{-1/2}\rho V^2 S \tag{2.13}$$

when flow is laminar. For a turbulent boundary layer the friction drag coefficient is

$$C_{Df} = 0.072(Re)^{-1/5},$$

and the corresponding drag becomes

$$D_{ft} = 0.036(Re)^{-1/5}\rho V^2 S. \tag{2.14}$$

These values agree well with experiments with smooth surfaces.

The areas used for profile and parasite drag differ, however, since the former refers to the wings and the latter to the body of the animal. The velocity of air meeting the wings and body in a flapping animal also is different, because the resultant velocity for the wings includes a flapping component (see below). Therefore, the resultant velocities of the wings and body have different directions and the drag components must be treated separately. The wing profile drag can be written as

$$D_{pro} = (1/2)\rho V^2 S C_{D,\,pro}, \tag{2.15}$$

where S is the wing area of the animal and $C_{D,\,pro}$ is the profile drag coefficient, and the parasite (body) drag as

$$D_{par} = (1/2)\rho V^2 S_b C_{D,\,par} = (1/2)\rho V^2 A_e. \tag{2.16}$$

S_b is the frontal projected area of the body and A_e is the area of a flat plate with the parasite drag coefficient $C_{D,\,par} = 1$, which gives the same drag as the body. A_e is conventionally named the *equivalent flat plate area* (see further Sect. 9.5.3).

The total aerodynamic drag of a flying animal thus becomes

$$\begin{aligned}D &= D_{ind} + D_{pro} + D_{par} \\ &= 2\,k(Mg)^2/\rho V^2 \pi b^2 + (1/2)\rho V^2(S C_{D,\,pro} + A_e).\end{aligned} \tag{2.17}$$

2.3.2 Power Required to Fly

The flight muscles of a flying animal do mechanical work when the point of application of a force is moved. The work is the distance moved multiplied by the force component in the direction of motion (a force of 1 N moving 1 m does 1 joule (J) of work). The rate at which this work is done is the mechanical power required to fly, and is $P = DV$ (measured in watts, W). Because drag is due to different things, the *aerodynamic power* is required for different purposes, and we recognize three main components:

1. *Induced power* P_{ind} — the rate of working required to generate a vortex wake whose reaction generates lift and thrust;
2. *profile power* P_{par} — the work needed against form (pressure) and friction drag of the wings; and
3. *parasite power*, P_{par} — the work needed against form and friction drag of the body (trunk).

The *inertial power*, P_{iner}, the work needed to accelerate the wings at each stroke (= oscillate the wings) also must be considered.

Multiplying the aerodynamic drag components in Eq. (2.17) by speed V and adding inertial power P_{iner}, the total flight power then becomes

$$P = 2k\, Mg^2/\rho V\pi b^2 + (1/2)\rho V^3(SC_{D,\,pro} + A_e) + P_{iner}. \tag{2.18}$$

The power components are associated with the muscles' mechanical efficiency, η, the ratio of mechanical power output, P (flight power or rate at which mechanical work is done), to metabolic power input, P_i (rate at which chemical, metabolic energy is consumed) (see Sect. 3.2). In addition to these components is the cost of internal body functions, or resting metabolism, P_b (see Sect. 3.3), so the power input is

$$P_i = (P_{ind} + P_{pro} + P_{par} + P_{iner})/\eta + P_b. \tag{2.19}$$

Tucker (1973) estimated the extra mechanical power required during locomotion for circulation of the blood and for ventilation of the lungs each to be about 5% of the total power for other purposes. Therefore, the sum of flight power and power for the standard metabolic rate (power input) should be multiplied by a circulation and ventilation factor of 1.10 (i.e. a 10% addition), so Eq. (2.19) can be written as

$$P_i = 1.10\,[(P_{ind} + P_{pro} + P_{par} + P_{iner})/\eta + P_b]. \tag{2.20}$$

The total power can then be expressed as

$$P_{tot} = 1.10\,[2k\, Mg^2/\rho V\pi b^2 + (1/2)\rho V^3(SC_{D,\,pro} + A_e) + P_{iner} + \eta P_b]. \tag{2.21}$$

Pennycuick (1968a, 1969) developed a model for estimating the various power components required to fly using classical aerodynamic theory including blade-element theory combined with the induced fluid velocities predicted for a momentum jet (see Table 9.1). Tucker (1973) and Pennycuick

(1975) further elaborated the theory, and most studies (e.g. Tucker and Parrott 1970; Tucker 1973; Greenewalt 1975; Norberg 1976a, b, 1985b) used this aerodynamic theory, which assumes that steady-state or quasi-steady-state aerodynamics prevail.

But any type of flapping flight also involves nonsteady periods, particularly at the reversal points where active pronation (forward-downward rotation) and supination (upward-backward rotation) of the wings occur. Nonsteady effects are especially important when the stroke amplitude and rate of twisting are large, for example, at low speeds and when an animal is hovering with stroke plane tilted and wings flexed during the upstroke, providing little or no useful force (= asymmetrical hovering). For flight modes where nonsteady aerodynamic conditions are important, the momentum jet approach underestimates the induced power, but this theory is often useful because it gives a *minimum* limiting value. The induced flow (downwash) fluctuates because of the nature of the wing stroke, so the induced velocity must be higher during some phases of the wing stroke to compensate for the lower velocity when the stroke is reversed. This means that more power is required than with a constant jet. Ellington (1978, 1980, 1984e) and Rayner (1979a,b,c) used vortex theory (see Sects. 8.5 and 9.6 and table 9.1) for calculating the induced power, and this probably gave more realistic values for hovering and slow flight. It is, however, important to note that the profile and parasite powers can be calculated with conventional (blade-element) theory, as outlined in this book.

The *induced power* is the main power in hovering and slow flight. In hovering flight it is $P_{ind} \propto Mg^{3/2}/b$, and in forward flight $P_{ind} \propto Mg^2/b^2V$ [cf. Eqs. (2.10) and (2.18)] so that to minimize induced power the weight should be low and the wingspan long [see also Eq. (2.48), Sect. 2.4.3].

The *profile power* increases with wing area (wing length and width) and with increasing speed; $P_{pro} \propto b^3S/T^3$ in hovering flight (Rayner 1979c) and $P_{pro} \propto SV^3$ in forward flight [cf. Eqs. (2.15) and (2.18) and Table 9.1], where T is wingstroke period and about proportional to b (Pennycuick 1975; see also Chap. 10); hence $P_{pro} \propto S$ in hovering.

The *parasite power*, the power required to overcome the drag of the body, is proportional to the frontal (cross-sectional) area of the body, and to the cube of the forward speed, $P_{par} = D_{par} V = (1/2)\rho V^3 S_b C_{D,par}$ [cf. Eq. (2.16)]. To minimize the parasite power, the body should be streamlined, which is especially important for fast flying animals, since the parasite power rises with the cube of the flight speed.

The *inertial power* is considered to be low in medium and fast flight, since wing inertia is convertible into useful aerodynamic work at the bottom of the stroke (Pennycuick 1968a; Norberg 1976a). It may be neglected for medium and fast speeds, but may be of some importance in hovering and slow flight. The inertial power is

$$P_{iner} = I \omega/T, \tag{2.22}$$

where I is the moment of inertia of the wings and ω the angular velocity of the wings. The moment of inertia depends on the distribution of the mass along the wing, and is

$$I = \int_{r=0}^{r=I_w} m_w' \, r^2 \, dr, \qquad (2.23)$$

where I_w is wing length and m_w' the mass of a wing-element at distance r from the fulcrum (shoulder joint). The moment of inertia can be estimated by strip analysis, where the wing is divided into a number of chordwise strips and each strip is weighed, and l is calculated for each strip (Fig. 2.5). The spanwise position of the centre of mass of a strip can be approximated to be at its spanwise mid-point. During the upstroke the radii are usually reduced by wing flexion, strongly reducing the moment of inertia since the radius appears squared in the equation.

The kinetic energy of a pair of wings in the flapping plane, relative to the body, is $I\omega^2$ (i.e. the wings' "internal kinetic energy" with flight speed disregarded). The angular velocity reaches its maximum value, $\omega_{max} = 2\pi n\gamma$, in the middle of the downstroke and in the middle of the upstroke, or twice during a wingbeat cycle (n = 1/T is stroke frequency and γ is the wing's positional angle). The total work done accelerating the wings is $2I \, (2\pi n\gamma)^2 = 8\pi^2 In^2\gamma^2$, and the power required is n times this amount,

$$P_{iner} = 8\pi^2 In^3\gamma^2. \qquad (2.24)$$

For geometric similarity, the wing mass m_w is about proportional to b^3, and $r \propto b, \omega \propto b^{-1}, T \propto b$; hence the inertial power is about proportional to wingspan squared, $P_{iner} \propto b^2$.

Osborne (1951) suggested that acceleration forces could play a significant role in flapping flight aerodynamics, and Ellington (1984b,f) accounted for these forces together with his vortex theory of hovering flight. The wing acceleration and deceleration during flapping set the surrounding air in motion, with an apparent increase in wing mass, the *wing virtual mass,* or added mass, m_v. This added wing mass is found only during accelerations of unsteady wing motions, and therefore it is not important in steady and quasi-steady motions. Flow theory shows that this mass should be equal to the mass of air in an imaginary cylinder around the wing with wing chord as diameter.

$$m_v' = \rho\pi c^2/4. \qquad (2.25)$$

The mean wing chord can be obtained by dividing the area of one wing with the length of the wing, $c = S_w/l_w$, so the total virtual mass of the wing becomes

$$m_v = \int_0^{I_w} m_v' \, dr = (\rho\pi/4) \int_0^{I_w} c^2 \, dr, \qquad (2.26)$$

which should be added to the wing mass for the calculation of the total moment of inertia. The total inertial power, or acceleration power, for a wing in unsteady motion therefore becomes

$$P_{iner, u} = I_t\omega/T, \qquad (2.27)$$

where

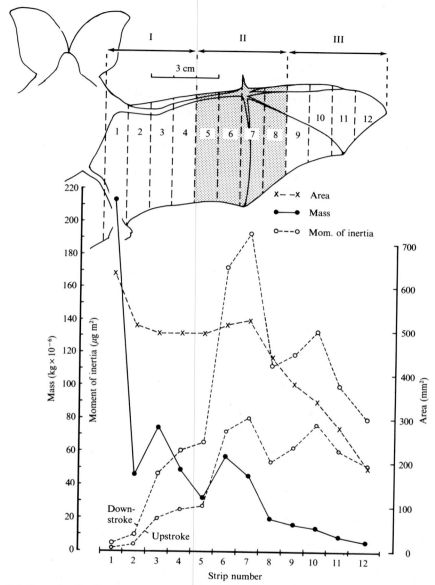

Fig. 2.5. Spanwise distribution of area, mass, and moment of inertia of the wing of the long-eared bat (*Plecotus auritus*). Flexion of the wing during the upstroke accounts for the difference between the downstroke and upstroke curves for the moment of inertia. (Norberg 1976a; by courtesy of The Company of Biologists Ltd)

27

$$I_t = \int_{r=0}^{r=I_w} (m_w' + m_v') \, r^2 \, dr \qquad (2.28)$$

is the total moment of inertia.

Figure 2.6 shows the total flight power P plotted against flight speed of a bat. The sum power curve typically has a shallow U-shape, and its bottom point defines the *minimum power speed* V_{mp} at which the animal can fly the longest time on a given amount of energy. The *maximum range speed* V_{mr} where the power/speed ratio (or energy/distance ratio) reaches its minimum is obtained by multiplication with time t; P/V = (energy/time)/(distance/time) = energy/distance. It can be found by drawing a tangent to the curve from the origin, when metabolic power is added to the flight power in the diagram, and should be used for maximization of flight distance on a given amount of energy

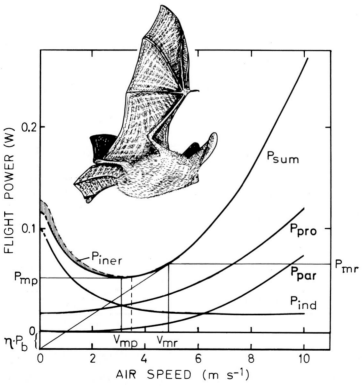

Fig. 2.6. Power curves for a bat the size of *Plecotus auritus* (mass 0.009 kg, wingspan 0.27 m, wing area 0.0123 m²). The extra drag of the large ears is disregarded. The various powers for flight are the mechanical power components over and above the resting metabolic rate P_b, which is set at zero in the diagram and marked below the origin, along the vertical axis. P_{ind} is induced power, P_{par} is parasite power, P_{pro} is profile power, P_{iner} is inertial power, P_{sum} is sum of the aerodynamic power components, P_{mp} is minimum power, P_{mr} is maximum range power, η is mechanical efficiency, V_{mp} is minimum power speed, and V_{mr} is maximum range speed. (After Norberg 1987)

28

(Pennycuick 1969, 1975). A low power curve can be obtained with high aspect ratio AR (long, narrow) wings and is advantageous for enduring flight. The higher the aspect ratio with a constant wing area, the higher the lift to drag ratio L/D of the wings.

The minimum power speed can be found by differentiating Eq. (2.18), assuming that $P_{iner} = 0$, and setting the first derivative equal to zero,

$$dP/dV = -2k(Mg)^2/(\rho\pi b^2 V^2) + (3/2)\rho V^2 (SC_{D,pro} + A_e) = 0,$$
$$V_{mp} = [4k(Mg)^2/(3\rho^2\pi b^2 SC_{Df})]^{1/4}. \tag{2.29}$$

The power required at this speed (minimum power P_{mp}) is obtained by substituting V by V_{mp} in Eq. (2.18) and neglecting the inertial power.

To estimate the maximum range speed V_{mr} and power P_{mr}, we have to find the maximum value of the distance Y travelled per unit work done, where Y = V/P. In the same way as V_{mp} we find the speed at which $1/Y$ is a minimum by differentiating $1/Y = P/V$ and setting the first derivative equal to zero,

$$(d/dV)(1/Y) = -4k(Mg)^2/(\rho\pi b^2 V^3) + \rho V^2(SC_{D,pro} + A_e) - P_b/V^2 = 0,$$
$$V_{mr} = [4k(Mg)^2/(\rho^2\pi b^2 SC_{Df})]^{1/4}. \tag{2.30}$$

Ignoring the metabolic power term (which decreases with speed) the maximum range speed V_{mr} is thus 1.32 times the minimum power speed V_{mp}. The aerodynamic power required at V_{mr} (maximum range power P_{mr}) is obtained by substituting V by V_{mr} in Eq. (2.18), so P_{mr} is 1.14 times P_{mp}. But the real V_{mr} and P_{mr} (when the metabolic power is included) will become somewhat higher (Fig. 2.6).

On the basis of morphologic and kinematic data in a quasi-steady analysis, Weis-Fogh (1972, 1973) derived analytical expressions for the average lift coefficient, the aerodynamic power, the moment of inertia of the wings, and the dynamic efficiency in animals that perform hovering with horizontal stroke plane (symmetrical hovering, see Sect. 8.2 and 8.4.2), such as hummingbirds and various insects. Norberg (1976a,b) devised equations based on Pennycuick (1968a) and Weis-Fogh, which, together with kinematic and morphologic data, allow the exact calculation of the mean lift and drag coefficients (as averaged over the whole wing and the entire wing stroke) for forward flight and asymmetrical hovering. This study is an analytic method, and differs from the trial and error used by Pennycuick (see Sect. 9.5.6 for details).

As Ellington (1980) noted, Norberg's model can be used as a proof-by-contradiction of the applicability of the steady-state theory for some animals in some types of flight. If the calculated coefficients are within the range of experimental steady-state values, then steady-state aerodynamics may entirely explain the flight but the quasi-steady assumption cannot be excluded. But if the calculated values are greater than the maximum observed coefficients for steady-state conditions, unsteady effects will be important. The lift and drag coefficients and the onset of stall cannot be described with quasi-steady theory when unsteady effects are significant. In the long-eared bat (*Plecotus auritus*), the estimated coefficients for slow forward flight are consistent with steady-state conditions whereas they are not for hovering flight (Norberg 1976a,b).

2.4 Vortex Theory of Flight

Vortex theories have been derived for animal flapping flight by Cone (1968), Betteridge and Archer (1974), Ellington (1978, 1980, 1984d-f) and Rayner (1979a,b, 1986), and their essence has been clearly summarized by Rayner (1979c, 1986, 1988) and Ellington (1984d-f).

With a vortex theory of flight the coefficients of lift and drag need not be calculated as in the blade-element theory, for it considers how an aerofoil influences the airflow to generate lift and can be used for unsteady flows. The vortex system forces air to move downwards behind the wing, and the reaction of this momentum flow is experienced by the wing as lift. *Vorticity*, or intensity of vortex motion, is measured by the rotational velocity of the air. The total strength of all vortices together is called *circulation*. The stronger the vortex the greater the lift generated, but with some energy loss to drag.

2.4.1 Bound, Trailing and Starting Vortices

Vortex motion occurs in the boundary layers along the surfaces of both sides of the wing and in the wing's wake. The pressure difference between the top and bottom surfaces of the wing has to disappear gradually towards the wingtips, and because of the greater pressure below the wing, some air will flow upwards around the wingtips. Streamlines run along the wings, with an outward direction on the bottom surface and inward direction on the upper surface (Fig. 2.7a). These line vortices are called the *bound vortices* of the wing, and are formed by the boundary layer or vortex sheet which surrounds the surface. The bound vortex can be regarded as a spanwise vortex line concentrated at a quarter chord distance behind the leading edge, but it is just a useful theoretical construct and not a physical reality.

However, the actual vortex system is more complicated because the spanwise circulation is not constant across the wingspan. It decreases to zero at the wingtips, and a vortex cannot end freely in the interior of the air. Whenever the strength of the bound vortex changes by some amount $d\Gamma$ an equal, opposite, amount of circulation $-d\Gamma$ must be shed from the trailing edge into the wake (*Kelvin's circulation theorem*), so the sum of circulations about a wing section and all vortices shed into the wake is zero.

The free vortex system will usually consist of a sheet of vortices leaving downstream from the trailing edge of the wing. Interactions between them generally cause the wake to roll up into a pair of distinct line vortices (along the direction of flight) which also convects downwards. These line vortices are called the *trailing vortices* of the wing. The distance a between the two trailing vortices behind the wings is somewhat shorter than the wingspan, and is approximately represented by $a = b\pi/4$. In this way a "horseshoe" vortex system is obtained (Fig. 2.7b).

A downward velocity and momentum are developed between the two vortex systems, and the vertical velocity component of this induced downwash is the

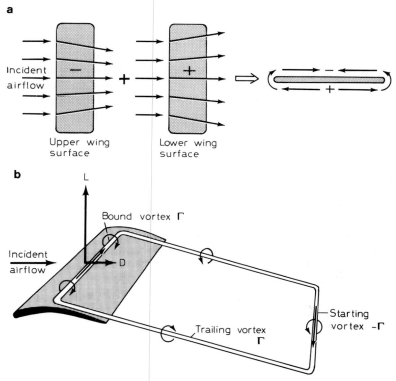

Fig. 2.7. *a* Airflow direction over the upper surface of the wing (*left figure*) and past the lower surface of the wing (*middle figure*) as well as the flow in the spanwise direction as seen from in front (*right figure*)— defines subpressure and + defines superpressure. *b* Simplified vortex distribution of an aerofoil in steady flight. The vortex system consists of a *bound vortex* on the wing and two *trailing vortices*, one in each wake behind the wingtips. A transverse *starting vortex* is shed whenever the strength of the bound vortex (= circulation Γ) changes. *L* is lift and *D* is drag

induced velocity. The lift on the wing can be interpreted as the reaction to the time rate of change of this momentum, while the horizontal component of the wake vector is the induced drag, which is required to generate the wake vortices.

The trailing vortices follow along the path of the wings and *starting vortices* are shed at the place where the force generation began or where the circulation changes. The starting vortex is formed along the whole trailing edge of the wing and appears to be unstable, separating from the wing and carried along by the general forward motion of the animal. The circulation of the starting vortex is equal but opposite to the circulation around the wing. The two trailing vortices are formed from the air particles that are driven from the high pressure (lower) surface of the wing to the low pressure (upper) surface, and make the connection between the wingtips and the starting vortex (Fig. 2.7b); the bound vortex of the wings completes the vortex loop. The trailing vortices can be seen clearly as condensation trails behind aircraft wings. The circulation produces centrifugal forces that tend to move the air outwards in the vortex leaving a low-pressure core made visible by the condensation of water vapour (depending on the air pressure and temperature). In "dust devils" dust and sand are sucked up in this central low-pressure "tunnel", while tornadoes over water readily suck up water.

Circulation has an important technical advantage because it permits the measuring of the integrated vorticity, even in infinite vortices as in a vortex sheet. Circulation increases with the angle of incidence to a maximum value beyond which the circulation can no longer be sustained and the wings stall and the lift rapidly falls away. In hovering flight (Chap. 8) the circulation is zero at mid-span, since the wing base does not move.

2.4.2 Steady Motion

The strength, or circulation Γ', of the bound vortex of a wing section (Fig. 2.2b) is a measure of the velocity difference in the circulating air associated with the pressure difference across the section. The lift per unit wingspan due to this circulation in steady motion is given by the Kutta-Joukowski theorem as

$$L' = \rho V' \Gamma', \tag{2.31}$$

and the total lift experienced by the wings of span b is

$$L = \rho V b \Gamma. \tag{2.32}$$

The value $\rho V b \Gamma$ is equivalent to $2\rho VV'S$ in Eq. (2.10), since $b = S/c$.

Lift can be understood only by assuming a circulation flow superposed on the translational airflow past the wing. Circulation Γ is measured by the rate of rotation about the vortex lines, and is a function of speed V, chord c, and angle of incidence α'. In steady motions there is only one value of Γ that places the rear separation point of the wing at the trailing edge (Kutta condition for finite fluid condition),

$$\Gamma_t = \pi c V \sin \alpha' \approx \pi c V \alpha'. \tag{2.33}$$

It is the *translational circulation* for linear motion, which is briefly the circumference of the aerofoil (about twice the wing chord; Prandtl and Tietjens 1957b) times the velocity V', $\Gamma_t = 2cV'$. The Kutta condition is fundamental and extremely important in aerodynamic analyses.

Combination of Eq. (2.8), (2.32) and (2.33) gives the lift coefficient

$$C_L = 2L/\rho V^2 S = 2\pi \sin\alpha', \tag{2.34}$$

which measures the circulation, and increases with the angle of incidence up to stall. However, experiments have shown that this value of the lift coefficient may seem of little use in an analysis of animal flight, even for gliding flight. Nachtigall (1979) found, for three wing profiles from the pigeon in a two-dimensional flow, that at Reynolds numbers from 27,000 to 80,000, the lift coefficient was about twice as high as predicted by theory. For a dragonfly model in steady flow at lower Re the lift coefficient is compatible with theory (Newman et al. 1977), but the effective profile cannot be predicted by theory (Ellington 1984d), so that the lift coefficient must be measured experimentally. The circulation around a wing can then be calculated by inserting the above expression for the lift coefficient (2.34) in Eq. (2.32):

$$\Gamma = (1/2)cVC_L. \qquad (2.35)$$

2.4.3 Lifting-Line Theories

Lifting-line theories are comparable with the blade-element theory and can be used for steady flow as in gliding, but they also give good approximations for flapping flight, where quasi-steady and unsteady effects prevail. In the 1920's, Prandtl worked out a theory concerning the influence of the aspect ratio on the forces acting on an aerofoil (e.g. von Mises 1959), and this theory is summarized below. The problem of *Prandtl's lifting-line theory* is simplified by a number of approximations:

1. The free vortex sheet extending from the trailing edge downstream is assumed to have the form of a plane ribbon extending from the trailing edge downstream to infinity.
2. The resulting vortex lines on the sheet are assumed to be straight lines parallel to the velocity (Fig. 2.8a).
3. The wing is supposed to be infinitely thin so that the vortex lines on the top and bottom surfaces can be replaced by resultant vortex lines on a simple sheet (Fig. 2.8a).

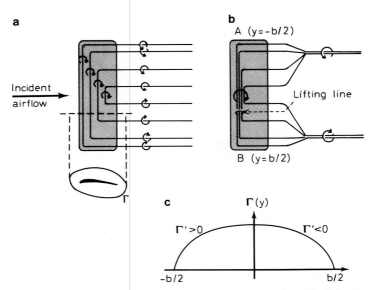

Fig. 2.8. a, b Simplified vortex distribution according to assumptions (1) to (4) in Prandtl's lifting-line theory. *c* Lift (or Γ–) distribution over wingspan *b*. AB is lifting line. See text for further explanation. (After von Mises 1959)

4. Each vortex line is supposed to consist of three straight parts; further-more, the "horseshoe vortices" are supposed to lie on one straight segment AB (Fig. 2.8b) which represents the whole wings.

AB, a non-constant vortex line of finite vorticity, is known as the *lifting line* which is zero at A, increases towards the middle plane, and decreases again to zero at B. The circulation Γ along the circuit around the wing profile (Fig. 2.8a) equals the integral of the vorticities of these vortex strips that pierce the plane of the circuit. These can also be described as the vortex strips that lie to the side of the plane of the circuit on the free vortex sheet. The circulation Γ thus increases from zero at the wingtips to a maximum at the mid-span (Fig. 2.8c). The function $\Gamma(y)$ represents also the concentrated vorticity for any point y of the lifting line, and is called the Γ-*distribution* or *lift distribution over the wingspan*.

In Prandtl's wing theory the downward-induced (downwash) velocity w at points near the lifting line plays an important role; w is normal to the wingspan and to the direction of motion and is small compared to the velocity V of the general airstream (Fig. 2.9). The effect of the induced velocity is a reduction of the (geometric) angle of incidence of the wing section; if α' is the *angle of incidence* (the angle between the direction of flow and the zero lift direction of the wing section), then the *effective angle of incidence* will be

$$\alpha_o = \alpha' - \tan^{-1}(w/V) \approx \alpha' - (w/V) \tag{2.36}$$

(Fig. 2.9).

The induced velocity varies along the chord of the wing section, resulting in a change of effective camber. But the theory can be developed accurately enough by

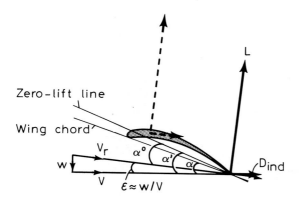

α = angle of attack

α' = angle of incidence

α^o = effective angle of incidence

$\alpha' - \alpha^o \approx w/V$

Fig. 2.9. Definitions of various angles of incidence of the airflow. V is the velocity of the general stream, w is the downwash velocity, and V_r is their resultant

assuming that the chord is small and by assuming a constant value of the induced velocity along the chord. The component of the induced velocity along the span is also neglected since it is small and unimportant, except possibly at the wingtips.

The lift force L is inclined backwards at the angle $\tan^{-1}(w/V) \approx (w/V)$ to the vertical (for horizontal relative motion of the wing). The drag force is caused by the induced velocity of the trailing vortices, and its horizontal component is the *induced drag*. The induced drag coefficient is

$$C_{Dind} = [\tan^{-1}(w/V)]C_L \approx (w/V)C_L, \tag{2.37}$$

so the total drag coefficient of the wing section becomes

$$C_D = C_{Dpro} + (w/V)C_L. \tag{2.38}$$

The work done on the air by the induced drag of the wings appears as the kinetic energy of the trailing vortex system.

The typical wing section experiences a lift force at the effective angle of incidence $\alpha_o = \alpha' - (w/V)$ so the direction of the lift force is rotated backwards through the angle $\tan^{-1}(w/V)$. The normal induced velocity at any point y_1 of the span is the sum of the effects of all the trailing vortices of the vortex sheet extending across the span

$$w(y_1) = (1/4\pi) \int_{-b/2}^{b/2} \Gamma(y)/(y_1-y) \, dy. \tag{2.39}$$

If the circulation Γ is assumed to be constant across the wingspan, the lift on the wings can be expressed in the alternative forms

$$L = \rho V \int_{-b/2}^{b/2} \Gamma(y) \, dy = \rho b V \Gamma = (1/2)\rho S V^2 C_L \tag{2.40}$$

[cf. Eq. (2.8) and (2.32)], and hence

$$\Gamma = (1/2)SVC_L/b = (1/2)cVC_L, \tag{2.41}$$

where $c = S/b$ is the mean chord of the wing.

The induced drag is

$$D_{ind} = \rho \int_{-b/2}^{b/2} w(y) \, \Gamma(y) \, dy. \tag{2.42}$$

When circulation and downwash velocity are known, the lift and induced drag are obtained from the first expression in Eq. (2.40) and from Eq. (2.42). The downwash velocity can be found with the momentum jet theory (giving a minimum value; see Sect. 8.3) or with vortex theory (Sect. 8.5).

The wings of flying animals are tapered towards the tip, and some birds have a more or less elliptic wing form. An elliptical wing with invariable profile and without twist experiences an *elliptic lift distribution*, meaning that the magnitude of the circulation at any point of the wingspan is proportional to the ordinate of an ellipse with the span as the major axis (Fig. 2.8c). The vorticity distribution is given by

$$\Gamma(y) = \Gamma_o[1 - (2y/b)^2]^{-1/2}, \tag{2.43}$$

where Γ_o is a constant. The lift distribution curve is the upper half of the ellipse with the equation $(\Gamma/\Gamma_o)^2 + (2y/b)^2 = 1$; Γ has the maximum value Γ_o in the median plane of the wingspan and drops to zero at the tips of the wings ($y = \pm b/2$). For a given aspect ratio, the wing that has the smallest drag for a given lift is the wing with an elliptic Γ-distribution. This is usually considered optimal for subsonic fixed-wing aircraft, but other circulation patterns can be more useful in animal flapping flight.

With elliptic loading the downwash velocity has the value

$$w = V C_L/\pi\, AR \tag{2.44}$$

across the span, and the induced drag coefficient has the value

$$C_{D\,ind} = C_L^2/\pi\, AR \tag{2.45}$$

[cf. Eq. (2.11)]. For a given C_L (or weight Mg) $C_{D\,ind}$ becomes smaller as AR increases. The geometric and effective angles of incidence then have the following relations

$$\alpha' = \alpha_o + C_L/\pi\, AR. \tag{2.46}$$

The induced drag is

$$D_{ind} = (1/8)\pi\rho\Gamma_o^2, \tag{2.47}$$

where Γ_o is equal to the circulation of the vortices in the wake, Γ. The induced power is the work done against the induced drag, equivalent to the rate of increase of wake kinetic energy per unit time, and is

$$P_{ind} = D_{ind}V = (1/8)\pi\rho\Gamma_o^2 V, \tag{2.48}$$

where V is the freestream velocity.

2.4.4 Quasi-Steady Assumption

Under the quasi-steady assumption, the instantaneous forces on a wing in unsteady motion are assumed to correspond to steady motion at the same instantaneous velocity and angle of attack. The instantaneous lift on a wing section depends on the circulation satisfying the Kutta condition [Eq. (2.33)]. In addition to translational motion, Ellington (1984d) considered the quasi-steady aerodynamic mechanism that includes the effects of profile rotation on lift production (Fung 1969).

The rotation of a wing section with chord c occurs about some axis located at distance l_x behind the leading edge, and has angular velocity ω. The circulation of this rotation is

$$\Gamma_r = \pi\omega c^2(3/4 - l_x/c), \tag{2.49}$$

and it will affect the airflow around the wing section. So the total quasi-steady circulation Γ_q around the section is the sum of the circulation for translation Γ_t [Eq. (2.33)] and for rotation Γ_r, and is

$$\Gamma_q = \Gamma_t + \Gamma_r = \pi c V \sin\alpha' + \pi \omega c^2 (3/4 - l_x/c)$$
$$\approx \pi c V[\alpha' + (\omega c/V)(3/4 - l_x/c)]. \tag{2.50}$$

The ratio $\omega c/V$ represents the angle through which the wing rotates during one chord of translation, while the expression in square brackets expresses the angle of incidence at the three-quarters chord point due to translation and rotation about the axis l_x/c. Because flapping velocity varies linearly along the wing, the ratio $\omega c/V$ must decrease with the distance from the wing base, so the rotational effects will be largest near the wing base and at the start and end of the half-strokes. In hovering, $\omega c/V$ would become infinitely large if the flapping velocity were zero at the turning points (which generally is not the case since the wing moves some distance at the turn). On the other hand, the circulation due to translation will be more important for the outer parts of the wing and in the middle of each half-stroke.

In bats, the wing membrane is highly flexible and the camber of the wing can be altered. Therefore, in slow flight and during the upstroke, the trailing edge lags behind, reducing the camber and rotation at this part, an effect also found in insects (Ellington 1984d). Ellington suggested that "the effect of this differential rotation may be interpreted as a profile with camber inversely related to the translational velocity".

According to Eq. (2.31) the quasi-steady lift per unit wingspan on a wing in translation and rotation is

$$L_q' = \rho V' \Gamma q'. \tag{2.51}$$

So sectional lift will be proportional to the first power of velocity. The quasi-steady lift coefficient, given from Eq. (2.34), is

$$C_L = 2\pi[\sin\alpha' + (\omega c/V)(3/4 - l_x/c)] \approx 2\pi[\alpha' + (\omega c/V)(3/4) - l_x/c)]. \tag{2.52}$$

2.4.5 Unsteady Effects

The flow over an animal's flapping wing is unsteady compared with the flow over the fixed wing of an aircraft, and distortion of the bound vortex arises mainly from movements of the wings, and increases with the flapping speed of the wing. The induced downwash varies during the wingbeat cycle and controls the bound vortex. There is no variation in the induced downwash during gliding, and in fast flight it is small. However, in hovering and slow flight the variation is more significant and can thus no longer be neglected. The energy loss to induced drag rises when circulation increases.

In slow flapping flight and hovering these unsteady aerodynamic effects to force generation are probably significant, and a highly complex vortex theory must be used for a satisfactory result. The variation in wing movement and circulation pattern introduce large complexities, and no simple approximation for unsteady effects is available for flapping wings. A flapping animal may be able to apply chordwise pitching moments to the wing sections to control circulation and wingstroke performance (Rayner 1986), and introducing some simplifications may give rough estimates.

The unsteadiness stems mainly from two phenomena associated with flapping: *"lift-delay"* and *"stall-delay"*. The former occurs when the angle of incidence increases rapidly below the stall, and the latter when the angle increases rapidly from below to above the stalling angle.

Lift-delay is best described in terms of the circulation around the wing, for when the angle of incidence increases rapidly, a strong vortex is shed at the trailing edge of the wing (Fig. 2.10a). This abrupt change in circulation was first analyzed by Wagner (1925), and is known as the *Wagner effect*. The circulation of this vortex is equal and opposite to that of the bound vortex which generates the lift increment on the wing, and in turn is proportional to the increase in lift. The shed vortex affects the airflow by delaying the build-up of the bound vortex circulation until the wing has left it about three chord lengths in its wake. After the wing has travelled this length, the lift increment reaches about 80% of its eventual value so the unsteady effects of the wake depend on the distance between the wing section and a particular vortex in the wake. The number of chord lengths travelled by a wing section at distance r from the wing base during a half-stroke is given by

$$n_c = \emptyset r/c, \tag{2.53}$$

where $\emptyset$ is the stroke amplitude in radians (see, for example, Ellington 1984d).

Stall-delay. Stall occurs when the angle of incidence becomes high enough to cause large-scale flow separation (about 15° for aircraft wings, but higher for cambered animal wings) which results in loss of lift, increased drag, and flight instability. But when the angle increases rapidly from below to above the stalling angle, the lag in the accompanying flow separation (Fig. 2.10b) means that the wing travels a few chord lengths with higher lift than can be attained in steady flight before drag increases and instabilities occur (= stall delay).

Lift-delay and stall-delay have opposing effects on the lift generated by a flapping wing whose angle of attack exceeds the stalling angle: sudden large changes below the stall produce with lift-delay an initial delay in lift generation, while stall-delay produces an increase in lift. Because of this, it is generally believed that aerofoils perform better when operated under unsteady conditions than under steady-state conditions. For animal wings, estimated lift coefficients for flapping wings exceed those possible for fixed wings under steady-state conditions (Norberg 1975, 1976b).

Unsteady theory simply adds the effects of wake-induced circulation to the quasi-steady analysis, and Ellington (1984d) derived simple equations to demonstrate this. The downwash decreases the effective angle of incidence α_0 from α' by a small amount $\varepsilon = \tan^{-1}(w/V) \approx w/V$ [cf. Eq. (2.36)]. The Kutta circulation for translation, given by Eq. (2.33), gives

$$\Gamma = \Gamma_q + \Gamma_i = \pi c V(\alpha' + \varepsilon), \tag{2.54}$$

which can be regarded as the sum of the quasi-steady circulation Γ_q and the circulation induced by the wake vorticity Γ_i, assuming that all wake vorticity is located at the starting point. Ellington noted that this may be a reasonable

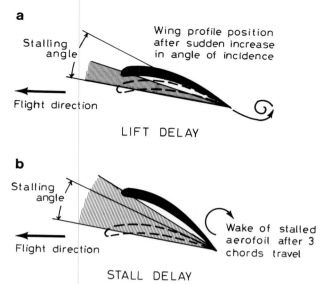

a

Stalling angle

Wing profile position after sudden increase in angle of incidence

Flight direction

LIFT DELAY

b

Stalling angle

Flight direction

Wake of stalled aerofoil after 3 chords travel

STALL DELAY

Fig. 2.10 a,b. Lift and stall delays as the effective angle of incidence increases. See text for explanation. (After Kuethe 1975)

approximation after the wing section has moved several chord lengths, because most wake vorticity is concentrated at that point.

The two-dimensional unsteady lifting-line theory for a flapping wing (von Kármán and Burgers 1935) shows that thrust can be obtained only if circulation changes in time. The trailing vorticity is, however, necessarily ignored in the theory, so it becomes of only limited value to flapping flight.

A number of three-dimensional lifting-line models have been proposed (for example, Schmeidler 1934; von Holst and Küchemann 1941; Cone 1968; Betteridge and Archer 1974; Archer et al. 1979; Phlips et al. 1981), but only Phlips et al. (1981) consider a non-planar wake, and all of the models assume that the circulation is as in a steadily moving wing.

Rayner (1979b) introduced a model describing fast flapping flight with constant circulation and with flexed or swept upstroke. He used a lifting-line theory and took account of the non-planar wake and of time variations in induced downwash. The vortex theories given by Ellington (1978, 1980, 1984d-g) and Rayner (1979a,b,c) for flapping flight are further discussed in Chapters 8 and 9.

Chapter 3

Physiology of Flight

3.1 Introduction

The foregoing has presented the theoretical grounds for estimating the energetics of flight. These theoretical methods predict the mechanical power required to fly, that is, the power output, which also is treated for different flight modes in Chapters 8 and 9. A large part of the metabolic rate appears as heat production in the body. The actual, metabolic, energy cost of flight (chemical power output) depends on the mechanical power requirement (chemical power input) and the mechanical efficiency of the muscles (power output/power input) in converting chemical, metabolic, energy into mechanical work. The power input can be measured by various methods, but, like the theoretical approaches, each method has its advantages and disadvantages. Energy expenditure during flight, therefore, can best be understood when several, theoretical and experimental, approaches are used concurrently.

Flight physiology is complex and includes several functions, such as respiratory mechanics and gas exchange, circulation, temperature regulation and water loss which are briefly treated in the following sections. The physiology and energetics of flight have been reviewed and discussed by several workers, e.g. Duncker (1971), Lasiewski (1972), Berger and Hart (1974), Calder (1974), Calder and King (1974), Berger (1981), and Butler and Woakes (1985). Calder (1974, 1984) extensively analyzed and reviewed the influence of body size on avian and mammalian energetics.

3.2 Energy and Mechanical Efficiency

A body is said to possess *potential energy* if the energy can be used to move it, while a moving body is said to possess *kinetic energy* because it can be made to do work against a force that resists its motion. There are two distinct types of energy — work and heat. In the work category are included mechanical work, electrical energy and chemical free energy, and all these forms are equivalent to each other. A muscle produces mechanical work from chemical energy. Muscles that are active and consume metabolic energy can operate in any of three modes (e.g. Alexander 1983): (1) They shorten and perform mechanical work (positive work, (2) their length is unchanged as they exert force and stabilize joints (zero work), or (3) they are stretched while exerting tension, tending to resist their lengthening, thus dissipating mechanical energy (negative work). When muscles do positive work, chemical energy is converted to work and heat,

chemical energy = work + heat.

This work is thus done *by* the muscles. When muscles do negative work, they consume more chemical energy than if they were inactive, and the balance is

work + chemical energy = heat.

This work is done *on* the muscles.

The *efficiency value* gives the proportion of chemical energy converted into external work, and is the ratio of the mechanical work which is done and the quantity of metabolic energy needed.

The mechanical efficiency η can be expressed as the ratio that relates the change in rate of work performed, ΔP, to the change in metabolic rate or power input, ΔP_i, as

$$\eta = \Delta P / \Delta P_i. \tag{3.1}$$

[cf. Eq. (2.19)]. The efficiency has experimentally been determined in a number of birds and bats flying at different speeds in a wind tunnel — Tucker (1972) outlined the theoretical basis and general method. The oxygen consumption of an animal flying horizontally at speed V in the wind tunnel is determined and then the wind tunnel is tilted through an angle φ so the effect on the animal is the same as though it were climbing at φ. This means that the animal will have to increase its thrust by the same value as the force component of its weight along the inclined flight path. The increment of power output ΔP required is

$$\Delta P = Mg \, V \sin\varphi, \tag{3.2}$$

which is the same as the increment of thrust of the animal's wings times the speed along the inclined path [cf. Eq. (5.2)]. The corresponding increment of power input ΔP_i (the difference between the metabolic rates during level flight and during flight when the tunnel is tilted) is recorded. The efficiency can be estimated using linear regression when power input is plotted as a function of the change in power output. It then equals the reciprocal of the slope at any given air speed. The following values have been reported: 0.19–0.28 for the laughing gull (*Larus atricilla*) (Tucker 1972), 0.20–0.29 for the fish crow (*Corvus ossifragus*) (Bernstein et al. 1973), 0.32–0.40 for the white-necked raven (*Corvus cryptoleucus*) (Hudson and Bernstein 1983), and 0.13–0.34 and 0.24–0.27 for the spear-nosed bat *Phyllostomus hastatus* (Microchiroptera) and the black flying fox *Pteropus alecto* (Megachiroptera) (Thomas 1975).

The efficiency of level flight probably varies with flight velocity and wingbeat kinematics (Rayner 1979c). When the mechanical efficiency is unknown, it is commonly taken to be about 0.20–0.25, so that power input is four to five times the power output.

3.3 Metabolic Rates

Metabolic rate (MR) is the rate at which energy is obtained from food for maintenance and for all the normal functions of the living animal. Kleiber (1961) called metabolism "the Fire of Life", and the metabolic rate is a measure of the intensity of this fire. As already mentioned, the metabolic rate is actually the same as power input (P_i).

Resting metabolic rate (RMR) is the rate of energy use by fasting animals at rest. It is the maintenance metabolism, including circulation of the blood and ventilation of the respiratory system, and is the minimum value within thermoneutrality. The term *basal metabolic rate* (BMR) is often used as maintenance metabolic rate, but it excludes the costs for circulation and ventilation. It is the near-minimum estimate of the metabolic rate of an inactive and fasting (postabsorptive) animal within its thermo-neutral zone (in which there are no costs for thermoregulation). BMR is often replaced by the *standard metabolic rate* (SMR), which is not necessarily minimal, since it need not be measured in the thermo-neutral zone, (see further definitions in Paynter 1974, p. 295 and Peters 1983, p. 27). The metabolic rate at rest and the change of rate of circulation and respiration metabolisms must be added to the metabolism associated with any locomotor activity, and is denoted P_b in Eq. (2.19). SMR or BMR have been used for this purpose.

The resting and standard metabolic rates vary allometrically with body mass and have been measured in various animals by determining oxygen consumption or carbon dioxide production. Kleiber (1932) found that the slope of the metabolic regression line for mammals was 0.75, and not 0.67 expected from the "surface role" (see also Schmidt-Nielsen 1984), but Heusner (1982) found the exponent 0.67 when calculating the regression equation for individuals of each of seven species of mammals. He suggested that the 0.75 slope is an artefact (see example in Fig. 10.1) and defined an overall mean regression line for intraspecific lines of slope 0.67.

Lasiewski and Dawson (1967) compiled a large amount of data on energy metabolism in birds and calculated the regression equations for the standard metabolic rate during 24 h, which when recalculated with power in watts and body mass in kg are

$$SMR = 6.25 \, M^{0.724}$$

for 48 passerine birds of 36 species, and

$$SMR = 3.79 \, M^{0.723}$$

for 72 non-passerines of 58 species, but

$$SMR = 4.19 \, M^{0.668}$$

for all birds taken together. Later, Aschoff and Pohl (1970) calculated the standard metabolic rates separately for days (activity metabolism) and nights (resting metabolism) and obtained the following relations between metabolic rate and body mass (here recalculated in W and kg): for 14 passerines (11 different species of different sizes) the activity equation was

$$SMR = 6.83 \, M^{0.704}$$

and the rest equation was

$$RMR = 5.55 \, M^{0.726},$$

and for 17 non-passerines (seven species) the activity equation was

$$SMR = 4.41 \, M^{0.729}$$

and the rest equation was

$$RMR = 3.56 \, M^{0.734},$$

Calder (1974) averaged together the data of Aschoff and Pohl and their recalculations of those from Lasiewski and Dawson to be comparable with the activity equations:

$$SMR = 6.61 \, M^{0.72} \tag{3.3}$$

for passerines, and

$$SMR = 4.29 \, M^{0.73} \tag{3.4}$$

for non-passerines.

The equations suggest that passerines operate at higher metabolic levels than do non-passerines of similar size. However, Prinzinger and Hänssler (1980) found that the rest equation for 24 non-passerines of 24 different species within the same mass range was $RMR = 5.39 \, M^{0.716}$, which is very similar to the corresponding equation of Aschoff and Pohl for passerine birds ($RMR = 5.55 \, M^{0.726}$), indicating that there are no significant differences in metabolic rates between similar-sized passerines and non-passerines.

The metabolic rate for *general activity* was guessed by Pennycuick and Bartholomew (1973) to be $2.5 \times SMR$ in wild lesser flamingos (*Phoeniconaias minor*), an estimate confirmed by experiments.

Bennett and Harvey (1987) used collected data from published literature and compared the allometric scaling of active metabolic rate (AMR) with that of resting metabolic rate for 14 bird species of different families. They found that the slope for AMR was significantly shallower than that for RMR (recalculated in W and kg as $AMR = 10.6 \, M^{0.61}$ and $RMR = 4.02 \, M^{0.68}$), meaning that small birds expend greater amounts of energy above resting levels than active large birds. The ratio AMR/RMR thus is not constant for different-sized birds; the equations give that the ratio for a 10-g bird is 3.6 and for a 1-kg bird 2.6.

The energy cost of general activity is equivalent to that of free existence (or average daily metabolic rate) used by Hails and Bryant (1979) for the house martin (*Delichon urbica*). They used the doubly labelled water method (see Sect. 3.5.3) and found that the cost of free existence for house martins feeding young was $2.22–5.27 \times SMR$ with mean $3.9 \times SMR$. Utter (1971) found similar measures for the purple martins (*Progne subis*) and mockingbirds (*Mimus polyglottos*) from 2.7 to $3.0 \times SMR$.

Nagy (1987) summarized and measured allometrically the total energy a wild animal expends during the course of a day (the *field metabolic rate*, FMR, that is synonymous with the rates for general activity and free existence) in 25 species of birds analyzed with the doubly labelled water method. The regression equation for passerine birds is (here recalculated in W and kg)

$$FMR = 18.2 \ M^{0.749}, \tag{3.5}$$

and for non-passerines

$$FMR = 2.35 \ M^{0.749}. \tag{3.6}$$

Thus, the fied metabolic rate for a 100 g bird is about 2.58 × SMR for passerines and 2.35 × SMR for non-passerines (cf. Eqs. 3. 3 and 3.4).

3.4 Oxygen Uptake

Estimates of flight metabolism are provided from measurements of oxygen consumption, $\dot{V}_{o_2}$, during flight. Oxygen uptake from the environment must balance the oxygen consumption of metabolism as averaged over a long time, and in a steady state system the rate of flow is the same at any level:

$\dot{V}_{o_2}$ respiratory uptake = $\dot{V}_{o_2}$ transported in blood = $\dot{V}_{o_2}$ consumed.

Oxygen is delivered to the tissues by the respiratory and circulatory systems in turn, but is incompletely removed from each preceding stage.

3.4.1 Respiratory Mechanics

The avian respiratory system consists of a pair of lungs and a number of air sacs connected to the lungs. During inspiration air passes through the trachea, primary, secondary and tertiary (or para-) bronchii and lungs into the air sacs, and during expiration back through the parabronchi in the same direction. Gas exchange occurs in the parabronchi which contain a dense network of air capillaries and blood capillaries. The cross-current system with bulk air flow at right angles to bulk blood flow through the parabronchi is more effective at exchanging gases (CO_2 in particular) than the mammalian lung (Scheid 1979). The mammalian lung is composed of bronchial and bronchiolar airways terminating in alveolar airspaces where gas exchange occurs. It is only slightly lighter than the lung of a bird of similar size, and because it is less than half as dense, it occupies about twice the volume. Despite the large morphological differences between bird and bat respiratory systems, the performance of the bat respiratory mechanics is more similar to that of birds than to that of non-flying mammals.

The respiratory movements of the rib cage that induce airway pressure change and the pectoral muscle system that controls the wingbeat movements

44

can work independently of one another. Although not obligatory, coordination between wingbeats and lung ventilation usually occurs in birds and bats. Wingbeat frequency (f_w) generally exceeds respiration frequency (f_r) during flight in most species, but a one-to-one coordination has been recorded in crows and pigeons (Hart and Roy 1966; Butler et al. 1977) and in five species of bats (Suthers et al. 1972; Thomas 1981; Carpenter 1986). The most common coordination observed in birds is $f_w:f_r = 3:1$, but ratios as high as 5:1 occur (Berger and Hart 1974). The ratio can be changed within one individual during one flight, for wingbeat and respiration frequencies vary with flight speed, body mass and other factors; sudden changes of wingbeat frequencies occur for instance during turns and take-offs and landings. Wingbeat frequency decreases with increasing body mass (Greenewalt 1960) and with decreasing wing loading. Berger and Hart (1974) showed that the value of the relative wing loading $M^{2/3}/S$ strongly influences the coordination ratio, with the highest ratios in species coinciding with the highest relative wing loadings. Calder (1984) gave the following allometric equations for the coordination ratio and body mass:

$$f_w:f_r = 1.42 \, M^{-0.19} \qquad\qquad (3.7)$$

for birds of low wing loadings (passeriformes), and

$$f_w:f_r = 2.76 \, M^{-0.07} \qquad\qquad (3.8)$$

for birds of high wing loadings (ducks and fowl).

In most bird species the beginning of inspiration occurs at the middle or end of the upstroke, while the beginning of expiration occurs at the end of the downstroke. Furthermore, the inspiratory tracheal air flow during the downstroke is lower than during the upstroke (e.g. Berger and Hart 1974).

In flying budgerigars (*Melopsittacus undulatus*), the maximum respiratory rate occured at low flight speeds and the minimum at intermediate speeds; at higher speeds it increased again (Tucker 1968b). This is apparently correlated with the changes in oxygen consumption with flight speed and mirrors the U-shaped power curve predicted from aerodynamic theory.

3.4.2 Respiratory Gas Exchange

Assuming completely aerobic metabolism, the metabolic rate can be determined by measuring the oxygen consumption. A steady state condition must then prevail such that the rate of oxygen consumption equals the rate of oxygen delivery by the respiratory and circulatory systems in series. Thus, by the conservation of volume (Stahl 1962), we find that

$$\begin{pmatrix} \text{pumping} \\ \text{frequency} \end{pmatrix} \times \begin{pmatrix} \text{stroke or} \\ \text{tidal volume} \end{pmatrix} \times \begin{pmatrix} \text{fractional removal of O}_2 \\ \text{from cycled air or blood} \end{pmatrix}$$

$$= \begin{pmatrix} \text{oxygen consumption} \\ \text{rate} \end{pmatrix}, \qquad\qquad (3.9)$$

which is called the Fick equation. An allometric equation can be substituted for each variable; an excellent analysis and review are given in Calder (1984) of the allometry of the respiratory and circulatory systems in this respect for birds and mammals in rest and during exercise. Because most existing allometric relationships for flying birds and bats are based on relatively few species, they have relatively low predictive reliability. Data for more species would possibly change both the y-intercept and the slope (exponent) in the power equations.

Referring to the respiratory system, the rate of oxygen consumption is the product of the respiratory rate f_r, tidal volume V_T (volume of air respired during one breathing), and concentration difference between inspired ($F_{I_{O_2}}$) and expired ($F_{E_{O_2}}$) air (oxygen extracted):

$$
\left(\begin{array}{c}\text{frequency,}\\ \text{breaths min}^{-1}\end{array}\right) \times \left(\begin{array}{c}\text{tidal or stroke}\\ \text{volume, ml}\end{array}\right) \times \left(\begin{array}{c}\text{oxygen extracted,}\\ \text{ml O}_2 \text{ (ml air)}^{-1}\end{array}\right)
$$

$$
f_r \quad \times \quad V_T \quad \times \quad (F_{I_{O_2}} - F_{E_{O_2}})
$$

$$
= \left(\begin{array}{c}\text{oxygen consumption,}\\ \text{ml O}_2 \text{ min}^{-1}\end{array}\right)
$$

$$
= \quad \dot{V}_{O_2} \tag{3.10}
$$

(cf. Table 5.3 in Calder 1984). Lasiewski and Calder (1971) predicted that $(F_{I_{O_2}} - F_{E_{O_2}}) = 0.05 \, M^{-0.05}$ and estimated the following relationships for resting non-passerine birds for the parameters in Eq. (3.10): $17.2 \, M^{-0.31} \times 13.2 \, M^{1.08} \times 0.05 \, M^{-0.05} = 11.3 \, M^{0.72}$.

In resting birds respiratory frequency decreases with increasing body mass as between $f_r \propto M^{-0.28}$ and $M^{-0.33}$ (Berger and Hart 1974), and was estimated to be on average

$$
f_r = 17.2 \, M^{-0.31} \text{ breaths min}^{-1} \tag{3.11}
$$

for a sample of non-passerine birds (Lasiewski and Calder 1971). During flight f_r is 3–19 times higher than during rest (Fig. 3.1a), with the higher values referring to larger birds. Carpenter (1986) compiled data for flying birds and bats, and found that the number of breaths per minute for birds (n = 18 observations of five species) is given by the formula

$$
f_r = 160 \, M^{-0.15}, \tag{3.12}
$$

and for bats (n = 16 observations of four species) by the formula

$$
f_r = 171 \, M^{-0.30}. \tag{3.13}
$$

But when the regression is based on mean values for the same bat species (Carpenter's three species, the straw-coloured fruit bat *Eidolon helvum*, the hammer-headed fruit bat *Hypsignathus monstrosus*, the grey-headed flying fox *Pteropus poliocephalus*, and the spear-nosed bat *Phyllostomus hastatus* from Thomas and Suthers 1972) the least-squares regression becomes

$$
f_r = 127 \, M^{-0.65} \tag{3.14}
$$

(n = 4, correlation coefficient = 0.995), which differs from Eq. (3.13).

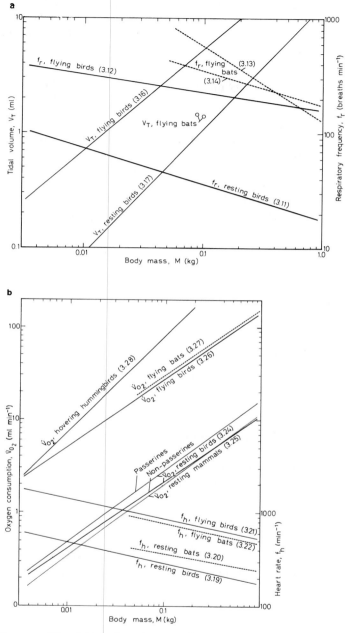

Fig. 3.1. *a* Respiratory frequency (f_r) and tidal volume (V_T) plotted against body mass in resting and flying birds and in flying bats. *b* Heart rate (f_h) and oxygen consumption ($\dot{V}_{O_2}$) plotted against body mass in various animals during rest and in flight. The equations used for calculation of the regression lines in *a* and *b* are identified by their numbers in the diagrams.

Tidal volume can be calculated when respiration rate and ventilation minute volume $\dot{V}$ are known, or ventilation rate can be obtained when tidal volume and respiration rate are known, since $V_T = \dot{V}/f_r$; the ventilation rate can be determined by measuring the respiratory water loss (Sect. 3.4.5). Ventilation rate in birds during flight varies with body mass as

$$\dot{V} = 6505 \ M^{0.73} \ \text{ml min}^{-1} \tag{3.15}$$

(Berger and Hart 1974). The resting value for ventilation was derived from respiration rate and tidal volume, and is given by $\dot{V} = 227 \ M^{0.74}$ for non-passerines (Lasiewski and Calder 1971). The ventilation during flight is thus 23 times that in resting birds the size of a pigeon. In flying pigeons ventilation was found to be 20 times the resting value (Hart and Roy 1966); the comparable value for the budgerigar was 12 times (Tucker 1968b). Variation of $\dot{V}$ was smaller than that for f_r in birds, but V_T was very variable since f_r and V_T tended to show compensatory shifts (Berger and Hart 1974). Tidal volume in flying birds can be calculated from the ratio of ventilation rate $\dot{V}$ and respiration frequency f_r [Eq. (3.15) and (3.12)],

$$V_T = 40.7 \ M^{0.88} \ \text{ml breath}^{-1}. \tag{3.16}$$

The allometry for the tidal volume in resting birds is

$$V_T = 13.2 \ M^{1.08} \tag{3.17}$$

(Calder 1984; Fig. 3.1a).

Flight data on ventilation rate and tidal volume are available for one bat species only. In two spear-nosed bats (mass M = 0.087 and 0.101 kg) $f_r = 576$ breaths min^{-1} in both specimens, $\dot{V} = 863$ and 827 ml min^{-1}, and $V_T = 1.50$ and 1.43 ml breath^{-1} (marked by circles in Fig. 3.1a), respectively (Thomas and Suthers 1972). Both ventilation rate and tidal volume are much lower, but respiratory frequency is higher, than predicted from the bird equations (Fig. 3.1a). The respiratory frequency is also much higher than predicted from Carpenter's Eq. (3.13) for bats.

3.4.3 Circulation

The circulatory system serves as an intermediary between diffusion of oxygen from the parabronchi and alveoli into the working tissues. The primary function is the delivery of oxygen, the rate of which can be described by the Fick equation [cf. Eq. (3.9)]:

$$\underbrace{\overset{\begin{pmatrix} \text{heart rate} \\ \text{or frequen-} \\ \text{cy, per min} \end{pmatrix}}{f_h} \times \overset{\begin{pmatrix} \text{blood volume per} \\ \text{beat or stroke} \\ \text{volume, ml} \end{pmatrix}}{V_s}}_{\text{cardiac output, ml min-1}} \times \overset{\begin{pmatrix} \text{concentration change,} \\ \text{arterial to mean venous} \\ \text{blood, ml } O_2(\text{ml blood})^{-1} \end{pmatrix}}{(Ca_{O_2} - Cv_{O_2})}$$

$$= \begin{pmatrix} \text{oxygen uptake,} \\ \text{ml } O_2 \text{ min}^{-1} \end{pmatrix} \tag{3.18}$$

$$= V_{O_2}$$

where Ca_{O_2} is the oxygen content of arterial blood and Cv_{O_2} is the average oxygen content of mixed venous blood [cf. Eq. (5.19a) in Calder 1984].

Heart rate and oxygen consumption in flying bats show more agreement with birds than with other mammals during exercise (Berger and Hart 1974; Butler 1981; Butler and Woakes 1985), and heart sizes in bats are also more similar to those in birds than to other mammals (Hesse 1921). Heart rate during flight in birds is usually more than twice the resting value (up to six times the resting value in the pigeon; Butler et al. 1977), but is higher during the first seconds of flight, at take-off and just after take-off. Heart rate also slightly diminishes with increasing body mass, in resting birds scaling as $f_h = 156 \, M^{-0.23}$ for non-passeriforms (Calder 1968) and as $f_h = 264 \, M^{-0.244}$ for the larger passeriforms (Calder 1984). Berger et al. (1970a) gave the following relationship for resting birds of different sizes,

$$f_h = 174 \, M^{-0.21} \tag{3.19}$$

(Fig. 3.1b). Heart rates for resting bats have been reported for three species (Studier and Howell 1969; Thomas and Suthers 1970; Carpenter 1985), and give the least-squares regression

$$f_h = 240 \, M^{-0.17} \tag{3.20}$$

($n = 3$, correlation coefficient = 0.998; Fig. 3.1b) which differs little from the allometry for resting mammals ($f_h = 241 M^{-0.25}$; Stahl 1967).

Carpenter (1986) compiled data on heart rates per minute for flying birds and bats and obtained the following regressions:

$$f_h = 534 \, M^{-0.145} \tag{3.21}$$

for birds, and

$$f_h = 454 \, M^{-0.224} \tag{3.22}$$

for bats (Fig. 3.1b).

The cardiac stroke volume (V_s) can be calculated from direct measurements of heartbeat frequency (f_h) and cardiac output ($\dot{V}_b$) where the cardiac output is the product of heartbeat frequency and the stroke volume, $\dot{V}_b = f_h \times V_s$ ml stroke^{-1}. Cardiac output is greater in birds than in similar-sized mammals (Grubb 1983), perhaps reflecting a necessary adaptation to high flight metabolism. In resting mammals the stroke volume scales as $V_s = \dot{V}_b/f_h = 187$

49

$M^{0.81}/241\ M^{-0.25} = 0.78\ M^{1.06}$ ml min^{-1} (Stahl 1967), and in resting birds as $V_s = 1.72\ M^{0.97}$ (Grubb 1983); in both cases stroke volume is almost independent of body mass. There are only a few records of the stroke volumes in flying birds. Berger and Hart (1974) compiled data on the minimum stroke volume for four bird species during flight and obtained the equation

$$V_{s,\,min} = 1.43\ M^{0.87}\ \text{ml min}^{-1}. \tag{3.23}$$

They pointed out, however, that the estimates involved uncertain assumptions and would hardly be comparable to those for resting values or any other activity.

Stroke volume data from one bat species during flight fit the bird Eq. (3.23) very well. Heart rate for two spear-nosed bats (M = 0.0875 and 0.101 kg) was f_h = 780 min^{-1}, stroke volume V_s = 0.168 and 0.217 ml (Thomas and Suthers 1972), and cardiac output $\dot{V}_b$ = 131 and 169 ml min^{-1}, respectively.

In flying pigeons (V = 10 m s^{-1}) the oxygen content of arterial blood, Ca_{O_2}, is maintained slightly below the resting value, but Cv_{O_2} is halved, and (Ca_{O_2} − Cv_{O_2}) becomes 1.8 times the resting value (Butler et al. 1977).

3.4.4 Oxygen Consumption Versus Body Mass

The oxygen consumption in flying animals is higher than in any non-flying, exercising animals (Fig. 3.1b). Even the highest metabolic values obtained in short tests with non-flying mammals do not approach the values for long flights in birds. Oxygen consumption varies allometrically with body mass and with flight speed (see Sects. 3.3 and 3.7). The resting values for birds, giving the RMR equations from Aschoff and Pohl (1970) in Sect. 3.3 where 1 W corresponds to 2.985 ml O$_2$ min^{-1} for STPD (standard temperature and pressure), fit the regressions

$$\dot{V}_{O_2} = 16.6\ M^{0.726}\ \text{ml O}_2\ \text{min}^{-1} \tag{3.24a}$$

for passerines and

$$\dot{V}_{O_2} = 10.6\ M^{0.734}\ \text{ml O}_2\ \text{min}^{-1} \tag{3.24b}$$

for non-passerines. The resting value for non-flying mammals varies with body mass as

$$\dot{V}_{O_2} = 11.6\ M^{0.76}\ \text{ml O}_2\ \text{min}^{-1} \tag{3.25}$$

(Stahl 1967).

The effects of ambient temperature, T_a, on oxygen consumption in resting bats have been estimated for a number of species. The least squares regressions for the specific oxygen consumption in the phyllostomid bats *Macrotus californicus* (M = 0.013 kg), *Leptonycteris sanborni* (M = 0.22 kg) and *Artibeus hirsutus* (M = 0.044 − 0.053 kg), and the pteropodid bat *Rousettus aegyptiacus* (M = 0.146 kg) were $\dot{V}_{O_2}$ = $(174 - 4.5\ T_a)$ ml O$_2$ kg^{-1} min^{-1} (Bell et al. 1986), $\dot{V}_{O_2}$ = $(154 - 4.2\ T_a)$ ml O$_2$ kg^{-1} min^{-1} and $\dot{V}_{O_2}$ = $(93.3 - 2.0\ T_a)$ ml O$_2$ kg^{-1} min^{-1} (Carpenter and Graham 1967), and $\dot{V}_{O_2}$ = $(51.7 - 1.22\ T_a)$ ml O$_2$ kg^{-1} min^{-1} (Noll 1979), respectively. Their respective RMR values in thermal neutrality were 0.27, 0.61, 1.9 and 2.0 ml O_2 min^{-1}. Only the value of *L. sanborni* fits Stahl's equation (3.25; giving the values 0.42, 0.64, 1.2 and 2.7, respectively for the four species).

The regression for seven species of birds flying in wind tunnels at minimum power speed is

$$\dot{V}_{O_2, min} = 150 \, M^{0.73},$$ (3.26)

(recalculated from equation for power input — cf. Eq. (3.35) — in Fig. 2 in Butler and Woakes 1985), and for five species of bats

$$\dot{V}_{O_2, min} = 155 \, M^{0.71}$$ (3.27)

($r = 0.9898$; calculated from empirical data from Carpenter 1986 and Thomas 1975). Those values are much lower than those for hovering hummingbirds, where

$$\dot{V}_{O_2} = 857 \, M^{1.03}$$ (3.28)

($n = 3$; $r = 0.9998$; calculated from empirical data from Lasiewski 1963b, Hainsworth and Wolf 1969, Berger and Hart 1972), reflecting the higher power output required for hovering flight. The close similarity in birds (other than hummingbirds) and bats suggests convergent evolution that has led to optimization in flight physiology.

Most of the experimental results on oxygen consumption during flight fit the theoretical U-shaped power-versus-speed curve. An increase of oxygen consumption occurs at low and high speeds in the budgerigar (Tucker 1968b) and in three megachiropteran bats (Carpenter 1986), and with increasing speeds in the white-necked raven (Hudson and Bernstein 1983), in the fish crow (Bernstein et al. 1973) and laughing gull (Tucker 1972). However, no significant differences were found over a large range of flight speeds in the starling (*Sturnus vulgaris*; Torre-Bueno and Larochelle 1978).

3.4.5 Mass Loss

Animals lose mass during exercise, for different reasons. The rate of body-mass loss m_b can be calculated from the equation

$$\dot{m}_b = \dot{m}_{CO_2} - \dot{m}_{O_2} + \dot{m}_f + \dot{m}_{w1},$$ (3.29)

where $\dot{m}_{CO_2}$ is the mass loss by CO_2 release, $\dot{m}_{O_2}$ is the mass gain by O_2 consumption, $\dot{m}_f$ is the mass loss by fuel consumption, $\dot{m}_d$ is the mass loss due to defecation, and $\dot{m}_{w1}$ is mass loss of water.

The components $\dot{m}_{CO_2}$ and m_{O_2} depend on the fuel or fuel mixture used (indicated by the RQ value; see definition on p. 55) and can be calculated using the results of gas exchange measurements and the specific masses of O_2 and CO_2 (for details, see Biesel and Nachtigall 1987 and Rothe et al. 1987). The difference between $\dot{m}_{CO_2}$ and $\dot{m}_{O_2}$ is zero ($\dot{m}_{CO_2} - \dot{m}_{O_2} = 0$) at RQ = 0.723, but > 0 (mass loss) at RQ > 0.723, and < 0 (mass again) at RQ < 0.723.

The water loss $\dot{m}_{w1}$ depends on the mass of the cutaneous and respiratory evaporative water loss $\dot{m}_e$ and the mass gain by metabolic water production $\dot{m}_{wp}$, and is $\dot{m}_{w1} = \dot{m}_e - \dot{m}_{wp}$. The rate of evaporative water loss $\dot{m}_e$ depends on the ambient temperature and the flight speed (Carpenter 1969). Assuming no defecation, the total mass loss of the animal equals $\dot{m}_e$ only during pure fat combustion (RQ = 0.72) and maintained water homeostasis (i.e. metabolic water production $\dot{m}_{wp}$ = evaporative water loss $\dot{m}_{w1}$). In pigeons, the decrease of mass loss in time reflects the change-over from a carbohydrate-rich fuel to a fat-rich one (Biesel and Nachtigall 1987). Carbohydrates (e.g. glucose) are the major energy source at the beginning of flight and give 17.6 kJ and 0.6 g of metabolic water per gram oxidized. Fat (e.g. triolein) gives 39.8 kJ and 1.06 g of water g^{-1} fat oxidized.

3.4.6 Heat Loss and Exchange

Outside the thermoneutral zone, heat production for thermoregulation during rest is determined by the ambient temperature. A large part of the metabolic power input during flight is lost as heat and only 20–30% can be converted into mechanical work for flight. The heat produced is lost through the skin and by expired air, by evaporation of water and as dry heat. Some of the heat may be stored in the body tissues resulting in an increase of the body temperature, but after reaching a steady state the excess heat must be dissipated. At low ambient temperatures most of the excess heat production during flight is dissipated as conduction to the air and as heat radiation, not by evaporation, but at temperatures above 20–25°C there is a significant increase in evaporative water with increased temperature (Tucker 1968b; Berger et al. 1971; Berger and Hart 1972; Bernstein 1976; Torre-Bueno 1978). In long-distance flights, evaporation as a cooling mechanism causes dehydration when evaporative water loss exceeds metabolic water production, so migrating birds and bats should fly at nights and at altitudes with appropriate air temperatures to dissipate heat by non-evaporative routes.

The various components of heat exchange have been described in detail by Biesel and Nachtigall (1987) and are summarized in Fig. 3.2. Under steady-state conditions the heat production $\dot{H}_p$ is the sum of the evaporative heat loss $\dot{H}_e$ and the heat loss by non-evaporative means $\dot{H}_{ne}$,

$$\dot{H}_p = \dot{H}_e + \dot{H}_{ne}. \tag{3.30}$$

Heat production during flight may be computed from the measured oxygen consumption, using the energy equivalent of oxygen for the substrate metabolized and a conversion efficiency of metabolic energy to external work. Assuming an efficiency η of 25% as in mammals (Kleiber 1961), and using an average energy equivalent e of 0.335 W (20.1 J) per ml O_2 (STPD), the heat production in W can be written as

$$\dot{H}_p = (1 - \eta) \times e \times \dot{V}_{o_2} = 0.75 \times 0.335 \times \dot{V}_{o_2}. \tag{3.31}$$

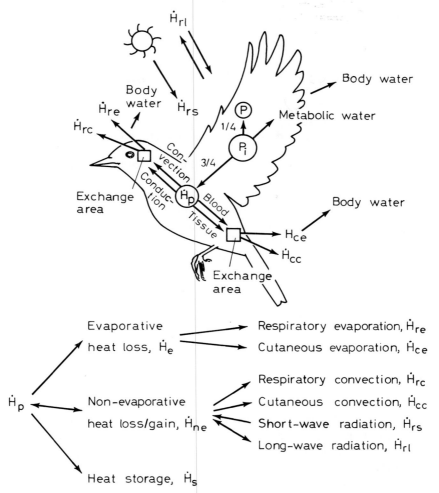

Fig. 3.2. Heat production $\dot{H}_p$ and heat exchange in a flying animal. P_i is power input and P is power output. (After Biesel and Nachtigall 1987)

Evaporative heat loss of respiratory air can be calculated from water loss assuming that 40.4 W per g of evaporated water is lost:

$$\dot{H}_e = 40.5 \text{ W g}^{-1} \text{H}_2\text{O}. \tag{3.32}$$

Knowing $\dot{H}_p$ and $\dot{H}_e$, the non-evaporative heat loss $\dot{H}_{ne}$ can be calculated from Eq. (3.30). Evaporative heat loss is not the main means of heat loss during flight in birds, but its importance grows with increasing ambient temperature, and it makes up 50% only in extreme cases.

The *thermal conductance* C characterizes the ability to emit heat (the reciprocal of thermal insulation) and is the non-evaporative heat flow Q per unit temperature difference of mean body temperature T_b and ambient temperature T_a (in W m^{-2} °C^{-1});

$$C = Q (T_b - T_a)^{-1} \tag{3.33}$$

(Kleiber 1967). The heat flow is (in W m^{-2})

$$Q = \dot{H}_{ne} \times S_{bs}, \tag{3.34}$$

where S_{bs} is the body surface area, generally taken as proportional to $M^{2/3}$. High C means high heat loss per unit body area and temperature difference (that is, bad insulation properties).

Heat loss and evaporative water loss during flight have been estimated for only a few birds (Tucker 1968b; Berger et al. 1971; Lasiewski and Calder 1971; Bernstein 1976; Berger 1978; Hudson and Bernstein 1981; Biesel and Nachtigall 1987) and bats (Carpenter 1985, 1986). In the pigeon at constant speed (V = 12 m s^{-1}), the evaporative heat loss $\dot{H}_e$ accounts for the loss of 10–29% of the heat produced, where $\dot{H}_p = 0.75 \times$ SMR, and increases with the ambient temperature (Biesel and Nachtigall 1987). At constant T_a of 15°C, $\dot{H}_e$ has a minimum at flight speeds of 10–14 m s^{-1} (about minimum power speed) but is higher at both lower and higher speeds, whereas $\dot{H}_{ne}$ and C increase with speed. The non-evaporative heat loss $\dot{H}_{ne}$ decreases with increased T_a, whereas the efficiency of $\dot{H}_{ne}$ (that is, the thermal conductance C) is raised.

3.5 Altitudinal Changes

There are only a few investigations on the physiological reactions to flight under high-altitude (reduced pressure) conditions. Migrating animals should fly at high altitudes during migration for water economy and temperature regulation, but what are the physiological responses to the reduced pressure?

The decreasing partial pressure of oxygen may reduce the capability for oxygen uptake. House sparrows (*Passer domesticus*) could not fly for more than a few seconds at simulated altitudes of 6100 m without artificial oxygen supply (Tucker 1968a). Experiments with mammals and birds exercising (not flying) on treadmills show no notable differences in oxygen uptake with declining partial oxygen pressure between the two groups (Segrem and Hart 1967; Berger and Hart 1974).

Berger (1974a,b, 1978) studied the effects of air pressure reduction (simulated altitude) in hovering hummingbirds and found a slight (6–8%) increase in oxygen consumption up to 4000 m height (0.6 atm) compared to sea level. Hyperventilation at high altitudes resulted from increases in respiratory frequency (15%; but no panting-like frequency was observed) and tidal volume (18%). Berger (1974b) concluded that the oxygen supply limits high-altitude capacities. Aerodynamically the hummingbirds can hover up to at least 8000 m, and to maintain lift at the lower air density the birds increased wingbeat amplitude, angular velocity and the geometric angle of attack.

3.6 Estimates for Cost of Flight

Flight metabolism can be deduced from oxygen consumption, carbon dioxide production and nitrogen excretion (Kleiber 1961). Nitrogen excretion has generally not been measured in studies on flying animals, but an energetic equivalent of oxygen can be estimated from the total RQ (respiratory quotient, $\dot{V}_{CO_2}/\dot{V}_{O_2}$) with negligible error (King and Farner 1961). Provided that the animal is in a steady state, the RQ value permits a reasonable estimate of the proportions of fat and carbohydrate fueling flight. During flight, however, the quotient $\dot{V}_{CO_2}/\dot{V}_{O_2}$ would increase above the resting metabolic RQ if the anaerobic threshold is exceeded, producing an increase in lactic acid and excess in CO_2 elimination. The *respiratory exchange ratio* RE is used to separate the quotient during exercise from the resting metabolic RQ, but in a steady-state system with aerobic metabolism, RE is the same as RQ. As RQ is really a metabolic term (i.e. part of the internal metabolism), the term RE is often used to describe $\dot{V}_{CO_2}/\dot{V}_{O_2}$ in the whole animal.

In a steady-state system the oxygen consumed equals the metabolic power required divided by a conversion factor k:

$$\dot{V}_{O_2} \text{ consumed } = \text{ metabolic power}/k. \tag{3.35}$$

By measuring oxygen uptake and assuming that metabolism is aerobic, it is possible to convert this to metabolic power input, P_i, which is expressed in watts with 1 ml O_2 min^{-1} = 0.335 W for STPD (standard temperature and pressure) when the RQ is 0.8. The power input, therefore, is $P_i = k\dot{V}_{O_2} = 335\dot{V}_{O_2}$.

Recorded respiratory exchange ratios (RE) in flying birds range between 0.66 and 0.85 (Tucker 1968b, 1972; Berger and Hart 1972; Torre-Bueno and Larochelle 1978), and in bats between 0.77 and 0.9 (Thomas 1975; Carpenter 1975, 1985, 1986). The values are generally higher at the beginning of a flight and in the first flights of consecutive experiments (see, for example, Butler et al. 1977 and Rothe et al. 1987).

Flight costs also have been calculated from fat loss in migrating chaffinches (*Fringilla coelebs*; Dolnik et al. 1963) and from fat loss and CO_2 production in homing pigeons (LeFebvre 1964) and with the doubly labelled water method in purple martins (Utter and LeFebvre 1970, 1973).

Most estimates on flight metabolism usually refer to measurements of the total amount of energy consumed, while the animal performed short flights at various speeds and made occasional manoeuvres and rests. The calculation of power output from such results involves many assumptions under the most favourable circumstances, and such measurements cannot be used for comparison with predictions made from any mechanical theory (Pennycuick 1989).

3.6.1 Direct Measurements of O_2 Uptake and CO_2 Production

Pearson (1950) and Lasiewski (1962, 1963a,b) estimated the rate of oxygen consumption in hovering hummingbirds by measuring the partial pressure of oxygen in a closed system. Hainsworth and Wolf (1969) and Berger and Hart

(1972) used an open flow system where the hummingbirds drank from a small funnel and the expired air was collected and analyzed. Head masks have been used for the collection of expired air by sucking air through the mask, past the animal's head, at a volume per unit time large enough to prevent expired air from escaping from the mask; data have been collected from a number of birds and bats flying in free air in flight cages or wind tunnels (birds: Tucker 1966, 1968a,b, 1969, 1972; Berger et al. 1970b; Bernstein et al. 1973; Bernstein 1976; Butler et al. 1977; Nachtigall and Rothe 1978; Torre-Bueno and Larochelle 1978; Hudson and Bernstein 1983; Rothe et al. 1987; bats: Thomas and Suthers 1970, 1972; Carpenter 1975, 1985, 1986; Thomas 1975). The basic methods used for power consumption were outlined in the previous sections.

Although the animals are flying under controlled conditions, wind tunnels may not be ideal because of the restricted space, the loud noise from the motor and fan, and the stationary surroundings. The flying speeds are also usually slower than in free flights (Rothe and Nachtigall 1980) and the flight patterns sometimes different (Butler et al. 1977). In spite of this and corrections for the aerodynamic drag of the masks, trailing tubes and leads that are attached to the animals, wind tunnels are the only practical way of simultaneously studying changes in several physiological and flight parameters.

In most species the power input increases with flight speed above the minimum power speed, and with the flight angle when it has been changed. In several cases the power input versus speed curve is U-shaped (Fig. 3.3), as predicted in theoretical versions (Pennycuick 1968a, 1975; Rayner 1979c).

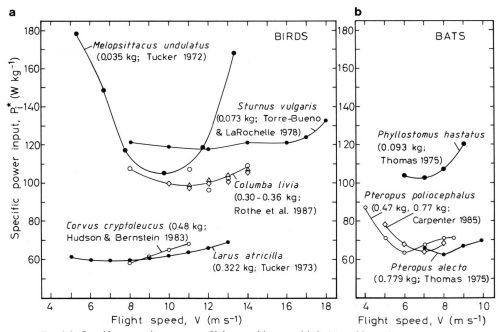

Fig. 3.3. Specific power input versus flight speed in some birds (*a*) and bats (*b*)

3.6.2 Mass Loss

Birds and bats feeding in flight may gain weight during long flights, but many migratory birds lose considerable amounts of weight during long-distant migrations, much of it by evaporative water loss. Mass loss equals evaporative water loss only during pure fat combustion (RQ = 0.72) and maintained water homeostasis (Sect. 3.4.5), assuming that no feeding or defecation have occurred during free flight. In these cases the mass loss equals the mass of fat consumed.

Nisbet et al. (1963) used loss of weight obtained by mean values from different species to estimate the metabolic rate during a long flight in the blackpoll warbler (*Dendroica striata*). They assumed that fat constituted the major part of this weight loss, and used the caloric value of fat (39.8 kJ g^{-1}), and assumed that the evaporation of the metabolic water required about 2.5 kJ g^{-1} (giving the net energy gain 37.3 kJ g^{-1} to the bird). They calculated the power consumption as approximately 1.16 W, which was only about 40% of that predicted by allometric formulae (Butler 1985). Weight-loss estimates for barn swallows (*Hirundo rustica*) and house martins (*Delichon urbica*) (Lyuleeva 1970) were about 50% of the estimates derived from oxygen consumption measurements in flight (Berger and Hart 1974), while those for other birds during migration (Pearson 1964; Raveling and LeFebvre 1967; Dolnik and Blyumental 1967; Hussell 1969) agreed well with values derived from oxygen consumption (Hart and Berger 1972). The low weight loss in martins and swifts may be associated with their gliding habits and/or feeding during flight. Because water loss is strongly dependent on ambient temperature while total energy output is not, the water loss and so the mass loss may greatly exceed the production of metabolic water, particularly at high temperatures (Berger and Hart 1974). This makes mass loss a questionable way of determining power input (Berger and Hart 1974), particularly for short flights when change-over to fat has not yet taken place, and at high ambient temperatures when the dependence of evaporative water loss on temperature is significant.

3.6.3 Doubly Labelled Water Method

An animal's total energy expenditure over a given time period (e.g. a day) can be estimated by the doubly labelled water method (DLW), which was developed by LeFebvre (1964) and Lifson and McClintock (1966). When the time budget for various activities is continuously recorded in detail by direct observation, by biotelemetry, or otherwise, the DLW method can be used to separate the metabolic costs of the various activities by multiple regression.

The procedures of the method are described in detail by, for example, Bell et al. (1986) and Kunz and Nagy (1987). The method relies on the fact that the oxygen in respired CO_2 is in isotopic equilibrium with the oxygen in the body water, via action of carbon anhydrase in blood (Lifson et al. 1949). An animal is given intraperitoneal injection of isotopes of hydrogen and oxygen in the form of water (D_2O^{18} or H_2O^{18}), and after isotope equilibration in the body (30–35 min for bats weighing 7–90 g, but longer when animals are dehydrated, Richmond et al. 1962), a reference blood sample is taken from a vein, for example, in the leg (birds) or tail membrane (bats). A second blood sample is taken after the period of activity for which the energy expenditure should be measured (hours or days). Speakman and Racey (1988) pointed out the importance of recapture intervals being as close as possible to 24 h (or a multiple thereof). When estimates made over periods deviating from 24 h but corrected to a daily, or hourly rate, and when CO_2 production is temporally variable, an error is introduced. The error varies between 2 and 7% for each hour the inter-sample period deviates from 24 h.

The hydrogen isotope is lost from the body as water and the oxygen isotope is lost as water and as respiratory CO_2. The CO_2 production is proportional to the difference between the oxygen turnover in body water and the hydrogen turnover, which is a measure of the metabolic rate. The CO_2 production (mM) can be calculated from the formula

$$CO_2 = (N/2.08)(K_O - K_D) - 0.015\ K_D N, \tag{3.36}$$

where K_O and K_D are the fractional turnover rates of ^{18}O and D, respectively, and N is the mean body water content (mM) (for details, see Lifson and McClintock 1966; Nagy 1975, 1980).

Comparisons with CO_2 production measured directly with respirometers showed an overestimation of 3–4% with the doubly labelled water method for the pigeon (LeFebvre 1964) and house martin (Hails 1979). LeFebvre used the DLW method on pigeons flying at least 480 km and the value on power input obtained (25.6 W, or about 64 W kg^{-1}) was similar to the estimates from fat loss, and also to direct measurements of V_{O_2} in a wind tunnel (Butler et al. 1977). Estimated CO_2 production of resting house martins (17 W kg^{-1}; Hails and Bryant 1979) fits the equation of Aschoff and Pohl (1970) for passerines (cf. Sect. 3.3). Tatner and Bryant (1986) estimated the flight costs for robins (*Erithacus rubecula*) with the DLW method and obtained a value of 7.1 W (or 381 W kg^{-1}), which is about twice that predicted by various allometric equations for flight cost versus body mass. The flights were, however, short and brief and more costly because of the large numbers of accelerations, decelerations and, possibly, manoeuvres. The value obtained better fit the equation of Teal (1969) for the cost of brief flights in small birds based on direct measurements of CO_2.

Costa and Prince (1987) measured the flight costs of six grey-headed albatrosses (*Diomedea chrysostoma*) and obtained the lowest cost of flight yet measured, 36.3 W or 9.8 W kg^{-1}, which is 3.2 times the predicted BMR. The low flight cost is consistent with the economic dynamic-soaring flight of albatrosses (see Sect. 6.2.4).

3.6.4 Radio Telemetry

Short-range radio transmitters have been used to record oxygen uptake and respiratory variables in flying birds. Hart and Roy (1966) mounted head-masks and transmitters on flying pigeons and recorded a large amount of respiratory data of the birds just after take-offs, when values generally were higher than during steady-state flights. Kanwisher et al. (1978) used a long-range (80 km) transmitter attached externally to herring gulls (*Larus argentatus*) to record heart rates during longer flights.

One must use other devices to avoid the effects of head-masks and other apparatuses for physiological estimates. Torre-Bueno (1976) implanted a small temperature-sensitive transmitter into starlings and measured their temperatures when flying in a wind tunnel. Butler and Woakes (1980) used a two-channel transmitter in free-flying barnacle geese (*Branta leucopsis*) to record heart rate and respiratory frequency before, during, and after flights of relatively long duration. But to obtain a greater number of physiological variables larger multichannel transmitters would be necessary.

3.7 Flight Duration, Flight Range and Cost of Transport

Berger and Hart (1974) derived allometric expressions for flight duration and flight range from physiological measurements. The maximum continuous flight power can be given by $\dot{V}_{O_2} = 146\ M^{0.72}$, which can be transformed into watts as

$$P_i = 48.8\ M^{0.72}, \tag{3.37}$$

but it is likely that this equation overestimates energy expenditure on long flights.

Fat is assumed to constitute the most important and concentrated energy reserve for migratory animals, and in small migratory birds fat content may be 30–50% of total body mass. The potential flight duration t (in hours) is given as the ratio of available energy E and energy expenditure P_i, and is given by

$$t = E/P_i = 367\ M_1^{0.28}f/(2-f)^{0.72} \approx 249\ fM^{0.28}, \tag{3.38}$$

where M_1 is the initial body mass and f the initial fractional fat content. The flight range Y is determined as t × flight speed V,

$$Y = tV = EV/P_i \tag{3.39}$$

[cf. Eq. (7.2)]. Given a fat content of 40%, a 4-g bird could fly for 21 h and a 40-g bird for 37 h. A flight time of 37 h was recorded for the 150-g lesser golden plover (*Pluvialis dominica*) (25% fat; Johnston and McFarlane 1967), which is 20 h less than the maximum possible time predicted from Eq. (3.38). Assuming a reasonable flight speed of 14 m s^{-1}, a 40-g golden plover with a 40% fat content would be able to fly 1900 km.

Pennycuick (1975) gave more complicated theoretical models for estimation of the flight range and flight duration, which take into account the weight loss during a long flight as fat reserves are mobilized. Weight decrease also leads to a lower maximum range speed, and both factors can increase the maximal flight distance (see Sect. 7.3 for theoretical grounds).

Observations by Smith et al. (1986) on migrating Swainson's hawks (*Buteo swainsoni*) and broad-winged hawks (*B. platypterus*) indicated that their migration from southern US to northern South America could be accomplished without feeding. Smith et al. collected migratory hawks from their night roosts in Panama and noted no feces or pellets or food in the birds' guts. Based on the assumptions that the birds use their fat reserves, rely mainly on soaring flight during migration (10 h day^{-1}), roost at nights, and that soaring costs 2 × daily BMR and roosting 0.8 × daily BMR, the authors modelled the maximum flight ranges for the hawks. They concluded that a 0.6-kg hawk could deposit 0.2 kg of fat, and at a flight speed of 240 km day^{-1} the maximum flight range would be about 5800 km. The assumption that soaring would cost 2 × BMR comes from values for gliding in the herring gull (*Larus atricilla*), which cost 2.17 × BMR (Baudinette and Schmidt-Nielsen 1974). But the assumed soaring costs are very low compared to albatrosses in dynamic soaring (3.2 × BMR) estimated by the DLW method (Costa and Prince 1987).

The cost of transport is the metabolic energy cost of moving an animal's weight a given distance, $C = P_i/MgV$, and is dimensionless. The minimum cost of transport should be attained at the maximum range speed. The relation between cost of transport and body mass in insects and birds was calculated by Tucker (1970, 1973) as

$$P_i/MgV = 0.896 \, Mg^{-0.227}. \tag{3.40}$$

Carpenter (1985, 1986) measured the endurance of four species of fruit bats and found that, as expected, flight times were significantly longest at the airspeed of minimum power input (V_{mp}).

60

Chapter 4

Morphological Flight Parameters

4.1 Introduction

Wing shapes vary among flying animals, and while some species have broad
wings of large areas, others have narrow wings of high aspect ratio, some with
rounded wingtips, others with more pointed ones. The geometry of a wing
determines local vortex strength and flight manoeuvrability. This chapter
reviews a number of morphological characters useful for flight analyses.

Morphological parameters describing the morphology and shape of a flying
animal can be divided into two groups (Ellington 1984b), the first one including
the primary measures, such as wing lengths, wing areas, body and wing masses.
These are used to derive *aspect ratio* and *wing loading*, powerful measures for
interpreting the adaptive function of the overall wing morphology, but which do
not describe the shape and size of the component portions of the wing, particularly the wingtip. Also included in the first group is the virtual wing mass, or
added mass, that is the mass of the air accelerated and decelerated together with
the wing at the top and bottom of the wingstroke. The second group of
morphological parameters provides a second-order description of the morphology that characterizes the overall shape of the distribution of a quantity.

4.2 Lengths, Areas, Masses

Useful linear measures of wings and their components include wingspan, wing
length, lengths of arm wing and hand wing, all expressed in metres (m). The
wingspan, b, is the distance between the wingtips of an animal with extended
wings and their leading edges held along a straight line (normal to the long axis
of the body; Fig. 4.1). The *wing length*, l_w, is the distance from the shoulder joint
to the wingtip of the extended wing, the *arm wing length*, l_{aw}, the distance from
the shoulder joint to the wrist, and the *hand wing length*, l_{hw}, the distance from the
wrist to the wingtip.

The total *wing area*, S, should be measured as the area of both wings and the
intercepted body area expressed in square metres (m²). In bats, the tail membrane, *uropatagium*, should be included as well, for it is moved up and down with
the wings and contributes some useful forces during flight. The area of one wing,
S_w, is measured from the body and hind leg (in bats) and does not include any part
of the trunk or uropatagium. The division line between the *arm wing area*, S_{aw},

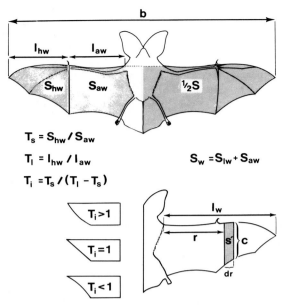

Fig. 4.1. Definitions of morphological quantities to describe the bat wing, most of which are used also for the bird wing. The wingspan, b, is measured from tip to tip of the extended wings; S is wing area, including the area of the body in between the wings in birds and bats and the tail membrane in bats (when present) but excluding the projected area of the head. S_{hw} is the area of the handwing, that is, the area distal to the fifth digit in bats and distal to the wrist and including the primary feathers in birds. S_{aw} is the area of the armwing, that is, the area between the fifth digit, the body and legs in bats and the area between body and wrist including the secondary feathers in birds. l_{hw} and l_{aw} are the corresponding lengths, their sum being equal to wing length l_w. l_{hw} and l_{aw} are used to define the tip length and tip area ratios, T_l and T_S, and the wingtip shape index, T_i. S' is the area of a wing strip with chord c and at distance r from the wing base. (The *upper figure* is from Norberg and Rayner 1987; by courtesy of The Royal Society)

and the *hand wing area*, S_{hw}, lies between the primary and secondary feathers in birds and along the fifth digit of the extended wing in bats.

The overall size of the animal is measured by the total mass, or *body mass*, M, expressed in kilograms (kg). Weight, Mg (mass times gravitational acceleration g = 9.81 m s^{-2}), is given in Newtons (N).

The *wing loading* Mg/S (given in N m^{-2}) is the weight divided by the wing area, and the mean pressure on the wings is dependent on the wing loading. Since the aerodynamic force elicited by an aerofoil varies as the square of the velocity, any characteristic flight speed (defined as any characteristic point on the power-versus-speed curve, such as the minimum power speed and maximum range speed) is proportional to the square root of the wing loading, which cannot be used as a measure of wing shape. But the wing shape may also affect any characteristic flight speed.

4.3 Wing Shape

The overall shape of the wing can be described by the *aspect ratio*, AR, a measure of the wingspan divided by the mean wing chord, c, AR = b/c, and an indication of the narrowness of the wing. The taper of most animal wings makes it difficult to calculate the mean wing chord, but by a simple mathematical manipulation, the mean chord can be circumvented. Since wing area S equals bc, then the aspect ratio can be written as

$$AR = (b/b) \times (b/c) = b^2/S. \tag{4.1}$$

Higher aspect ratios mean greater aerodynamic efficiency and lower energy losses in flight, particularly at low flight speeds.

Wings with rounded tips tend to have low aspect ratios, but shape variation cannot be expressed by aspect ratio alone. Norberg and Rayner (1987) introduced three wingtip indices to describe the shape of the wingtip (Fig. 4.1); the *tip length ratio*, T_1, which is the ratio of the length of the hand wing l_{hw} to the length of the arm wing l_{aw},

$$T_1 = l_{hw}/l_{aw}, \tag{4.2}$$

the *tip area ratio*, T_S, which is the ratio of the hand wing area S_{hw} to the arm wing area S_{aw},

$$T_S = S_{hw}/S_{aw}, \tag{4.3}$$

and the *tip shape index*, T_i, which is determined by the relative size of the hand and arm wings and so is a measure of wingtip angle and hence of wingtip shape independent of the extent of the hand wing. The tip shape index is given by

$$T_i = T_S/(T_1 - T_S). \tag{4.4}$$

High tip shape indices indicate rounded wingtips and low values indicate pointed tips. According to the model a hypothetical value of infinity would indicate a rectangular wing and when $T_i = 1$ the wingtips would be triangular. Wingtip shapes for T_i values smaller and larger than 1 are shown in Fig. 4.1.

4.4 Weis-Fogh's and Ellington's Shape Parameters

Area, mass, and virtual mass of the wing can be considered as variables defined over the wing length. The moments about the wing hinge (wing base) of these characters can be used as parameters of the distribution of these variables along the wing (Weis-Fogh 1973; Ellington 1984b). For a given wing length and area, the moments of area depend only on the shape of the wing. Let S' = c dr be the area of a wing strip at distance r from the wing base (Fig. 4.1). The total area of the wing of length l_w then is $S' = \int_0^{l_w} c\, dr$, and the mean chord is $\bar{c} = S_w/l_w$.

Ellington (1984b) introduced non-dimensional, normalized, chord $\hat{c}$ and radius $\hat{r}$, which are defined as $\hat{c} = c/\bar{c} = l_w c/S_w$ and $\hat{r} = r/l_w$, or $c = S_w \hat{c}/l_w$ and $r = l_w \hat{r}$. The kth moment of wing area then is defined by

$$S_{w,k} = \int_0^{l_w} c \, r^k \, dr = S_w \, l_w^{\ k} \int_0^1 \hat{c} \, \hat{r}^k \, d\hat{r}. \tag{4.5}$$

Weis-Fogh (1973) showed that, for hovering flight, the mean lift force is proportional to *the second moment of area* and the aerodynamic torque and mean profile power are proportional to the *third moment of area*.

If the mass per unit wing length is m_w', and $m_w' dr$ is the mass of a wing element, then the normalized mass per unit length is $\hat{m}_w' = m_w' l_w / m_w$, where m_w is the mass of one wing. The kth moment of wing mass is then defined by

$$m_{w,k} = \int_0^{l_w} m_w' r^k \, dr = m_w \, l_w^{\ k} \int_0^1 \hat{m}_w' \, \hat{r}^k \, d\hat{r}. \tag{4.6}$$

The *first moment of mass* is proportional to the resultant inertial force on the flapping wing, and *the second moment of mass* is equal to the moment of inertia of the wing [Eq. (2.23)] and is proportional to the inertial torque of the flapping wing.

In the same way, the kth moment of wing virtual mass is defined by

$$m_{v,k} = \int_0^{l_w} m_v' r^k \, dr = m_v \, l_w^{\ k} \int_0^1 \hat{r}^k \, d\hat{r} = (\rho \pi c^2 l_w / 4) \int_0^1 \hat{r}^k \, d\hat{r}. \tag{4.7}$$

So the shape of the wing and the spanwise mass distribution determine the magnitude of the moment of inertia, whereas the shape of the wing determines the wing virtual mass. An increase of the chord near the wingtip (the tip being more rounded) will cause an increase of the virtual mass with ensuing increase of the inertial forces.

Chapter 5

Gliding Flight

5.1 Introduction

Gliding flight is very cheap compared with active flight. Physiological measurements indicate that gliding in the herring gull (*Larus argentatus*) costs only 2.17 times BMR (Baudinette and Schmidt-Nielsen 1974), for the flight muscles do not perform any mechanical work but produce only static forces to keep the wings down on the horizontal plane, opposing the aerodynamic force. The gliding performance, aerodynamics, and stability and control of movements will be treated in this chapter. Gliding is the main component in soaring flight, which will be treated in the next chapter.

5.2 Gliding Performance

When gliding, the animal's wings leave behind a continuous vortex sheet that rolls up into a pair of vortex tubes (wingtip vortices), as in fixed-wing airplanes. This effect has been demonstrated in a gliding kestrel (*Falco tinnunculus*) by using helium bubbles (Spedding 1987a). Since a gliding animal does not produce any mechanical energy, none is transferred from the animal to its environment. The lift force produced approximately balances the weight of the animal, but potential energy must be used to overcome the total drag.

Stationary aerodynamic theory is appropriate for explaining passive flight (gliding and parachuting). Therefore, an animal gliding at steady speed descends at an angle θ to the horizontal (Fig. 5.1a), and the lift:drag ratio, L/D, establishes the gliding angle, θ ($L/D = 1/\tan\theta$), and is called the *glide ratio*. The transition between gliding and parachuting is conventionally taken to occur at the 45° inclination of the glide path, where lift and drag contribute equally to weight support. At shallower paths $L/D > 1$ an animal glides, and at steeper paths $L/D < 1$ it parachutes (Fig. 5.1b).

When an animal begins to glide, either after flapping or after taking off from some elevation, the glide path steepens first and the speed increases to the equilibrium speed (when this has not been attained by flapping). The glide path flattens as the equilibrium speed is approached until the force F (now vertical) becomes equal to the weight, Mg. The resultant force, F, can be resolved into a drag component, D, backward along the glide path, and a lift component, L, perpendicular to it. In equilibrium (stable) gliding, the lift and drag equal

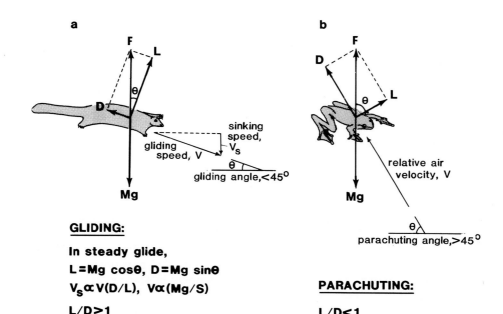

GLIDING:

In steady glide,

L=Mg cosθ, D=Mg sinθ

$V_s \propto V(D/L)$, $V \propto (Mg/S)$

L/D>1

PARACHUTING:

L/D<1

Fig. 5.1. Aerodynamics of *a* gliding flight and *b* parachuting. In parachuting the relationships between the lift:drag ratio and the equilibrium angle of descent are as for gliding, but because the area supporting weight is so small, lift becomes small in relation to the combined drag of the body and the gliding surfaces. Therefore, the overall lift to drag ratio, *L/D*, is lower, resulting in a correspondingly steeper glide path. (Norberg 1985a; by courtesy of Harvard University Press)

$$L = F\cos\theta = Mg\cos\theta, \text{ and} \qquad (5.1)$$
$$D = F\sin\theta = Mg\sin\theta. \qquad (5.2)$$

The total drag D is the sum of the wing drag D_w (induced and profile drags) and the body drag D_b (parasite drag). The L/D ratio for the wings alone is better than that determining the glide angle θ, because the wings' L/D ratio depends on the relative magnitudes of the wing and body drag. Body drag is proportional to the speed squared and makes up 12%–20% of the total drag at minimum power speed in modern birds (Pennycuick 1975; Rayner 1979c). In equilibrium gliding, the horizontal component of the body drag $D_{b,h}$ equals the horizontal thrust component T of the wings' resultant force R_w.

In steady gliding at small angles and in level powered flight with no vertical acceleration, the lift force as averaged over a whole wing and wing stroke equals the weight, $L \approx F = Mg$. Insertion of $L = Mg$ in Eq. (2.8) gives the animal's speed, or gliding speed,

$$V_g = (2Mg/\rho SC_L)^{1/2}. \qquad (5.3)$$

So, with higher lift coefficients you have lower gliding speeds. The minimum gliding speed, $V_{g,min}$, occurs at the maximum obtainable lift coefficient, $C_{L,max}$, which means that the minimum stalling speed is approximately

$$V_{g,min} = (2Mg/\rho SC_{L,max})^{1/2}.$$ (5.4)

But the minimum gliding speed also depends on the wing loading. Maximum lift coefficients of 1.5 to 1.6 have been reported from gliding birds and 1.5 from a gliding bat (Pennycuick 1971c, 1972a; Tucker and Parrott 1970). Values of 1.5 have been estimated for a gliding pterosaur (*Pteranodon*, Bramwell 1971).

An animal cannot glide more slowly than the stalling speed that can be reduced by increasing the wing area, which improves the aerofoil planform and profile, or by delaying stall by other structural means. Splaying of the primaries ("wingtip slots", Sect. 11.5.4) of vultures is thought to increase the maximum lift coefficient in gliding (Pennycuick 1971a). Tucker (1988) calculated the maximum lift coefficients to be 2.2 in landing white-backed vultures [*Pseudogyps (Gyps) africanus*], which probably use all their morphological high-lift devices for slow descents without stalling. High gliding speeds can be obtained by flexion of the wings, which reduces the wing area and increases the wing loading.

During stable gliding at speed V_g the animal loses height at a vertical sinking speed (or *rate of sink*) V_s

$$V_s = V_g \sin\theta.$$ (5.5)

From Eq. (2.17) it follows that $D = Mg \sin\theta = 2k \ Mg^2/\rho V_g^2 \pi b^2 + (1/2) \rho V_g^2 SC_{Df}$, where k is the induced drag factor and $C_{Df} = C_{Dpro} + C_{Dpar}$. Multiplying both sides by V_g and dividing by Mg gives

$$V_g \sin\theta = 2k \ Mg/\rho V_g \pi b^2 + \rho V_g^3 SC_{Df}/2 \ Mg,$$ (5.6)

an alternative expression for the sinking speed V_s. If V_g increases, one term on the right hand side of the equation (the induced drag component) decreases and the second term (form and friction drag component) increases. But as a gliding bird increases its speed by reducing its wingspan and wing area, it increases the induced drag and simultaneously lowers the form and friction drag. On the other hand, when speed increases, the induced drag decreases and the form and friction drag become more important. Minimum sinking speed is obtained at an intermediate, optimum airspeed, which is somewhat higher than the minimum speed. An increase of aspect ratio tends to reduce the sinking speed, but the optimum (with respect to minimum sinking speed) wing area ($S \propto b^2$) is larger at low speeds (= low induced drag) and smaller at high speeds (= large form and friction drag). An increase of aspect ratio usually will increase the overall lift coefficient because the gliding animal will then consist of relatively more aerofoil and less body (with poor lift capacity), resulting in larger potential speed reduction. At low and medium speeds an increase of span increases lift more than drag because of the increase in overall lift coefficient and the overriding importance of induced drag at these speeds. This produces an increase of L/D and a flattening of the glide path.

The glide polar is obtained by plotting sinking speed against forward speed and it can be used to summarize the gliding performance of a gliding animal. From Eq. (5.6) the glide polar can be written as a regression equation (Welch et al.) in the form

$$V_s = a_1/V_g + a_2V_g^3,$$ (5.7)

where $a_1 = 2k\,Mg/\pi\rho b^2$ and $a_2 = \rho S C_{Df}/2Mg$. If a_1 and a_2 can be determined empirically, k and C_{Df} are obtained from Eq. (5.6) and (5.7) as

$$k = a_1 \pi \rho b^2/2\,Mg \text{ and} \tag{5.8}$$
$$C_{Df} = 2\,Mg\,a_2/\rho S. \tag{5.9}$$

Assuming that $k = 1$ (as it is when there is an elliptical spanwise lift distribution), a_1 can be calculated from Eq. 5.8. The curve of glide polars for some gliding animals and a sailplane (Fig. 5.2a) extends down to the minimum gliding speed $V_{g,\,min}$ [Eq. (5.4)]. The speed for minimum sink, V_{ms}, corresponds to the minimum power speed, while the speed V_{bg} for best glide ratio corresponds to the

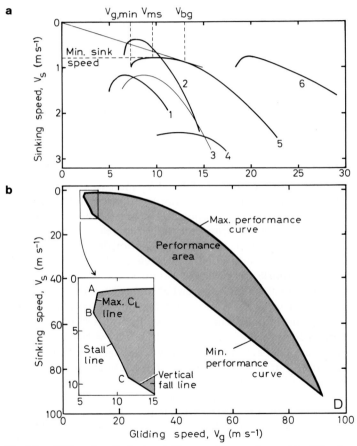

Fig. 5.2. *a* Glide polars displaying sinking speed versus air speed in gliding for some animals and for a motor glider. *1* Dog-faced bat *Rousettus aegyptiacus* (Pennycuick 1971c); *2* pterosaur *Pteranodon ingens* (Bramwell 1971); *3* fulmar petrel *Fulmarius glacialis* (Pennycuick 1960); *4* pigeon *Columba livia* (Tucker and Parrott 1970); *5* white-backed vulture *Gyps africanus* (Pennycuick 1971a); *6* motor glider ASK-14 (Pennycuick 1971a). $V_{g,\,min}$ is minimum gliding speed, V_{ms} is minimum sink speed, V_{bg} is speed for best glide ratio. *b* Sinking speed versus gliding speed with the performance area of a hypothetical gliding bird. See text for explanation. (Slightly modified from Tucker 1988)

maximum range speed. The speed for best glide ratio can be found by drawing a tangent to the curve from the origin of the graph.

The glide angle θ depends upon the geometry of the wing, and because the induced drag D_{ind} of the wings decreases with increased span [cf. Eqs. (2.9)–(2.11)], wings having high aspect ratio have high L/D ratios, and allow shallow glide angles. The best glide ratios (which equal L/D ratios) range from 10:1 to 15:1 for vultures and birds of prey and reach 23:1 in the wandering albatross (Tucker and Parrott 1970; Pennycuick 1972a, 1982). The best ratio for the pterosaur *Pteranodon* was estimated to be 13:1 (Bramwell 1971). Modern gliders can achieve much higher ratios, for they can exceed 45:1 and correspond to a glide angle of 1.3°. Mean lift coefficients in straight glides of the magnificent frigatebird (*Fregata magnificans*), the brown pelican (*Pelecanus occidentalis*) and the black vulture (*Coragyps atratus*) range from 0.72 to 0.84 (Pennycuick 1983).

Many birds, especially large ones, are very good gliders and soarers (see Chap. 6). It follows from Eq. (5.4) that the gliding speed is proportional to the square root of the wing loading Mg/S, $V_g \propto (Mg/S)^{1/2}$, so that those with low wing loadings (large relative wing areas) can glide slowly with low sinking speed. Most of these birds possess separated primary feathers (wingtip slots) whose probable function is to reduce induced drag and increase lift, thereby increasing the overall L/D and reducing the glide angle with little change in the glide speed (see Sect. 11.5.4).

Gliding mammals range in size from 10 to 1500 g (as most bats), have low aspect ratios (1 to 1.5) and wing loadings similar to those of birds (Rayner 1981; Fig. 10.3). The marsupial *Petaurus breviceps* shows excellent control of gliding flight and use glide angles of 11 to 27° (glide ratios of 5:1 to 2:1) (Nachtigall 1979). Scott and Starrett (1974) observed that the glide angles of various flying frogs usually were around 30°, although *Hyla miliaria*, with a membranous flange along the hind edge of each limb, can glide at an angle of about 18° (Duellman 1970). Gliding rodents have cartilaginous spines that extend into the aerofoil. In sciurids these supports come from the wrist, and from the elbow in anomalurids, and in either case it stretches the membrane and somewhat increases the span. In flying geckos (*Ptychozoon*) broad flaps of skin on either side of the body are spread out by the limbs during flight, and they also have small flaps on either side of the neck and tail, webbed feet, and flattened body and tail (Fig. 1.1). The angles of descent of the geckoes are 45° or steeper (Russell 1979), making them parachutists and not gliders. A colubrid snake of genus *Chrysopelea* from Borneo has a glide angle of about 30°, and during gliding its body is held rigid in coils forming a composite triangle-shaped aerofoil. The transverse coils are close to one another, and Rayner (1981) suggested that this configuration could operate as some kind of "slotted delta wing". Bats *can* glide, but they rarely do, perhaps because they are not as able as birds to vary wing area and control gliding.

5.3 Effects of Change in Wingspan on Gliding Performance

In theoretical studies, Tucker (1987, 1988) demonstrated how changes in wingspan affect gliding performance in a descending bird, and he defined a "performance area" in a diagram where sinking speed V_s is plotted against gliding speed V_g (Fig. 5.2b). The lowest V_s in this area defines a "maximum performance curve" (curve AD), which is obtained if the bird changes the wingspan, wing area and profile drag coefficient to minimize profile drag. A bird gliding on this curve descends as slowly as possible for a given gliding speed, and covers the maximum distance.

The "minimum performance curve" (curve BCD) describes the maximum V_s at each V_g. Along this curve the bird maximizes drag at each speed and descends through the air as rapidly as possible. In the low-speed part of this curve (BC), which Tucker calls the "stall line", the bird has fully spread, stalled wings and descends at a steep angle but cannot generate enough drag to equal its weight at these low gliding speeds. At point C, wing area and profile drag coefficient have their maximum values and the bird parachutes, but at higher speeds (along curve CD = "vertical fall line") the bird progressively folds its wings completely and descends vertically, and $V_s = V_g$ and $D = Mg$. At point D, wing area and wing profile drag become zero and the bird experiences only parasite drag in a head-first dive.

Along the transition curve (AB) between the maximum and minimum performance curves the bird's wings have their maximum lift coefficient ("maximum lift coefficient line"). The bird moves from A to B by increasing drag; lowering of feet and tail increases the parasite drag, and twisting of the wings increases the induced drag factor k (Pennycuick 1971b; Tucker 1987).

White-backed vultures were observed to continually adjust their body configurations as they moved between these points during descents at landings. Tucker (1988) found that most observations of gliding vultures fall within the boundaries of the theoretical performance area.

5.4 Lifting Line Theory

The aerodynamics of gliding can be explained by lifting-line theory, which assumes that the wings operate in the same fashion as conventional aerofoils. In all types of flight there are various degrees of unsteadiness, which are typically measured by the *reduced frequency*, the ratio of flapping velocity to the mean flight velocity (see also Sect. 8.1). In gliding flight the reduced frequency is zero, which means that unsteady effects are absent or of minor importance.

When an elliptically loaded wing is replaced by a lifting line with circulation as in a horseshoe system, the circulation can be expressed as $\Gamma = L/\rho bV$ [cf. Eq. (2.40)]. Since the spanwise spacing a between the two trailing vortices is $\pi/4$ times the wingspan (see Sect. 2.4.1) and because the lift is assumed to balance the weight of the animal, the circulation can be written as

$$\Gamma = 4Mg/\rho b\pi V. \tag{5.10}$$

Spedding (1987a) used this equation to calculate the circulation of a 0.21 kg kestrel gliding at 7 m s^{-1} to be 0.476 m^{2}s^{-1} and his measured value from estimations on wake velocities was 0.496 m^{2}s^{-1}, which is close to the theoretical value.

5.5 Flap-Gliding

Among seabirds, albatrosses (Diomedeidae), stormpetrels (Hydrobatidae) and the gannet (Sulidae) characteristically use flap-gliding in cruising flight (see dynamic soaring, Sect. 6.2.4), by alternating short glides with short bursts of flapping. Diving petrels (Procellariidae) and most other seabirds flap most or all of the time; however, a bird that flap-glides must flap efficiently to produce the speed near that for the best glide ratio in gliding flight, for otherwise gliding would be inefficient (Pennycuick 1987a).

Pennycuick made an extensive analysis of the flight mode, wing morphology and aerodynamics of seabirds. To investigate the flap-gliding he tried to determine the value of the lift coefficient where the gliding speed would be equal to the minimum power speed in powered flight. The minimum power speed can be expressed as

$$V_{mp} = 0.76k^{1/4}(Mg)^{1/2}\rho^{-1/2}A_e^{1/4}S_d^{1/4} \tag{5.11}$$

(Pennycuick 1975), where A_e is the equivalent flat area of the body (Sect. 9.5.3) and S_d the wing disk area ($S_d = \pi b^2/4$). Assuming that the induced drag factor is $k = 1.2$ and body drag is 0.25, Eq. (5.11) can be expressed as $V_{mp} = 4.19$ $M^{1/3}g^{1/2}\rho^{1/2}b^{-1/2}$ which should equal the gliding speed expressed in Eq. (5.3). The coefficient of lift at the minimum power speed can then be solved for and equals

$$C_{L1} = 0.114 \, M^{1/3}bS^{-1}. \tag{5.12}$$

When b is proportional to $M^{1/3}$ and S to $M^{2/3}$ as for geometric similarity, C_{L1} becomes proportional to M^0, and so would be independent of body mass. But if isometry did not prevail, the correlation could be different. Pennycuick (1987b) plotted the wingspan and wing area of procellariiform seabirds and for 22 species the wingspan varied as the 0.39 power of body mass; for 47 species S varied as the 0.63 power of the mass. With these correlations the lift coefficient would vary as $C_{L1} \propto M^{0.09}$.

The lift coefficient for Procellariiformes during the gliding phase might be expected to vary with the 0.09 power of the body mass, but this was not so. In eight species the lift coefficient varied with the 0.24 power of the body mass (Pennycuick 1987a). Furthermore, plotting the lift coefficient corresponding to mean airspeed in flapping and flap-gliding and the lift coefficient for gliding at

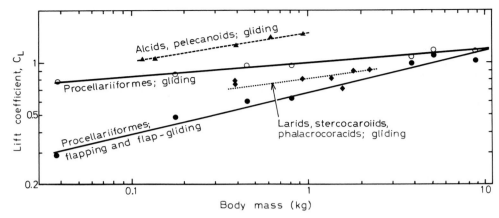

Fig. 5.3. Lift coefficients for gliding (at calculated minimum power speed), flap-gliding and gliding (at mean airspeeds). (Data from Pennycuick 1987a)

the calculated V_{mp} for the same species against body mass produces lines that converge at a body mass of 10.9 kg and a lift coefficient of 1.21 (Fig. 5.3). If an 11-kg albatross flew at V_{mp} during the flapping phase it would have to glide at a lift coefficient of 1.2, and this speed is too low and the coefficient too high for efficient gliding. The wandering albatross (*Diomedea exulans,* M = 8.7 kg) deviates below the line for flapping and flap-gliding because it flies faster than predicted and at a lower lift coefficient. The largest species probably do not have enough muscle power to fly much faster than V_{mp}, which is too low to be efficient, so they avoid powered flight and use soaring whenever possible. The smallest species in Pennycuick's analysis, the Wilson's storm petrel (*Oceanites oceanicus,* M = 0.034 kg), whose actual lift coefficient is far below that predicted for gliding at V_{mp}, is the only species that flew at a speed near its calculated maximum range speed (Pennycuick 1982).

Using wings under water has a drastic effect on flight adaptation in the seabirds (Pennycuick 1987a,b). Auks (Alcidae) and diving petrels (Pelecanoididae) use their wings also for swimming under water, and they have shorter spans and smaller wing areas than "normal" procellariiforms, such as albatrosses and storm petrels (Pennycuick 1987a), and thus lower aspect ratios and higher wing loadings. In penguins (Spheniscidae) and other flightless wing swimmers the wings are further reduced and optimized for swimming. The swimming motions of auks and diving petrels are penguinlike, the flapping frequency is much lower than in air and the wings are held with the wrist and elbow joints flexed.

The value of C_{L1} is much higher in auks and diving petrels than it is in other procellariiforms of similar mass. The high lift coefficients in the guillemot (*Uria aalge*; $C_{L1} = 1.46$) and the razorbill (*Alca torda*; $C_{L1} = 1.39$) would make these birds stall if they attempted to glide at the minimum power speed. Pennycuick noted that their high flapping speeds are too slow for flap-gliding, so they proceed by wing flapping. But auks do glide when slope-soaring along cliffs in strong

72

winds. The lift coefficients for the diving petrels are about the same as for albatrosses, but the former are much smaller and can fly faster than their minimum power speed. Unlike albatrosses, which may not have enough muscle power to fly much faster than V_{mp}, diving petrels do not flap-glide but flap all the time.

Gulls (Laridae), skuas (Stercorariidae), and cormorants (Phalacrocoracidae) in Pennycuick's (1987a) analysis have lower values of C_{L1} than the other seabirds, meaning that their gliding speeds are lower than their cruising speeds in flapping flight, so they tend to flap continuously in cruising flight. Gannets and boobies (Sulidae) have wings of very high aspect ratio and are similar to the "normal" procellariiform birds and they flap-glide in cruising flight (Pennycuick 1987b).

5.6 Stability and Control of Movements

In his article on prototypes in nature, Kuethe (1975) reviewed a passage from Wilbur Wright's first letter to Octave Chanute, dated 13 May 1900. I will use that passage here to introduce stability and control of movements: "My observation of the flight of buzzards leads me to believe that they regain their lateral balance, when partly overturned by a gust of wind, by a torsion of the tips of the wings . . .". This observation led the Wright brothers to invent the aileron for lateral control, which in turn led to man's first powered flight on 17 December 1903.

Animals in flight need high stability and control of movements, and even gliding animals must be capable of steady, controlled flight, for an animal gliding at a constant speed is in equilibrium with no residual forces or moments. Control means that the weight must be exactly balanced by the net aerodynamic force, which acts vertically upwards and through the centre of gravity (Fig. 5.1). If the equilibrium of forces in steady flight is disturbed, restoring moments tend to bring the animal back to its normal attitude.

Pitch is rotation about the transverse axis, running horizontally and transversely through the centre of gravity (Fig. 5.4a) in an animal flying horizontally. *Roll* is rotation about the median axis, running horizontally and longitudinally, and *yaw* is rotation about the vertical axis. The aerodynamic laws on control and stability in flight are documented in any book on classical aeronautical theory, and are discussed in great detail for birds by Pennycuick (1975). Here I give a brief summary of them.

5.6.1 Pitch

Pitch is closely related to the control of speed, and its control is best achieved by a long, sturdy, dorsoventrally flattened tail. Downward movement of the tail produces an upward lift and a nose-down pitching moment that results in faster

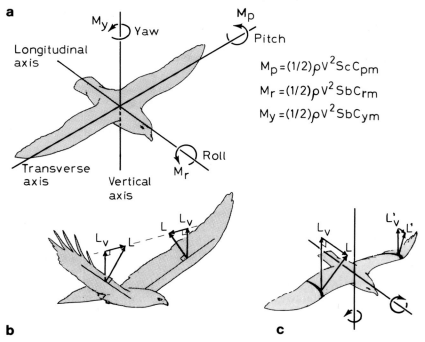

a Axes of rotation.

$$M_p = (1/2)\rho V^2 S c\, C_{pm}$$

$$M_r = (1/2)\rho V^2 S b\, C_{rm}$$

$$M_y = (1/2)\rho V^2 S b\, C_{ym}$$

b

c

Fig. 5.4. *a* Axes of rotation. M_p, M_r, and M_y are the pitch, roll and yaw moments about the transverse, longitudinal (median) and vertical axis, respectively, and C_{pm}, C_{rm}, and C_{ym} their respective moment coefficients. *b* Dihedral of the wings controlling roll moments, resulting in roll stability. A roll to the left (left wing down) would increase the lift L on the left wing and decrease the lift on the right wing because of increased angles of attack on the wing moving down and decreased angles of attack on the other wing. This force difference will lift the left wing again and restore the bird to the horizontal. L_v is the vertical lift force of one wing. *c* Partial retraction of the left wing decreasing left wing area and lift. L and L' are the lift forces and L_v and L_v' are the vertical lift forces of the two wings. Positions of wings in the middle of a downstroke

gliding speed. An upward movement of the tail produces a nose-up pitching moment and slower speed. But all gliding animals do not have a large tail and animals with no protruding tail achieve the same result by moving the legs (gliding mammals and bats) or by fore-and-aft movements of the wings relative to the centre of gravity (birds). In slow glides, downward deflection of the wingtips ("diffuser wingtips") in wings held straight or slightly forward is thought to give pitch stability. In fast glides stability is probably achieved by combining sweepback of the wings and twisting of the handwing in the nose-down sense (see further Pennycuick 1975).

5.6.2 Roll

A disturbance in roll leads to sideslip which can be stabilized by sweepback and/or by upward deflexion of the wings (V-attitude or *dihedral* about the body's longitudinal axis) (Fig. 5.4b). The control of roll moments can also be achieved by differential twisting of the wings, so that the angles of attack (= the amount of lift) become different on the two wings. A roll usually precedes a turn and can be used to change direction, while a quick turn (high rate of roll) can be effected by a partial retraction of one wing which gives a larger lift difference between the two wings (Fig. 5.4c).

5.6.3 Yaw

Yaw is controlled by the tail or by the twisting and flexing of the wings to give different drag coefficients on the two sides (Fig. 5.4c). Pennycuick (1975) suggested that deflected (diffuser) wingtips should provide some yaw stability.

Several of the characteristics needed for good control and stability probably occurred in early gliders as they do in modern gliding animals. The earliest forms of birds (*Archaeopteryx*) and pterosaurs (*Rhamphorhynchus*) had long tails that probably served for control and stability. In gliding mammals the flight membrane can be controlled by the limbs.

Chapter 6

Soaring

6.1 Introduction

Soaring costs a minimum of energy compared with other types of flight, for during soaring, energy is extracted from natural winds and converted to potential (height gain) or kinetic (speed gain) energy. Soaring birds usually glide using vertical and horizontal air movements, and they use only additional muscular energy to correct position and to hold the wings down in the horizontal position. Many large birds use soaring when searching for food, during migration, and sometimes for flights between roosting and foraging places.

6.2 Soaring Methods

Soaring can be classified in different ways. Cone (1962a) distinguished between *static soaring*, which depended on vertical movements of the atmosphere, and *dynamic soaring*, where energy was extracted from variations in horizontal wind speed. But most soaring includes both types, and to avoid confusion I have followed Pennycuick's (1975) classification of the soaring techniques according to the exploited meteorological process.

6.2.1 Slope Soaring

In *slope soaring* a bird flies in a region of rising air caused by upward deflection of the wind over a slope, forest edge, water wave, etc. (slope lift, Fig. 6.1a). Slope lift is used by many seabirds such as gulls (Laridae), for travel between roosting and foraging areas. Slope soaring along ocean waves is frequently used by petrels and albatrosses (Procellariiformes; Pennycuick 1982); but birds with high wing loadings, such as the auks (Alcidae), cormorants (Phalacrocoracidae) and the gannet (*Sula bassana*), can only slope soar in strong winds.

Alerstam (1985) described slope soaring by flocks of terns which flew inland in the afternoon to exploit thermal currents over land. The birds circled by gliding, mixed with flapping, and sometimes by almost pure soaring with fully spread wings and tail.

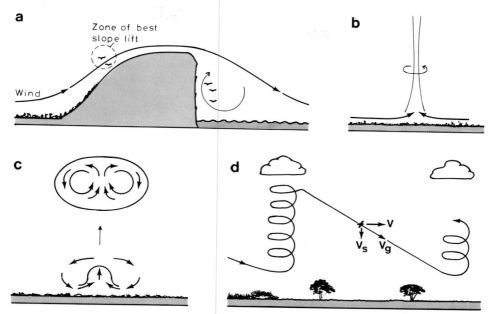

Fig. 6.1. a Slope lift. The best lift is found over a smooth slope (*left*), while vertical cliffs can produce more complicated flow patterns but sometimes give useable lift when facing downwind (*right*; Pennycuick 1975; by courtesy of Academic Press). *b* Dust-devil, triggered from solar heating of the ground. *c* Vortex-ring, triggered from heated ground and rising as a distinct bubble. *d* Cross-country soaring by climbing in a thermal, gliding off, and climbing in a new thermal again. *V* is forward speed, V_g is gliding speed, and V_s is vertical sink speed

6.2.2 Thermal Soaring

Thermal soaring occurs in thermals, which are rising volumes of warm air moving because of thermal instability in the atmosphere. The unstable vertical distribution of air of different temperatures results from differential heating from the ground or cooling from above which causes thermals to move. Thermals vary in form and structure but occur as two main types, the *dust devil* (columnar type, Fig. 6.1b) and the *bubble* or *vortex ring* (doughnut type, Fig. 6.1c).

Dust-devils are triggered by solar heating of the ground and consist of rapidly rotating columns of air with zones of reduced pressure up the middle caused by centrifugal force. A gliding bird can maintain or increase height by circling in the upward stream of air, but dust-devils last for only a few minutes and seldom provide lift beyond 500–1000 m above the ground.

Vortex-ring thermals may be triggered directly from the heated ground, and rise through the atmosphere as distinct bubbles. The lift is confined to a central core that is limited outwards by a zone of sinking air that turns inwards again at the bottom of the bubble. Lowered surrounding air pressure with increased height causes the rings to grow in size as they rise so that the cores can reach 1–2 km diameter 2–3 km above the ground. Circulation of air up in the middle and down at the outside of a vortex ring has a life of half an hour or more, so that birds may climb more than 2500 m in such a thermal. Vultures, eagles, buzzards, and storks use thermals for cross-country soaring by climbing in a thermal to some substantial height and then gliding off in the desired direction, losing height as they go (Fig. 6.1d; see Sect. 6.3.2).

6.2.3 Gust, Frontal and Wave Soaring

Gust soaring can occur in random eddies set up by the passage of the wind over a rough or wooded ground, gusts that can have substantial vertical and horizontal wind velocity components. Gust soaring has been observed in black kites (*Milvus migrans;* Pennycuick 1972b). *Frontal soaring* can be used in the narrow zone of lift (front) formed at the convergence of two air masses with different physical characters. Cold fronts and sea-breeze fronts are used for soaring by the European swifts (*Apus apus*). *Wave soaring* is used by some birds in the standing (lee) waves on the downwind sides of hills (Fig. 6.1a).

6.2.4 Dynamic Soaring

Wind-gradient, or *dynamic soaring,* does not depend upon vertical air movements but on variations in horizontal wind speed. Wind-gradient soaring might be possible in random turbulence but is more likely in the wind shear that occurs over a flat surface, such as the ocean. Near the surface the wind speed is slowed down by friction, forming a wind gradient (Fig. 6.2).

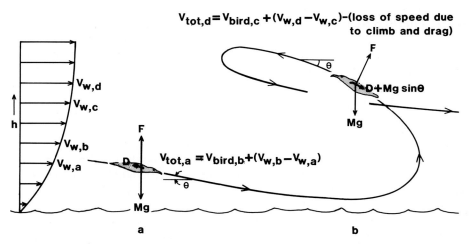

$$V_{tot,d} = V_{bird,c} + (V_{w,d} - V_{w,c}) - (\text{loss of speed due to climb and drag})$$

F

θ

$D + Mg \sin\theta$

$V_{w,d}$

$V_{w,c}$

Mg

h

$V_{w,b}$

F

$V_{w,a}$

D $V_{tot,a} = V_{bird,b} + (V_{w,b} - V_{w,a})$

θ

Mg

a   b

Fig. 6.2. Simplified analysis of the dynamic soaring of an albatross in a wind shear. The wind speed V_w increases with height h above the surface. The bird makes use of the wind gradient (change of wind speed with height), which is steep near the surface and decreases further up. Indices a-d denote height position, V_{bird} is the gliding speed of the albatross relative to the air, V_{tot} is the bird's airspeed at given heights. (Norberg 1985a; by courtesy of Harvard University Press)

The classical interpretation of dynamic soaring technique (e.g. Lighthill 1975 and Pennycuick 1982) is briefly summarized here. When a bird glides downward in a downwind direction (Fig. 6.2a) it gains airspeed (= kinetic energy). At sea level it soars in slope lift along the windward face of a wave and uses some kinetic energy for manoeuvering, and when it has used most of this energy the bird climbs into the wind (Fig. 6.2b). Although it tends to slow down (relative to the water) now because it is working against gravity, the bird gains airspeed (= kinetic energy) as it climbs into the wind because the wind is blowing progressively faster with increasing altitude. The process involves converting kinetic energy to potential energy, and when the wind gradient becomes too weak to allow further climbing, the bird turns downward and downwind again, using the gained potential and kinetic energy for manoeuvering and horizontal progression. Using only these techniques, albatrosses zigzag over the ocean, only occasionally flapping their wings.

Pennycuick (1982) found, however, that the main soaring method of albatrosses and large petrels was slope soaring along waves. Windward "pull-ups" in the dynamic soaring technique also occurred in large and medium-sized procellariforms, but Pennycuick concluded that most of the energy for the pull-up came from the kinetic energy gained by accelerated gliding along a wave in slope lift because the wind gradient would not be strong enough for the bird to sustain airspeed in a windward climb.

6.3 Soaring Performance and Flight Morphology

The large variations in foraging behaviour and in the way natural winds are used have produced variations in the structure and form of the wings according to different aerodynamic requirements. The aerodynamics of soaring have been treated in detail by Pennycuick (1971a, 1975), and in various papers (Pennycuick 1972b, 1982, 1983) he also related flight morphology to soaring performance.

6.3.1 Circling Performance

To climb in thermals a bird must fly in circles with low sinking speed, and a narrow thermal can be used only if the bird can make tight turns without losing height faster than the air rises. The turning radius can be calculated as follows. Consider first a bird flying steadily and horizontally with lift L equal to $L = (1/2)\rho V^2 S C_L$. A sudden increase of the angle of attack will increase the lift to $(1/2)\rho V^2 S C_L'$ and give the bird an upward acceleration a of

$$(1/2)\rho V^2 S(C_L' - C_L) = Mg\,a/g. \tag{6.1}$$

The bird will then begin to describe a curved path of radius of curvature r′ given by $a = V^2/r'$, where

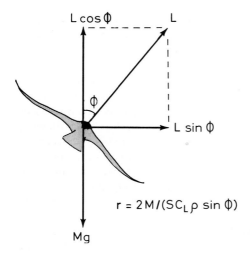

$$r = 2M/(SC_L \rho \sin \Phi)$$

Fig. 6.3. Mean forces acting on an animal in a balanced, banked turn. Mg is weight, L is lift, r is radius of a banked turn

$$r' = 2\, Mg/g\rho(C_L' - C_L). \tag{6.2}$$

The change in drag is then ignored. C_L' cannot exceed $C_{L,max}$, and if the speed is high, C_L is small. The absolute minimum value of r is given by

$$r_{min}' = 2\, Mg/g\rho C_{L,max} = V_{stall}^2/g, \tag{6.3}$$

where V_{stall} is the appropriate stalling speed. But this theoretical value cannot be attained, since $C_{L,max}$ is accompanied by a rather large drag.

Flying in circles means that the bird must bank its wings at some angle to the horizontal, and in a banked turn (Fig. 6.3) the direction of motion is longitudinal and without a side-force (which would produce *side-slip*). If Φ is the angle of bank, we have

$$L \cos \Phi = Mg, \; L \sin \Phi = Mg\, V_t^2/gr, \text{ and } \tan \Phi = V_t^2/gr, \tag{6.4}$$

where r is the radius of a banked turn and V_t the forward speed in the turn. At any particular angle of bank Φ the radius of turn is

$$r = Mg\, V_t^2/Lg \sin \Phi = 2\, Mg/SC_L \rho g \sin \Phi. \tag{6.5}$$

From this equation it follows that at any given lift coefficient and angle of bank the turning radius is directly proportional to the wing loading. Birds with low wing loading can make tight turns, glide slowly with low sinking speed, and are good at exploiting weak and narrow thermals. Birds with high wing loading can glide fast without excessive steepening of the glide angle, but they have larger turning radii. The more a bird steepens the bank angle, the more it reduces the radius of turn, and increases the sinking speed, so that if the lift coefficient is to be kept constant, the speed must be increased. The minimum radius of a banked turn, r_{min} takes the same value as r_{min}' in Eq. (6.3), because sin Φ cannot exceed unity. But r_{min} can never be reached because it would require infinite lift and vertical bank so thar r_{min} is approached asymptotically at very high angles of bank. The larger pterosaurs may also have used thermals to minimize flight costs, and r_{min} in *Pteranodon* (M = 18 kg) was estimated to be 5.2 m (Bramwell 1971).

Pennycuick (1971a) described a bird's circling performance in detail and constructed circling curves for various angles of bank for the white-backed vulture (*Gyps africanus*) (Fig. 6.4a). The relations between the forward speed V_t of a turn and the forward speed V in straight flight, and between the sinking speed V_{st} of a turn and the sinking speed V_s in straight flight at the same lift coefficient are

$$V_t = V/(\cos \Phi)^{1/2}, \tag{6.6}$$
$$V_{st} = V_s/(\cos^3 \Phi)^{1/2} \tag{6.7}$$

(Haubenhofer 1964). By combining Eqs. (6.4) and (6.6) it follows that the relation between the turning radius and the straight-flight forward speed is

$$r = V^2/g \sin \Phi. \tag{6.8}$$

The left-hand end of each curve in Fig. 6.4a is determined by the maximum lift coefficient, and the circling envelope of these curves defines the minimum sinking speed obtainable at any particular radius. The horizontal asymptote defines the minimum sinking speed in straight flight. A small increase in the angle of bank results in a large reduction of the turning radius with only a slight increase of sinking speed. At high angles of bank, however, a further steepening of the angle produces only a small reduction of the turning radius at the expense of a substantial increase in the sinking speed. Figure 6.4b shows the relation between radius of turn and angle of bank at three different flight speeds in the red-throated diver (*Gavia stellata*).

Figure 6.4c shows the profile of an arbitrary circular thermal, by plotting the upward air velocity against distance from its center. The rate of climb of a bird can be obtained by subtracting the bird's circling envelope from the thermal profile. In general, the maximum rate of climb occurs when the slope of the circling envelope is equal and opposite to that of the thermal (Goodhart 1965).

6.3.2 Cross-Country Soaring

The basic principle of cross-country soaring using thermals is that the bird gains height in a thermal, glides off through dead air until it finds a new thermal, and begins to climb again. Pennycuick (1971a) (see also Welch et al. 1968) showed how to determine the optimal speed V_{opt} for gliding between thermals. The average cross-country speed is the distance between the thermals divided by the sum of the times for the climb and interthermal glide. The optimal speed V_{opt} can be determined graphically by drawing a tangent from the achieved rate of climb V_{th} to the glide polar (Fig. 6.4d).

6.3.3 Wing Shape in Soaring Birds

The short soaring wing and the long soaring wing allow exploitation of different situations. Birds that use thermal soaring as their main means of locomotion

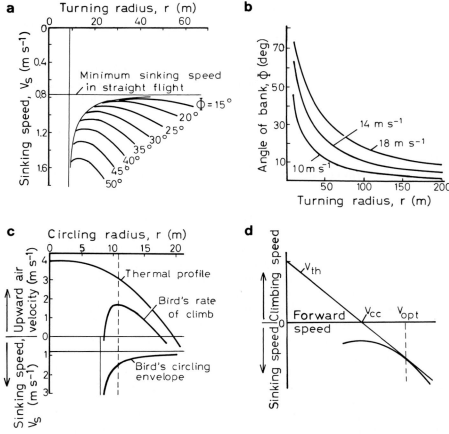

Fig. 6.4. *a* Circling curves for the white-backed vulture *Gyps africanus*. The radius of turn increases with increasing airspeed (*from left to right* not marked) at any particular angle of bank φ. The maximum lift coefficient determines the left-hand end of each curve, and the envelope at these ends defines the minimum sinking speed obtainable at any particular radius of turn. The *vertical* and *horizontal lines* are the asymptotes of the envelope. *b* Relation between angle of bank φ and radius of turn *r* at three different flight speeds for a bird the size of the red-throated diver *Gavia stellata*. *c* Profile of an arbitrary circular thermal as a function of vertical speed and circling radius. The thermal is strongest in the centre and the vertical upward air velocity decreases with increasing distance from its centre as shown. The bird's rate of climb is obtained by subtracting its circling envelope from its thermal profile. *d* Cross-country speed in thermal soaring. V_{th} is the achieved rate of climb in a thermal; the tangent from this point to the glide polar (cf. Fig. 5.2a) defines the optimum speed V_{opt} at which to glide between thermals in order to achieve the maximum cross-country speed obtainable. The tangent cuts the speed axis at the average cross-country speed V_{cc}. (*a,c,d* are from Pennycuick 1975, by courtesy of Academic Press. *b* is from R.Å. Norberg and U.M. Norberg 1971)

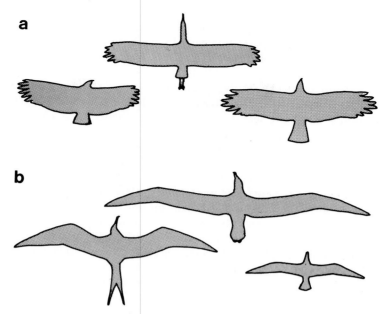

Fig. 6.5. *a* Short soaring wings in (*from left to right*) a vulture, a stork, and an eagle. *b* Long soaring wing in (*from left to right*) a frigate bird, an albatross, and a gull. (After Pennycuick 1987b and Herzog 1968)

typically have short soaring wings with very low aspect ratios (around 7 in many species; Pennycuick 1971b), large wing areas, and hence low wing loadings (Fig. 6.5a). The distal wing primaries are usually emarginated, forming a series of wingtip slots on the extended wing (see Sect. 11.5.4). This wingtype is typical of birds specialized for thermal soaring over land, such as vultures, eagles, storks and cranes (Pennycuick 1983). Long soaring wings are narrow with aspect ratios greater than 10 and they occur in albatrosses, frigate birds and gulls (Fig. 6.5b). Long soaring wings lack slots, have low camber and very low induced power. The gliding angle is small.

Albatrosses have high wing loading and very high aspect ratio (13–15; Pennycuick 1982) and are believed to use dynamic soaring. To explore why vultures do not have longer wings, Pennycuick (1971a) made circling performance estimates for an imaginary "albatross-shaped vulture", with the proportions of the wandering albatross (*Diomedea exulans*, body mass 8.5 kg) and the size of a 5.4 kg white-backed vulture (*Gyps africanus*). Pennycuick calculated the wing area of the albatross-shaped vulture as two-thirds of that in the white-backed vulture producing wing loading and turning radii 1.5 times greater. This means that although a white-backed vulture would climb better in narrow thermals up to 17 m radius, the albatross-shaped vulture would climb faster in wider thermals and have far higher average speeds when travelling across country. Once airborne, an albatross-shaped vulture would soar superbly in wide thermals, far outdistancing migrating storks and eagles.

Reality suggests that vultures must be able to soar in very small thermals and to be able to begin patrolling early in the morning when the thermals are weak and narrow (Pennycuick 1982). These factors would preclude an albatross-shaped wing.

Hankin (1913) observed that the order of initiating morning soaring by vultures correlated with increasing wing loading. The earlier in the day that the vultures can start searching for kills, the better their chances of successful foraging. This is particularly important, since the mammalian scavengers usually dominate the vultures, keeping them away from carcasses. On the other hand, high wing loading would give high cross-country speed, enabling a vulture to search larger areas and to transport food in its crop over larger distances.

Pennycuick went on to suggest that the low-aspect-ratio wing better reflected the requirements of take-off from the ground or trees, a task requiring large wing area, and one that might be hindered by long wings.

Pennycuick (1983) used field observations to compare the soaring performance of three contrasting species of gliding birds, the magnificent frigatebird (*Fregata magnificens*), the brown pelican (*Pelecanus occidentalis*) and the black vulture (*Coragyps atratus*). The frigatebird had the lowest wing loading (36.5 N m^{-2}) and highest aspect ratio (12.8), the pelican the highest wing loading (57.8 N m^{-2}) and rather high aspect ratio (9.8), and the vulture a rather high wing loading (55.7 N m^{-2}) and low aspect ratio (5.8). The mean circling radii were proportional to the wing loading and varied from 12 m (frigatebird) to 18 m (pelican). The low wing loading of the frigatebird gave it a low minimum sinking speed and a high rate of climb. The observed rate of climb was somewhat better in the frigatebird (0.48 m^{-1}) than in the black vulture (0.40 m s^{-1}), but contrary to expectation the pelican showed the highest mean rate of climb (0.57 m s^{-1}). The high aspect ratio and low wing loading in the frigatebird give low flight costs and are considered to be adaptations to long continuous flights when wing flapping cannot be avoided. In all species the angles of bank were 22.9 to 24.7°, and the mean circling lift coefficients 1.33 to 1.45.

6.4 Bats and Pterosaurs

Although bats cannot use soaring flight because convective air currents are absent at night, some bats slope soar. The largest pterosaurs were good gliders and soarers, and the wings of large pterosaurs appear to be direct natural counterparts of hang gliders with high aspect ratio and cylindrical camber (McMasters 1976). The largest hang glider of this type has a span of 11.8 m and a load mass of 100 to 120 kg, which probably is very similar to the largest known pterosaurs.

Chapter 7

Migration

7.1 Introduction

Many birds of the temperate zones migrate towards the equator during autumn when food becomes scarce, and several species from northern Europe travel to subsaharan Africa, a journey of >6000 km that can take several weeks. Non-stop flights of 2700 km by the lesser snow goose *Anser c. caeruleus* (Ogilvie 1978) and 3800 km by the Pacific golden plover *Pluvialis dominica fluva* (Johnston and McFarlane 1967) have been recorded, and an altitude of approximately 9000 m has been observed for bar-headed geese (*A. indicus*) over the summit of the Himalayas (Swan 1961). The most famous of all migrant birds is the arctic tern (*Sterna paradisaea*), which uses four to five long-distance flights to cover nearly 20,000 km from northern breeding grounds to Antarctica. The common tern (*S. hirundo*) makes migratory flights up to 15,000 km (Alerstam 1985). Both species have low wing loadings and high aspect ratio wings which makes flight inexpensive.

Most temperate bats hibernate in the winter, but some species make regular migrations of at least a few hundred kilometres to warmer places. Hibernating bats may sometimes also have to migrate shorter distances to their hibernaculae. The use of seasonal habitats by birds could never have developed without the evolution of prolonged powered flight.

Birds such as gulls, raptors, storks and cranes use cross-country soaring during migration or other long-distance flights, whereas terns, curlews (*Numenius arquata*), bar-tailed godwits (*Limosa lapponica*), arctic skuas (*Stercocarius parasiticus*), and cormorants (*Phalacrocorax carbo*) cross land mainly at night. These birds tend to migrate over sea and along coast-lines, using ground effect (discussed in Sect. 9.8.3).

7.2 Orientation and Navigation

A problem of long-standing interest is how migrating birds find their way, for some return to the same places year after year and use the same resting places on their flights. Animals use their different sense organs for orientation and navigation, where orientation is the ability to determine and keep a certain direction, and navigation the ability to decide position in relation to a definite destination.

Birds migrating by day are assumed to orientate and navigate by the sun and different landmarks for smell and taste are not particularly developed and may be of little value during long-distance migration. But smell can be used by homing pigeons, as a comparison of the orientation of pigeons subjected to different olfactory deprivation has shown (Wiltschko et al. 1987). This study compared performance in different areas and demonstrated a small olfactory effect near Ithaca NY (USA) and Frankfurt (Germany), but significantly larger near Pisa (Italy). The authors concluded that there are important regional differences in the strategies and cues pigeons use to navigate, suggesting a learned system which favours flexibility. Birds probably use any perceptible cues for navigation, with the optimal combination possibly different for each population.

Vision is well developed in birds; they have colour sense and homing pigeons (*Columba livia*) are even able to perceive the polarization of light. This allows them to estimate the approximate direction of the sun when the sky is partly overcast (Kreithen and Keeton 1974). But the pigeons need to see at least a part of the sky, since it is the blue sky that shows different patterns of polarization in different directions in relation to the sun. Birds also use star patterns for navigation (e.g. Emlen 1975).

The acoustic spectrum of hearing is wider in birds than in humans for they can detect infra-sounds (10^{-1} Hz). Infra-sounds frequently occur in thunderstorms, magnetic storms, earthquakes, sea waves, jet streams, and winds passing through gorges and around mountain tops, all of which could permit birds to obtain a rather detailed picture of their surroundings by listening (e.g. Quine and Kreithen 1981). Birds can also detect air pressure changes, permitting them to detect frontal changes and maintain constant flight altitudes.

The most effective sense "organ" for their orientation may be the magnetic sense, that allows them to use the geomagnetic fields around the earth (see, for example, Wiltschko and Wiltschko 1976; Gould 1982; Wiltschko 1983; Beck 1984). The basis for magnetic detection in birds appears to be small crystals of magnetite, which are found in microstructures between the skull bone and the dura mater and in the neck musculature. Each organ contains about 10^7–10^8 crystals, each about 10^{-4} mm large and with a total mass of about 10^{-4} mg, which is the amount needed for magnetic stability. Similar structures have been found in bacteria, bees, dolphins, mice, voles, and monarch butterflies, and may occur in bats and other far-moving animals as well. The magnetite crystals in bacteria permit perception of the inclination of the magnetic field.

Local geomagnetic anomalies from magnetic minerals may affect migrating birds in three different ways (Alerstam 1987):

1. Young birds hatched and raised near magnetically anomalous sites may develop aberrant compass/map sense. Alerstam and Högstedt (1983) experimentally shifted the magnetic field for young pied flycatchers (*Ficedula hypoleuca*) and found that those young exposed to a shifted magnetic field showed a temporary deviation from the normal orientation direction during the following autumn migration period. Similar results have been obtained for young pigeons (Wiltschko et al. 1983).

2. Birds may have difficulties in determining the right migratory or homing flight direction when leaving a magnetic anomaly. No studies have been done on departure directions from such areas, but studies on the behaviour of homing pigeons show disorientation at strong anomalies (Walcott 1978; Kiepenheuer 1982).

3. Magnetic anomalies may affect birds in migratory or homing flights, although migrating birds usually maintain their orientation when flying across a magnetic anomaly. Nevertheless, there are observations indicating that their orientation is affected (Richardson 1976; Moore 1977; Larkin and Sutherland 1977; Alerstam 1987). Alerstam (1987) radar tracked birds migrating across an area where the magnetic intensity exceeded the normal field by up to 60% at low altitude. As they crossed the area the birds changed altitude (in descents leading to height loss of about 100 m) more often than over areas with normal magnetic intensity, and flock formations were repeatedly broken up during these temporary descents. Alerstam noted that these findings support the theory that birds use the geomagnetic field and associated gradients during migration.

7.3 Flight Range

The distance a bird or bat can fly nonstop depends on its effective lift:drag ratio L'/D', the ratio of the weight to some average horizontal force needed to propel the animal along (Pennycuick 1972a, 1975). The mechanical power required to fly in steady horizontal flight at speed V is

$$P = T'V = Mg\,V\,(D'/L'), \tag{7.1}$$

so that

$$(L'/D') = Mg\,V/P, \tag{7.1a}$$

where T', D', and L' are the effective thrust, drag, and lift forces, respectively, as averaged over a number of wing-beat cycles.

In long migratory flights the situation is further complicated because the animal has to carry extra weight, e.g. before migration some passerines store up to half their lean body mass in fat (Odum et al. 1961), and as this fuel is consumed, the animal's mass decreases along with the power required to fly and the rate of fuel consumption. The actual distance Y flown is given by Pennycuick (1969, 1975) as

$$Y = (e\eta/g)\,(L'/D')\,\ln\,(M_1/M_2), \tag{7.2}$$

where e is the energy released on oxidizing unit mass of fuel, η is the mechanical efficiency, g the acceleration due to gravity, M_1 is the body mass at the beginning of the flight, and M_2 the mass at the end. The effective lift:drag ratio reaches its maximum at the maximum range speed V_{mr}, in turn giving the best distance travelled per unit work, and the power required is P_{mr}. From Eq. (7.1a) it follows that

$$(L'/D')_{max} = Mg \, V_{mr}/P_{mr}. \tag{7.3}$$

Pennycuick (1969) calculated that the P_{mr} required to fly at V_{mr} was proportional to $b^{-1.5}$, so that migratory birds should have long wings to cover as long a range as possible on a given amount of fuel.

For a bird with a typical L'/D' value of 6 and an M_1/M_2 ratio of 1.5, the maximum still-air range would be about 2000 km, which is regarded by Pennycuick (1972a) as a representative distance for a single stage flight by migrating small birds. Many passerines fly about this distance when crossing the Sahara (Moreau 1961), and this distance would be covered in 57 h at a cruising speed of 35 km h^{-1}.

The inverse of Eq. (7.1a) has been referred to as the *cost of transport*

$$C = P/Mg \, V \, (= D'/L'), \tag{7.4}$$

(cf. Sect. 3.7), for it represents the aerodynamic work done in transporting a unit of body weight over a unit of distance. The minimum cost of transport in migratory flights then becomes $C_{min} = P_{mr}/Mg \, V_{mr}$. The cost of transport is a non-dimensional quantity that is independent of body mass, for the dimensional units of $P/Mg \, V$ is (N m s^{-1})/(N) $\times$ (m s^{-1}). In theory, the cost of transport becomes independent of body mass and is therefore useful for comparisons between species of different sizes. However, based on data for Greenewalt's (1962) passerine birds, Tucker (1976) found a slight decrease with increasing body mass as

$$C_{min} = 0.60 \, M^{-0.20}, \tag{7.5}$$

while Pennycuick's (1975) theory yields approximately

$$C_{min} = 0.62 \, M^{-0.07}. \tag{7.6}$$

Rayner's (1979c) estimate on the same data set is

$$C_{min} = 0.21 \, (M)^{-0.08}. \tag{7.7}$$

Larger birds thus, have a slightly lower cost of transport. This results from differing scaling of body dimensions (Chap. 10), and all deviations in dimensional scaling may reflect the immense cost of lift generation in slow flight for larger birds (Rayner 1979c).

7.4 Cruising Speed and Flight Time

Radar observations of the cruising speeds of several migrating bird species agree rather well with the expected maximum range speed, V_{mr}, estimated with Pennycuick's (1975) equation (Alerstam 1982), although there are variations. The maximum range speed should be used whenever the longest distance should be covered on a given amount of energy (see also Sect. 7.5), but on a long flight the optimal cruising speed continually decreases as the bird's mass decreases because of mass loss. Pennycuick (1975) derived equations for the flight time and average speed that should be used by a migrating bird and in a series of geometrically similar birds, the average flight speed V would vary with the one-sixth power of the body mass, $V \propto M^{1/6}$, as would V_{mr} and V_{mp}. He also gave the procedures for estimating the flight range [using Eq. (7.2)], providing one method for small ($\leq$ 50 g) passerine birds and another for other birds.

Best flight economy for migrating birds would be obtained by a low wing loading, that is, slow flight (low V_{mr}), and high aspect ratio wings (see Chap. 12) if time is unimportant, but quite a few species have high wing loadings and high optimal speeds and save time. In some species with low wing loadings (such as the arctic tern, *Sterna paradisaea,* and the common tern, *S. hirundo*) the estimated flight speeds across land during migration can sometimes exceed the theoretical optimum because time saved will facilitate efficient restoration of energy reserves at the destination (Alerstam 1985). R. Å. Norberg (1981b) used the same argument to predict the optimal flight speeds in birds taking food to their young. Figure 7.1 shows the predicted relationship between the optimal speed and the rate of energy accumulation upon arrival at the destination. An energy accumulation rate of 1 W of mechanical power corresponds to the storage of about 0.4 g of fat per hour in the terns, which translates to a 4% daily increase of mass as fat, and Alerstam noted that this estimate is realistic in rich feeding waters. When the speed is increased above V_{mr}, the flight will become more expensive,

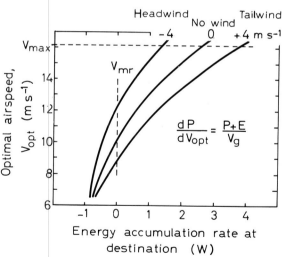

Fig. 7.1. Optimal airspeed versus energy accumulation rate upon arrival at the destination for arctic/common terns. An energy accumulation rate E equivalent to 1 watt of mechanical power P corresponds to the storage of about 0.4 g of fat per hour or a daily increase of mass due to fat storage in the terns of about 4% (assuming 12 h feeding and fat accumulation per day). V_g is ground speed, V_{max} is maximum speed, V_{mr} is maximum range speed. (Slightly modified from Alerstam 1985, which is based on R.Å. Norberg 1981b)

but as long as the birds can increase foraging time at their destination, the benefit remains. The faster the rate of energy accumulation at the destination, the greater the optimal flight speed should become.

7.5 Effect of Wind

The flight range of a migrating animal will be affected if it cruises in head- or tailwinds. If the animal flies at an airspeed V_a relative to the air and against a headwind $V_{w,h}$, its groundspeed V_g becomes $V_g = V_a - V_{w,h}$, and its flight range will be proportionately shorter. Similarly, flying in tailwind $V_{w,t}$, means that the animal's groundspeed becomes $V_g = V_a + V_{w,t}$, and its flight range will be longer. If the wind blows on the animal from some other direction, the relations between the speeds are found by the cosine theorem. Figure 7.2 shows that, knowing the animal's groundspeed, the windspeed V_w (either tailwind, sidewind or headwind) and the angle θ between these speeds, the animal's airspeed is given by

$$V_a = (V_g^2 + V_w^2 - 2V_g V_w \cos\theta)^{1/2}. \tag{7.8}$$

The time taken to fly a particular, shorter, distance (when speeds are assumed to remain constant) can be found by dividing the distance flown by the ground-speed. Conversely, the groundspeed, and hence airspeed, can be estimated when time, distance, windspeed and θ are known.

To obtain maximum range over the ground when affected by a wind, the animal should not fly at V_{mr}, the speed to attain maximum air range, but it should use a speed giving maximum ratio of groundspeed V_g to power. On the power

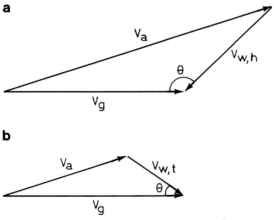

Fig. 7.2. Effect of headwind *a* and tailwind *b* on groundspeed and flight direction. V_a is the bird's airspeed, and V_g its resultant groundspeed, $V_{w,h}$ is the headwind, and $V_{w,t}$ is the tailwind

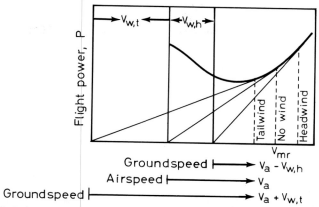

Fig. 7.3. When flying in a tailwind $V_{w,t}$ or headwind $V_{w,h}$, the maximum range speed (in relation to the surrounding air) is less and higher, respectively, than the maximum range speed V_{mr} in still air (when the bird's groundspeed equals its airspeed V_a). (After Pennycuick 1972a)

curve in figure 2.6 the bird's airspeed should be replaced by its groundspeed on the abscissa which will shift the y axis to the left for a tailwind or to the right for a headwind (Fig. 7.3). It follows that a tailwind reduces V_{mr} and a headwind increases it compared to the case with no wind.

Pennycuick (1969) estimated the effect of head- and tailwind components on specific range for eight different wind speeds for flying animals of different sizes and airspeeds (Fig. 7.4). The animals were assumed to fly at a constant airspeed regardless of the wind strength. For example, at a light headwind of 5 m s⁻¹ a small pipistrelle bat cruising at 7 m s⁻¹ would achieve only 25% of its still-air range, a pigeon cruising at 26 m s⁻¹ would achieve nearly 70%, and a swan cruising at 26 m s⁻¹ would achieve over 80%. Tailwinds have the opposite effect.

7.6 Evolution of Soaring Migration

The evolution of soaring migration and the migratory strategies in birds of different size were discussed by Pennycuick et al. (1979), and it is reviewed here.

Cross-country soaring depends on a more elaborate behaviour than flapping flight, and must be considered a more advanced flight mode developed from simple flapping flight. Specialization for soaring results in lower energetic cost per unit distance, but also in slower cross-country speed and longer time to reach the destination. Delayed migration due to unfavourable weather conditions would favour speed, while small fat reserves would favour economy. The advantages of flying fast or flying economically are different for large and small birds. Birds near the upper mass margin must fly economically and would benefit from soaring during migration for two reasons.

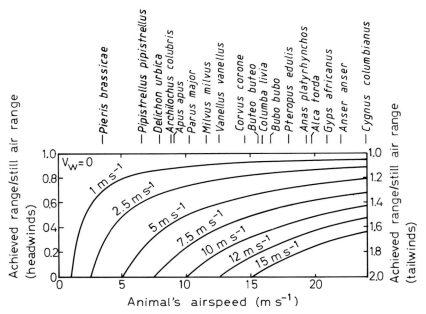

Fig. 7.4. Effect of head- and tailwind components V_w on flight range as compared with the achieved range in still air. The same cruising airspeed is assumed to be maintained, regardless of wind direction and strength. The wind speeds are indicated at the respective curve. Some birds, bats and insects are marked at their estimated maximum range speeds in still air. (Pennycuick 1975; by courtesy of Academic Press)

First, very large birds would have difficulty mustering sufficient muscle power to achieve the speed (at or near maximum range speed) required for maximum economy in powered flight. Second, soaring costs much less than horizontal flapping flight, for the energy cost arises from basal metabolism and keeping the wings outstretched. Since the basal metabolism per unit body mass is proportionately smaller in larger birds, the savings by soaring will become larger in heavier birds than in small ones, and soaring in large birds is beneficial in spite of the longer flight time, when additional energy is required. In smaller species, however, when migration speeds are slower in the first place, the longer migration time would eliminate any gain in flight economy by soaring and cost more than fast powered flight.

Therefore, soaring migration occurs in large and medium-sized birds, but is not universal even among large migratory species. Species like storks that rely heavily on soaring during migration must use particular times and routes to avoid flapping flight over sea. Cranes are partial soaring migrants using routes with no or slight updraughts and they fly across the Mediterranean at night. Swans, geese and some cranes are not known to use soaring during migration, and Pennycuick et al. (1979) suggested that the larger swans must have been subjected to strong selection for fast migration at the expense of economy. They instead save some energy by using formation flight (see below). Swifts may be the smallest birds using soaring during migration, but they differ from other small migrants because they do not land to feed or rest, and their high-aspect-ratio wings provide further flight economy.

7.7 Formation Flight

Some migrating birds, such as geese, swans and cranes fly in formation. Bird flocks have been categorised according to flight parameters and configurations, and Heppner (1974) distinguished between *line formations* and *cluster formations*. Line formations (Fig. 7.5a), flocks of birds flying in a line or queue that shows regularity in spacing and alignment, have been observed in larger birds, such as waterfowl and pelicans. There are different types of line formation, and line flocks may also be vertically spaced. Cluster formations (Fig. 7.5b) are three-dimensional and usually include large numbers of small birds, flying in close order. Birds in cluster formation can make very rapid turns without collisions.

A flying animal gets its lift by creating downward momentum within its span, and the wing leaves behind a vortex sheet which rolls up into a loop of concentrated vorticity (Fig. 7.6). The adjacent bird diagonally behind in the flock is thought to exploit this upwash and the same effect is obtained as flying in an upcurrent. This decreases the required strength of the self-induced downwash and the net induced drag of each wing. The induced power (generating lift moment) depends critically on downwash distribution and thus on spanwise loading, so that the aerodynamic effects are maximized when the spanwise loading is optimal. Minimum induced power occurs with an elliptical lift distribution (Sect. 2.4.3). The energy saved by each individual will depend on the geometry of the formation and is influenced by the relative positioning of the wingtips.

The optimal distance perpendicular to the flight path between the tips of two neighbouring birds (wingtip spacing) can be determined as a function of the wingspan. At the wing surface the distance between the two vortices of a pair is approximately equal to the wingspan, but the distance is somewhat shorter behind the wings. The distance, a, between the centre of the two vortices is approximately

$$a \approx \pi b/4 \tag{7.9}$$

(Fig. 7.5c; see also Sect. 2.4.1). The optimum wingtip spacing, s, based on theory is to overlap the outboard wing ahead by $(b-a)/2$ $(= 0.11b)$, so that $s = -(b-a)/2$ or $= (a-b)/2$. Hainsworth (personal communication) found that the median spacing of 55 Canada geese (*Branta canadensis*) from eight formations showed an overlap of 0.13b, closely approximating the predicted value, but there was considerable variation about the median (ranging from an overlap of 0.87b to a gap of 1.93b). In this study the depth between birds (distance along the flight path) was 1–3 wingspans.

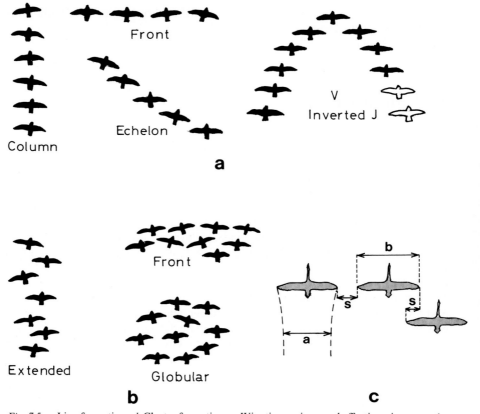

Fig. 7.5. a Line formations. *b* Cluster formations. *c* Wingtip spacing *s* and effective wingspan *a* (= distance between the two wingtip vortices at some distance behind the animal) of birds flying in flock formation. *b* is wingspan. (*a* and *b* after Heppner 1974)

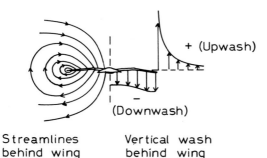

Fig. 7.6. Flow field of lifting wing seen from behind the bird. Flow streamlines are shown to the *left*, whereas the vertical velocity vectors of the induced flow are shown to the *right*. (After Lissaman and Shollenberger 1970)

Lissaman and Shollenberger (1970) calculated the energy savings in single birds flying in a *line-abreast formation* and assuming that the birds flew at optimal flight speed and L/D, had an elliptical loading, and that profile and parasite powers were unaffected by the formation flight. In their model they ignored wing movements and considered an essentially steady wing. Figure 7.7 shows the induced drag saving as a function of wingtip spacing in formations with 3, 9, 25 and infinite numbers of birds. The induced drag saving is expressed as the ratio of the induced drag of a bird in formation, $D_{ind, f}$, to that in solo flight, $D_{ind, s}$. Large drag savings would occur only when wingtip spacing was small and there were many birds. $D_{ind, f}/D_{ind, s}$ can be expressed as $1/e$, where e is the induced drag efficiency. Lissaman and Shollenberger estimated up to 71% savings, and concluded that the total power would be similarly reduced. Furthermore, in this situation flight range would increase by a factor of $e^{1/2}$, while the optimum cruise speed was reduced by a factor of $e^{-1/4}$. For a formation of 25 birds in tip-to-tip formation (s = 0), e = 2.9 gives a range increase of 71% with a cruise speed 24% below that of a single bird, so that the optimal flight speed in birds flying in formation would be lower than that for birds flying alone. A maximum of 51% was predicted for a formation of nine birds (Lissaman and Shollenberger 1970; Badgerow and Hainsworth 1981). But at the speeds used during migration (theoretically the maximum range speed, V_{mr}) the induced power is only a small part of the total power (about 20% in a pigeon of V_{mr}; Pennycuick 1975; Rayner 1979c), so the savings would be a small proportion of the total flight costs. If the induced power in the example of Lissaman and Shollenberger is assumed to be 20% of the total power (as estimated in the pigeon at V_{mr}), and if the profile and parasite powers were unaffected, then their estimate of the maximum savings would be only 14%, with a similar, smaller, increase of flight range.

Reducing the induced power means reducing the optimal speed (V_{mr}) so that birds flying in formation should move somewhat slower than birds in solo flight. But Alerstam (1982) gave several examples of waders in flocks that used similar speeds as single birds, and suggested that the birds use some of the drag savings

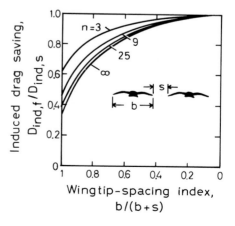

Fig. 7.7. Induced drag saving as a function of wingtip spacing. $D_{ind,f}$ is the induced drag of a bird in formation flight and $D_{ind,s}$ that in solo flight. (Data from Lissaman and Shollenberger 1970)

to increase speed above their V_{mr} to reduce flight time. Furthermore, at very slow flying speeds the vortices shed from the birds would become discontinuous and less useful to the neighbours in formation flights.

Lissaman and Shollenberger also considered the effect of *vee formations*. In a vee the drag can be more evenly distributed than in a line-abreast formation. But the leading bird experiences a weaker upwash on both sides from the birds behind, while the birds at the rear experience, on one side, a more fully developed field from the birds ahead. Leaders obtain relatively low savings, so their behaviour appears altruistic. They may also serve as a social or group orientation function (Hamilton 1967). Savings for the leading bird could, however, be enhanced with a swept vee. So the optimal vee shape, that gives a more equalized savings among the birds, is not with straight legs but with more sweep at the tip than at the apex, and with the depth from the leader to its neighbours less than the depth between other birds. If a bird in a vee moves ahead of the vee line, its induced power will increase and its speed decreases until the bird is back in line again. Behind the vee line less power is required. In this way the formation should be stable.

Using photographic data, Nachtigall (1970) demonstrated a phase relationship between wingbeats of neighbouring birds in flocks of geese flying in a vee formation. This would be expected according to the vorticity pattern, for shallow wingstrokes produce undulating vortex tubes behind the birds (Spedding 1987b). Berger (1972), however, found very little evidence for phase relations in a number of species, but this may vary with flight conditions.

Hainsworth (1987) filmed Canada geese (wingspan = 1.5 m) in vee formations during migration. The number of birds in each formation varied from 5 to 16, and most geese were clustered at depths between 1 m (0.67b) and 1.5 m (3.33b). Using the model of Lissaman and Shollenberger (1970), he estimated the induced power savings to be 36% at s = 0.13b by flying in a vee. This may correspond to a saving of 7% of the total power when flying at V_{mr}. The amount of savings depends on the bird's ability to fly with precision. Windy conditions during flight may make precision flight more difficult. Hainsworth observed that wingtip spacing variation depends importantly on the bird ahead. Any move by the bird ahead will change the wingtip spacing, and the trailing bird often actively adjusted its position in response.

Most migrating birds travel in *clusters* rather than in regular formations. Higdon and Corrsin (1978) used classical aerodynamic theory to estimate the total induced drag for two- and three- dimensional lattices of birds and compared them to data for the same number of birds flying individually. A simplified vortex model was used to estimate the relative upwash velocity at each bird in a formation confined to a single vertical plane. In their two-dimensional model each vortex pair was treated as a point doublet and then the array of point doublets as a continuous field of "doublet strength" (Fig. 7.8) producing a far-field approximation. The model shows that there is a decrease in total drag when the flock extends farther laterally than vertically, but a high, narrow flock will experience an increase in drag. Higdon and Corrsin assumed that a realistic flock shape is elliptical and the derived uniform induced vertical velocity within the ellipse then becomes

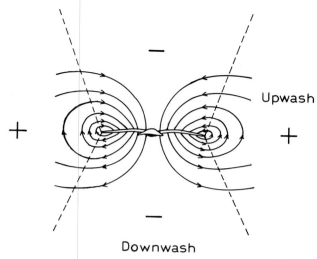

Upwash

Downwash

Fig. 7.8. Typical streamlines for a two-dimensional doublet (= vortex pair). The *dashed lines* separate regions of upwash (+) and downwash (−). Lift can be gained or lost by a bird in a flock depending upon where it is flying in relation to the other birds. In the positive flow regions of the adjacent bird ahead of it would gain lift, but it would lose lift in the downwash regions

$$w = \pi\mu (A - B)/(A + B), \tag{7.10}$$

where μ is a constant restriction implying that the birds are all the same size and weight, A is the length and B the height of the elliptic cross section of the flock. A circular flock would have $w = 0$ (since $A = B$). If the flock is high ($A < B$), there is a downwash ($w < 0$), and most members suffer increased drag. But in a shallow, wide flock ($A > B$) the model predicts drag-saving.

The constant μ can be expressed as a function of the distance a between the two vortices of the wings (Eq. (7.9)] and is

$$\mu = a\Gamma n/4\pi = b\Gamma n/16 \tag{7.11}$$

per unit area, n being the number of birds per unit area, and Γ is circulation. Since $L = \rho bV\Gamma = (1/2) \rho SV^2 C_L$ [Eq. (2.40)], substitution of this equation and Eq. (7.11) into (7.10) gives the following estimate for the upwash velocity at each bird due to the others:

$$w_u = (\pi^2 n C_L SV/16) (A - B)/(A + B) \tag{7.12}$$

It was shown in Eq. (2.44) that the self-induced downwash of each individual bird for the case of an elliptic spanwise lift distribution is $w_d = C_L SV/\pi b^2$. The relative change in downwash velocity, and hence the decrease in induced drag, is obtained by dividing the upwash velocity by the self-induced downwash,

$$w_u/w_d = (\pi^3 n b^2/4) (A - B)/(A + B) \tag{7.13}$$

The induced drag is found by integrating the product of the downwash and the circulation across the span according to Eq. (2.42). Higdon and Corrsin illustrated Eq. (7.13) by suggesting that if the flock is extended horizontally, say A =

97

5B, and the interbird centre-to centre spacing is 5b, then the reduction in induced drag is 5%.

A more detailed three-dimensional analysis used the horseshoe vortex pattern to separate each bird and neglected the disturbance due to flapping (Higdon and Corrsin 1978). The results from this model confirmed their findings on flock shape with the two-dimensional model when the interbird spacing is several wingspans, and they concluded that improved flight efficiency is not an important reason for migrating in large, three-dimensional flocks. This flock behaviour may instead have arisen for social reasons, such as security against predators.

Chapter 8

Hovering Flight

8.1 Introduction

For a flying animal, gliding and soaring are rather inexpensive, whereas flapping flight requires a substantial energy input. Hovering flapping flight is the most expensive, because it involves no forward speed component. This means that the surface area through which air is accelerated in a unit of time is much less than for forward flight. In spite of the costs it involves, hovering is performed by most insects and by many small birds and bats.

Hovering permits animals to forage in places otherwise inaccessible, for example, in front of flowers and fruits at plants that are too weak to support the animal's mass. Pyke (1981) observed that a hovering hummingbird (Trochilidae) can forage more quickly between flowers than one that perches, resulting in maximized net energy gain in spite of the higher flight costs. The distance the birds travel to reach a foraging place may influence whether they should perch or hover when they feed, but as distance decreases hovering becomes a more effective behaviour to maximize rate of net energy gain (Wolf and Hainsworth 1983; Hainsworth 1986). This was predicted in Norberg's (1977) foraging model, which says that more efficient and expensive foraging modes should be used for short flight or "search" times and at high food density. The size of the bird and its wingshape also dictate whether an animal should perch or hover (Norberg 1977; Norberg and Rayner 1987; see Sect. 12.3), as will floral architecture (Miller 1985).

The essence of hovering flight is the production of a vertical force to balance the weight of the animal. The main power drain in hovering and very slow forward flight is the induced power, for the effects of the induced velocity associated with the vortex wake are more significant than during forward flight, so large lift coefficients are required. Parasite power is minute since there is no forward speed. Only the induced downward speed gives some parasite drag, which can be neglected. Profile power is low since the resultant velocity arises only from flapping speed and downwash velocity. Inertial power may be of some importance.

Hummingbirds hover with fully extended wings during the entire wingbeat cycle and elicit lift also during the upstroke; this is what is known as *normal, true*, or *symmetrical hovering*. Other hoverers, such as many small passerine birds (Brown 1963; Norberg 1975), sunbirds (Nectarinidae, Zimmer 1943) and bats (Norberg 1970a, 1976b) flex their wings during the upstroke to reduce drag forces and any negative lift forces that may occur. This is called *asymmetrical hovering* (see further the next Sect. 8.2).

Hovering flight can be understood by the Rankine-Froude momentum theory (see Sect. 2.3.1) and by the vortex theory for induced power, but both must be supplemented by the blade-element theory for profile and inertial powers. The simple momentum theory, which gives a *minimum* value of the induced velocity, has frequently been used for force calculations in hovering birds, insects and bats (Pennycuick 1968a; Weis-Fogh 1972, 1973; Norberg 1975, 1976b). Ellington (1978, 1980, 1984d-f) and Rayner (1979a) developed vortex theories for hovering flight in insects and birds, which are summarized in Rayner (1979c) and Ellington (1984e,f).

The mechanics of hovering flight are easier to handle than those of forward flight, particularly for true hovering with symmetry in the wingbeat cycle. Weis-Fogh (1972, 1973) derived analytical expressions, using momentum and blade-element theories, for quantitative comparative studies of normal hovering in insects and birds. Norberg (1975, 1976b) derived Weis-Fogh's theory further to be applied to asymmetrical hovering in birds and bats. Weis-Fogh showed that normal hovering in hummingbirds usually is consistent with steady-state aerodynamics, whereas Norberg found that asymmetrical hovering in a bird (the pied flycatcher, *Ficedula hypoleuca*) and a bat (the European long-eared bat, *Plecotus auritus*) is not consistent with the quasi-steady conditions. Although asymmetrical hovering must thus be explained by non-steady-state phenomena, slow forward flight by bats can be explained by quasi-steady assumptions (Norberg 1976a).

But with his generalized vortex theory, Ellington (1984e,g) repeated Weis-Fogh's work on hummingbirds with more accurate data on morphology and kinematics, and, contrary to Weis-Fogh, he found that hovering flight in *Amazilia fimbriata* may not be consistent with quasi-steady conditions.

Problems with steady-state and non-steady-state mechanics of flapping flight are clearly summarized in Ellington (1984g). In the *quasi-steady assumptions* the instantaneous aerodynamic forces on a flapping wing is assumed to equal those which the wing would experience in steady motion at the same instantaneous velocity and angle of attack. The problem is whether or not the quasi-steady assumption is valid for large wingbeat amplitudes and high flapping frequencies, as in hovering and slow flights.

The validity of the assumption can only be tested in a *proof-by-contradiction*. If the mean forces, calculated according to the quasi-steady assumption, do not satisfy the net force balance of the flying animal, then the aerodynamics must be non-steady. If the mean forces are sufficient for flight, this only proves that the quasi-steady assumption could provide a satisfactory explanation. This does not prove that the wings work like conventional wings, because non-steady effects are always involved, more or less, in a flapping wing.

The quasi-steady mechanics usually apply reasonably well to fast flapping flight because of the smaller stroke amplitudes and the lower flapping frequencies. Not only is the contribution of wing flapping speed to the resultant relative air speed small in fast flights as compared with slow and hovering flight, but the angle of attack varies slowly. The mean flight velocity dominates the flow over the wings, which therefore have very low values of the *reduced frequency parameter*, the ratio of the flapping velocity to the mean flight velocity (= inverse

100

of the *advance ratio*). This parameter, first introduced by Walker (1925), is taken to indicate agreement with the quasi-steady assumption. It increases with decreasing flight velocities, and is thus high at slow flight where any unsteady effects due to flapping are important. Ratios below 0.5 are normally taken to indicate that unsteady effects are essential, and hovering with zero forward speed is the extreme case.

The lift coefficient, C_L, is a useful measure of the lift-generating capacity of the wing. If the mean lift coefficient required for hovering exceeds the maximum value ($C_{L,max}$) measured in steady flow, then the quasi-steady assumptions will be disproved. This maximum value of the lift coefficient is reached before stalling and its upper limit may be around 1.5 for flying animals (Pennycuick 1968b, 1971c; Ellington 1984g).

The lift coefficients attained by hovering animals may depend upon the orientation of the stroke plane. Animals hovering with a more or less horizontal stroke plane (many insects, hummingbirds; see below) are expected to use lift coefficients less than $C_{L,max}$ (Weis-Fogh 1972, 1973), but this apparently is not always the case. Passerine birds and bats (and dragonflies; R.Å. Norberg 1975) hover with an inclined stroke plane, and because their wings generate only relatively small forces on the upstroke these animals must rely on unsteady aerodynamics (Weis-Fogh 1973; Norberg 1975, 1976b). This is discussed in the next section (8.2).

Momentum jet, blade-element, and vortex theories for hovering flight will be described in Sections 8.3–8.6.

8.2 Kinematics of Hovering

The *normal hovering* in hummingbirds is defined by Weis-Fogh (1973) as active flight on the spot with wings moving through a large angle and approximately in a horizontal plane, or in a plane only slightly tilted to the horizontal, and with the long axis of the body strongly inclined to the horizontal (Fig. 8.1a). The wings describe a figure of eight with symmetrical halfstrokes, and the morphological downstroke is actually a forward stroke while the upstroke is a backstroke. During the backstroke the wings are rotated and twisted in the nose-up sense and the morphological upper surfaces face downwards with a slight, inverted camber of the wing. Lift is produced during the entire wingstroke except at the reversal points while drag is elicited during the forward stroke and an equally large thrust during the backstroke. Hummingbirds beat their wings through a large amplitude, usually about 120° (2.1 rad), and wingbeat frequencies during hovering vary with the size of the bird: 15 strokes s^{-1} for a 20-g *Patagona gigas* and 52 strokes s^{-1} for a 3.7-g *Archilochus colubris* (Weis-Fogh 1973). The angle of tilt of the stroke plane is 11° for a 5-g *Amazilia fimbriata* (Weis-Fogh 1972).

In hovering bats and birds other than hummingbirds the stroke plane is usually more tilted, and the wingstroke is rather similar to the one used in slow forward flight (Fig. 8.1b). These animals are anatomically unable to rotate and

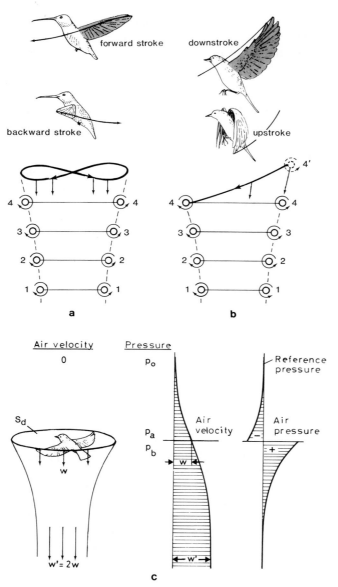

Fig. 8.1. Formation of vortex ring wake in four wingstrokes (*1–4* in chronological order) in symmetrical hovering *a* and in asymmetrical hovering *b*. By the time the wings have reached the bottom of the stroke, the first-formed part (*4′* in *b*) of the vortex ring has convected down to the same horizontal level; the complete rings, therefore, convect downwards in a horizontal orientation in spite of the inclination of the stroke plane. A continuous momentum jet past a hovering animal is shown in *c*. Air is sucked down from rest far above the animal, passes through the wing disk at the induced velocity w and reaches twice that velocity, $w' = 2w$, farther down in the most contracted zone. The air pressure changes from P_o far above the animal to P_a immediately above the wings, to P_b immediately below the wings, and to P_o again far below the animal. (After Norberg 1985a)

twist their wings to attain a positive or near zero angle of attack during the upstroke. To avoid large drag forces and negative lift forces, the animals flex their wings during the upstroke. In birds such as passerines each primary and secondary wing feather also rotates and twists in the nose-up sense, letting air through the wing (Norberg 1975; R.Å. Norberg 1985; see also Fig. 9.7 in the next chapter). In this *asymmetrical hovering* the upstroke is a recovery stroke, giving zero or slight negative lift, so that all useful lift must be produced during the downstroke. Large downstroke lift coefficients (larger than $C_{L,max}$) must compensate for the absence of upstroke lift, and these coefficients are usually not consistent with steady-state aerodynamics.

The long-eared bat (9 g) hovers with a stroke frequency of 11–12 strokes s⁻¹, a stroke amplitude of 120° (2.1 rad) and with the stroke plane tilted about 30° to the horizontal (Norberg 1970a, 1976b). The lift coefficient estimated with quasi-steady aerodynamics exceeds 3 (Norberg 1976b). The corresponding data for the pied flycatcher (12 g) are 14 strokes s⁻¹, 102° (1.8 rad), and 30°, respectively, and the estimated lift coefficient is about 5 (U.M. Norberg 1975). Dathe and Oehme (1980) estimated that the lift coefficient in the downstroke of the pigeon (*Columba livia*) was 2.8 and that of the black-headed gull 2.1 in hovering flight. All three values are higher than expected for steady-state conditions.

8.3 The Rankine-Froude Momentum Theory

This theory has been briefly described in Section 2.3.1 for forward flight. A hovering animal beating its wings horizontally in symmetrical hovering can be compared to a helicopter whose rotors drive the air downwards through the disk they sweep. The airflow through the disk is due entirely to the induced velocity w, which is assumed to be constant over the whole disk area. The airflow reaches this velocity when it accelerates downwards, first into the area of reduced pressure above the wings. Then the air continues to accelerate downwards to velocity w' below the wings because of a similar pressure gradient there (Fig. 8.1c). This pressure gradient occurs because the downward-flowing air diminishes below the flyer as the air jet area contracts.

In a real helicopter the air would be set swirling and not driven vertically, but it is simpler to discuss an imaginary (actuator) disk driving air downward with a purely vertical velocity distributed uniformly across the disk. This is assumed in the Rankine-Froude theory and it would require less induced power than any real rotor of the same radius. The wing disk area S_d is usually taken to be the area of a full circle whose diameter is the wingspan, although the wings mainly sweep out smaller sectors, making $S_d = \pi b^2/4$. There is a debate about whether the actual swept areas should be used or the entire disk area. Ellington (1978, 1984d-f) used the swept sector areas for his calculations on insect hovering, since they sweep out rather small proportions of the whole 360° disk, each being equal to $\emptyset b^2/4$, where $\emptyset$ is the stroke amplitude. He stated, however, that "this problem is quite tractable for an experimental study of the wake velocities using standard techniques . . .". The theory is also applicable to asymmetrical hovering so that if the stroke angle (angle between stroke plane and the horizontal) is β, then the horizontally projected disk area will equal $S_{d,proj} = S_d \cos\beta = \pi b^2 \cos\beta/4$.

The vertical force on the wings, which must equal the weight of the hovering animal, is given (by Newton's second law) as the rate at which downward momentum is imparted to the air. This rate of change of momentum is the product of the air velocity below the animal, w', and the mass flow of the air, f_m, through the wing disk. The mass flow is the rate at which mass passes through the disk, and equals $\rho S_d w$. The rate of change of momentum is then

$$w'f_m = \rho S_d ww' = Mg. \tag{8.1}$$

The air accelerates to the velocity $w' = 2w$ far below the animal which can be discovered by using the Bernoulli's theorem (cf. Sect. 2.2.1). Air is sucked down toward the disk from the ambient air pressure towards the low pressure region immediately above the disk. Then the air is driven by the high pressure immediately below the disk and accelerated down to the level where the pressure again equals the ambient pressure. Assume that the air approaching the wing disk changes its velocity from 0 and its pressure from p_o far above the animal to velocity w at the disk and pressure p_a immediately above the wing surface. Also assume that the air accelerates further downwards to velocity w' far below the animal with a pressure change from p_b immediately below the wing surface back to p_o further down (Fig. 8.1c). Then, by Bernoulli's theorem [Eq. (2.2)]

$$p_o/\rho = p_a/\rho + w^2/2$$

and

$$p_b/\rho + w^2/2 = p_o/\rho + w'^2/2,$$

which gives

$$p_b - p_a = \rho w'^2/2. \tag{8.2}$$

The pressure difference acting over the wing disk area must equal the weight of the animal, so that

$$Mg = (p_b - p_a)S_d = (1/2)S_d \rho w'^2 = (1/8) \rho \pi b^2 w'^2. \tag{8.3}$$

Combining Eqs. (8.1) and (8.3) gives the relations

$$\rho S_d ww' = (1/2)S_d \rho w'^2$$

and

$$w' = 2w. \tag{8.4}$$

The induced velocity can be obtained from Eqs. (8.3) and (8.4), and is

$$w = (Mg/2\rho S_d)^{1/2}, \tag{8.5}$$

where Mg/S_d is the *disk loading*. The induced velocity is therefore proportional to the square root of the disk loading, or to the inverse of the wingspan. It is referred to as $w_{R.F}$ (the Rankine-Froude model of induced velocity) in Sect. 8.4.

104

The momentum-jet-induced power, $P_{ind,RF}$, is the product of the weight of the animal (here equivalent to the induced drag) and the induced velocity, and is

$$P_{ind,RF} = Mg\,w = (Mg)^{3/2}/(2\rho S_d)^{1/2}. \tag{8.6}$$

This gives a *minimum* value of the induced power. The actual, effective, value involves an induced factor k [cf. Eq. (2.10)] which may be = 1.2 for animal wings (Pennycuick 1975). The value of k has also been estimated by the vortex theory, with similar results (see Sect. 8.5). Low weight and long wings will be the most effective way to reduce the induced power required for hovering.

8.4 Blade-Element Theory

8.4.1 Profile, Parasite and Inertial Power

The profile, parasite and inertial powers are all small components in hovering flight. In all power models the profile and inertial powers are estimated with blade-element theory, and the parasite power also can be calculated with conventional aerodynamic theory.

The *profile power* involves instantaneous values of the flapping speed and the nondimensional profile drag coefficient, and is given by

$$P_{pro} = (1/2)\rho V_r^3 S C_{D,pro} \tag{8.7}$$

[cf. Eq. (2.18)], where V_r is the resultant of the flapping speed V_f of the wing and the induced velocity w. Assuming that the wing's angular movement is sinusoidal, the flapping velocity can be expressed by Eq. (8.9) (see below). For *normal hovering* the induced velocity can be approximated by the momentum-jet theory and the profile drag coefficient can be assumed or measured in a wind-tunnel. Rayner (1979b) assigned a value of 0.02 for the zero-lift drag coefficient $C_{D,pro}$ (along the zero-lift line of the wing; see Fig. 2.9) for bird wings. Each quarter of the total wingstroke in normal hovering are similar to each other, and thus elicit equally large forces and require the same amount of power.

Rayner (1979b, c) calculated the profile power by a method derived from Osborne (1951), Pennycuick (1968a) and Weis-Fogh (1972), and for *asymmetrical hovering* it can be expressed as

$$P_{pro,hov} = 0.0166\mu\rho b^3 S C_{D,pro}\mathcal{O}/\tau^2 T^3. \tag{8.8}$$

The factor μ (*muscle ratio*) is defined as $(1 + m_s/m_p)$, where m_s is the mass of the muscles powering the upstroke and m_p of those powering the downstroke. Typically μ is about 1.1, so the upstroke muscles are about 10% of the mass of the downstroke muscles. The downstroke ratio τ is the time the downstroke takes in proportion to the time T for the whole wingstroke (downstroke duration is τT).

Parasite power can be neglected since the forward flight component is absent and the induced velocity is so low. *Inertial power* of the flapping wings may be of some importance in hovering flight for it is the power required for acceleration and deceleration of the wings at the start and end of each wingstroke, quite apart from the aerodynamic forces, and it can be calculated with Eq. (2.27), Section 2.3.2.

8.4.2 Weis-Fogh's Model for Normal Hovering

The primary aim of the analysis of large and small hovering animals by Weis-Fogh (1972, 1973) was to make quantitative estimates of force, work and power with the blade-element theory. His model used a compound drag coefficient to lump together the aerodynamic power components (induced and profile power). The model may give accurate values for normal hovering, when non-steady effects are assumed to be of minor importance, and it can be used as a test for the quasi-steady assumptions. The procedures of Weis-Fogh's quasi-steady approach are briefly summarized here.

The rationale of the method is to assume that steady-state aerodynamics prevail, to perform the calculations based on the theory for steady-state aerodynamics, and then to contrast the results against known values of the lift coefficient C_L. Based on this comparison one can determine whether or not the initial assumption can be upheld; in other words, whether or not steady-state aerodynamics provide sufficient explanation. When the estimated lift coefficient exceeds the value predicted by the wing-profile and the appropriate Reynolds' number (i.e. one > 1.5–1.6 for birds and bats), another approach, such as the vortex theory, should be used.

The wing's angular movement in hovering animals can conveniently be approximated with the equation for a simple harmonic motion, and calculated as

$$\gamma(t) = \bar{\gamma} + (1/2) \, \varnothing \sin (2\pi nt), \tag{8.9}$$

where $\gamma(t)$ is the positional, instantaneous, angle of the long wing-axis in the stroke plane, $\bar{\gamma}$ is the average angle, $\varnothing$ is the angle swept by the wings (stroke amplitude), n is wingbeat frequency, and t is time (starting at zero from the middle of the upstroke. In Fig. 8.2a these data are shown for a bat in horizontal flight. The instantaneous angular velocity is

$$d\gamma(t)/dt = \pi n \varnothing \cos (2\pi nt). \tag{8.10}$$

Assuming that the wing movement is sinusoidal, the flapping velocity V_f of a wing-element at distance r from the fulcrum, and along the long wing axis, is

$$V_f(r,t) = r \, d\gamma(t)/dt = r\pi \varnothing n \cos (2\pi nt). \tag{8.11}$$

The local, instantaneous, velocity $V_r(r,t)$ is the the resultant of the local, instantaneous, flapping speed and the induced velocity (Fig. 8.2b). According to the momentum theory for an ideal actuator disk, the induced velocity at the level of the disk itself is $w = (2Mg/\rho\pi b^2)^{1/2}$ [Eq. (8.5)]. The local, instantaneous, resultant aerodynamic force is

$$F(r,t) = (1/2)\rho V_r^2(r,t)S_r(C_L^2 + C_D^2)^{1/2} = (1/2)\rho \, V_r^2(r,t)S_r C_L/\cos \theta, \tag{8.12}$$

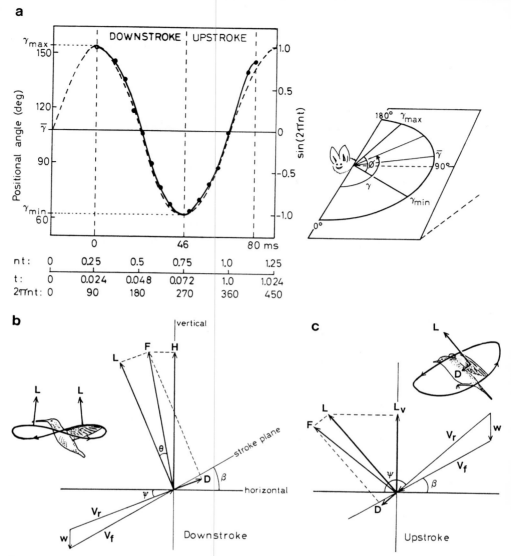

Fig. 8.2. *a* Angular movement of the long wing-axis of a long-eared bat (*Plecotus auritus*) in horizontal flight (*full curve* from high-speed cine film) showing that the wing performs essentially a sinusoidal movement (*dashed curve* calculated on the assumption that the wingstroke is a simple harmonic motion). Since the wings are not raised to the same level as in the beginning of the downstroke in this particular wingstroke, the period of the full curve is here somewhat shorter than that of the sinusoidal curve (which is twice the downstroke time). $\emptyset$ is wingstroke angle, γ is the wing's positional angle. (After Norberg 1976a) *b,c* The velocity and force systems acting on a wing-element in hovering flight. *b* A phase in the forward or backward stroke in symmetrical hovering and in the downstroke in asymmetrical hovering. *c* A phase in the upstroke in asymmetrical hovering. *H* is the vertical component of the resultant force *F* (used in Weis-Fogh's model for force equalization), L_v is the vertical component of the lift force (used in Norberg's model for force equalization), V_f is flapping velocity, *w* the induced wind, V_r the resultant velocity, and β is the angle of tilt of the stroke plane relative to the horizontal. The L/D ratio is taken to be 5. The *inset* bird figures show their wingtip paths during symmetrical *b* and asymmetrical *c* hovering and some elicited forces. (After Norberg 1975, 1976b)

where $\tan \theta = C_D/C_L$, whence the vertical component becomes

$$H(r,t) = F(r,t) \cos(\theta + \psi), \tag{8.13}$$

where ψ is the angle between the relative airspeed and the horizontal (Fig. 8.2b).

The kinematic data used in the calculations must be measured from cine films or from stroboscopic multiple-exposure still pictures of the hovering animal. The lift:drag ratio must be estimated, and in asymmetrical hovering it can be estimated by calculation (see Sect. 8.4.3). Weis-Fogh (1972) adopted the ratio of 6 for the hummingbird *Amazilia fimbriata*, while Pennycuick (1969) estimated ratios of about 5 for small passerine birds, a value also used by Norberg for the pied flycatcher (Norberg 1975). C_L furthermore is assumed to be constant during the entire wingstroke and along the whole span of the wing. The vertical component $H(r,t)/C_L$ must then be calculated for each of a number of chordwise wing strips and time-equidistant points during any quarter of a wingstroke, since in symmetrical hovering each quarter of a stroke is similar to the other. The average value of H/C_L can be found by graphically integrating the summed curves for all strips of both wings with respect to time for a quarter of a wingstroke, dividing by the corresponding time, and multiplying by 4 (Fig. 8.3). The vertical force H is averaged over a whole wingstroke and must equal the weight Mg of the animal. Hence the minimum average value of C_L can be estimated by finding what value C_L must take for the product of C_L and H/C_L to equal the weight Mg.

To calculate work and power for hovering hummingbirds, Weis-Fogh's quasi-steady analysis can be used, provided that the calculated average lift coefficient is consistent with such an approach. To estimate the power we must first calculate the angular acceleration of the wing, the inertial and the aerodynamic torques, and the work.

Derivation of the instantaneous angular velocity [Eq. (8.10)] gives the instantaneous angular acceleration

$$d^2\gamma(t)/dt^2 = -2\pi^2 n^2 \emptyset \sin(2\pi nt). \tag{8.14}$$

The bending moment caused by the acceleration of the wing-mass in the stroke plane, i.e. the inertial torque, then is

$$Q_i = I\, d^2\gamma(t)/dt^2 = -2I\pi^2 n^2 \emptyset \sin(2\pi nt), \tag{8.15}$$

where I is the moment of inertia of the wing-mass with respect to the wing-base [cf. Eq. (2.28)].

The aerodynamic torque attributed to a wing section is the distance (radius) from the fulcrum to the wing section under consideration times the aerodynamic force component in the stroke plane. This force component of the resultant force F_r is

$$M(r, t) = (1/2)\,\rho\, V_r^2(r, t)S_r\, C_L \sin(\theta + \psi + \beta)/\cos\theta, \tag{8.16}$$

and the aerodynamic torque is

$$Q_a(r, t) = rM(r, t). \tag{8.17}$$

The estimated average lift coefficient should be inserted in Eq. (8.16).

The total (aerodynamic + inertial) work produced by the wingstroke muscles during one wingstroke is the sum of the work done during the downstroke and that done during the upstroke, and is

$$W_t = \int_{\gamma_{min}}^{\gamma_{max}} (Q_a + Q_i)\, d\gamma + \int_{\gamma_{max}}^{\gamma_{min}} (Q_a + Q_i)\, d\gamma. \tag{8.18}$$

The total mechanical power imparted by the wing muscles to the wings is

$$P_{hov} = n\, W_t. \tag{8.19}$$

For comparisons between different-sized animals it is useful to estimate the *specific work* and *power*, which are found by dividing the obtained values by the total mass of the animal in question. Specific values are usually indicated by an asterisk after the symbol, i.e. the specific power is

$$P_{hov}{}^* = n\, W_t / Mg. \tag{8.20}$$

If polar diagrams from real wings are used to determine the lift:drag ratio, the induced drag is automatically taken into account. It is then meaningful to estimate the *aerodynamic efficiency of hovering*, which is the ratio between the power estimated from the momentum-jet theory [Eq. (8.6)] divided by the total aerodynamic power calculated from Eq. (8.19), $e_h = P_{ind}/P$. Weis-Fogh (1972) estimated the aerodynamic efficiency in a hovering hummingbird (*Amazilia fimbriata*) as 0.52, and similar values were found in hovering insects (Ellington 1984f).

8.4.3 Norberg's Model for Asymmetrical Hovering

In the asymmetrical hovering analyzed by Norberg (1975, 1976b), the upstroke differs greatly from the downstroke and has to be treated separately. In the pied flycatcher, the kinematics of the upstroke suggest that it contributes no useful forces (Norberg 1975), and the average value of $H(r, t)/C_L$ was calculated just for the downstroke. Integrated over the time for the whole wingstroke it must equal the weight of the bird. In the long-eared bat, the upstroke appears to give some useful forces and was treated separately (Norberg 1976b).

My analysis of the bat differs somewhat from that of Weis-Fogh, even though it is based primarily on his basic assumptions. The method is a further development of his and Pennycuick's (1968a) expressions for estimating the average lift coefficient in flapping flight. Pennycuick found the average resultant force coefficient C_F by setting the vertical components of the resultant force for the whole wing and wingstroke equal to the weight of the animal (pigeon). He then found the lift coefficient from his program by iteration and the ratio L/D from wind-tunnel measurements of a gliding pigeon. His calculations refer only to the downstroke. Weis-Fogh assumed an L/D ratio and also calculated the lift coefficient by setting the vertical components of the resultant force for the entire wing and wingstroke equal to the weight of the animal (see Sect. 8.4.2).

In my calculation (Norberg 1976b) I used the lift and drag forces (instead of the resultant force) for force equalization. The force coefficients and the lift:drag ratio are unknown, as are the directions of the resultant force at various wing positions. But the directions of the lift and drag forces (but not the magnitudes) can be found, since one can estimate the direction of the resultant airstream. The sum of the vertical components of the lift and drag forces themselves (rather than their resultant force), as integrated over the time of a whole wingstroke and over the whole wings, must equal the weight of the animal as integrated over the same time. In the same way the sum of the integral of the horizontal components in the direction of the flight path of the lift and drag forces, respectively, must equal the time integral of the body drag (Figs. 8.2c and 8.4). In hovering, the body is immersed in the induced wind, that probably is nearly vertical. The body therefore experiences a nearly vertical downward drag force, the horizontal component of which should be negligible. The horizontal body drag is set equal to zero. Actually, the vertical component of the body drag due to the induced wind should be added to the weight in the vertical force equation, but since it is very small this force also is omitted. With this equation system the lift and drag coefficients and their ratio can be calculated analytically. This model was also used to estimate the lift coefficients and power for forward flight in the bat (see Sect. 9.5.6), but it differs in the calculations of the resultant speeds and body drag.

The forces produced during a complete wingstroke can be found by dividing the wing into a number of strips and summarizing the calculated average speeds and forces for each strip. The wing movement of the long wing-axis (the axis between the humero-scapular joint and the wingtip) can be taken to be sinusoidal with respect to angular displacement (cf. Weis-Fogh's model, Sect. 8.4.1)
. The upstroke resultant airspeed of a wingstrip is different from that of the downstroke, since the flapping velocity component is directed upwards (Fig. 8.2c). Although induced velocity may fluctuate during the wingstroke, it was assumed to be constant in time throughout the entire wingstroke and uniformly

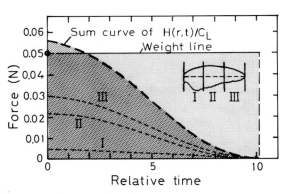

Fig. 8.3. The three wing-sections *I–III* in *Amazilia* (M = 5.1g, n = 35 s⁻¹) and the vertical dynamic pressure index $H(r,t)/C_L$ for each section as it varies with relative time t during a quarter stroke of symmetrical hovering (*broken curves*). The weight of the hummingbird is indicated by an *open circle* on the ordinate axis. The area under the summed curve (*heavy broken line*) must be multiplied by a value of C_L (here near to, but smaller than, 2) such as to make the product equal to the area under the straight weight-line during the time for a quarter of a stroke (the four quarter strokes being symmetrical). This is how the value of an average, minimum, lift coefficient is estimated by Weis-Fogh's quasi-steady analysis. (After Weis-Fogh 1972)

110

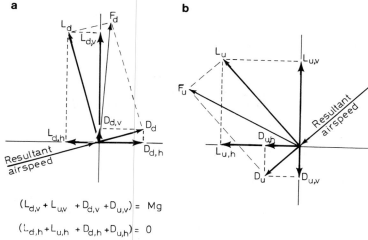

$$(L_{d,v} + L_{u,v} + D_{d,v} + D_{u,v}) = Mg$$

$$(L_{d,h} + L_{u,h} + D_{d,h} + D_{u,h}) = 0$$

Fig. 8.4. Average horizontal and vertical force components acting on a wing segment in the downstroke *a* and in the upstroke *b*. The sum of the horizontal components, as averaged over the whole wing and wingstroke, must equal the horizontal body drag D_b, which is around zero in hovering flight. The sum of the vertical components must balance the weight Mg of the animal. F is resultant force, L is lift, D is drag, T is wingstroke period; the force index d refers to the downstroke, u to the upstroke, h to horizontal and v to vertical

distributed over the wing disk. In reality, the induced velocity vector may be very small during the upstroke because of the partly flexed wings. During the downstroke it has to be large to compensate for the loss of lift in the upstroke. The induced velocity is, however, small compared to the flapping velocity and it will have rather a small effect on the direction and size of the relative air velocity.

In the asymmetrical hovering of the pied flycatcher and of the long-eared bat, the calculated lift coefficients exceeded those explainable by quasi-steady aerodynamics, so the vortex theory must be used for the induced power calculation.

8.5 Vortex Theory

The vortex theory relates the lift produced by the flapping wings to the pattern and velocity of the induced flow of the wake and to the power needed to generate that flow. It must be remembered that profile and parasite powers are not covered by vortex theories. Furthermore, because the induced power makes up such a small proportion of the total power (sum of parasite, profile, induced, inertial powers) in fast flights, the quasi-steady assumption is of little importance. In slow flight and in hovering, however, it will be significant. The induced power may comprise 80–90% of the total power in hovering, while at minimum power speed it is 25–75%, and at the maximum range speed 5–20% (Rayner 1979c).

The Rankine-Froude momentum theory gives a minimum value for the induced velocity, which may be a good approximation for normal hovering when the stroke plane is almost horizontal and the momentum jet is vertical. But this method does not take into account the effects of periodic pressure pulses produced by wing flapping, so it will underestimate the induced velocity, particularly when large unsteady effects are involved. For symmetrical hovering the momentum jet theory may underestimate the induced power by 10–15%, and for asymmetrical hovering by as much as 60% (Rayner 1979a; Ellington 1984e). Using the Rankine-Froude theory, the total mechanical power for symmetrical hovering may be underestimated by 8–18%, for asymmetrical hovering by 48–54%, for horizontal powered flight at medium speeds (near V_{mr}) by 15–45%, and for fast flight (near V_{mr}) by 3–12%.

The periodic pulses make the jet discontinuous, so the wake will consist of a chain of vortex rings (Fig. 8.1a). These rings are produced from the vorticity within the vortex sheet, which is continuously shed during each half-stroke in normal hovering and during each downstroke only in asymmetrical hovering. The ring convects downwards with the induced velocity field. The reaction of the momentum convected to the air is the lift experienced by the wings. The momentum of a vortex loop is equal to the product of the vortex sheet (loop) area and its circulation. This is the *impulse theory*, which is the time-average of the lift in the Kutta-Joukowski theorem [Eq. (2.31)]. The force impulse required to generate the sheet is derived from the vorticity of the sheet, and neither the impulse nor energy of the vortex change as the vortex rolls up. The mean wing lift is equal to the impulse divided by the period of generation. The induced power is calculated as the rate of increase of wake kinetic energy. Impulse analysis is a very general method for estimating the mean circulatory lift, and it is especially useful for the study of unsteady effects. It has been used, for instance, by Maxworthy (1979), Rayner (1979a, 1986), Ellington (1984d-f), and Spedding (1986) in their studies on animal flight.

Ellington (1978, 1980, 1984d-f) and Rayner (1979a,c) independently developed vortex theories for hovering flight in animals. Both combine suitable forms of the momentum theory with the vortex approach. Their models provide methods of calculating the mean lift and induced power for a given circulation profile during the wingstroke, for unsteady as well as for quasi-steady aerodynamics. Once the circulation profile is known during a wingbeat, the mean lift and induced power can be calculated. But their theories necessarily rely on assumed circulation profiles, where the circulation is a function of spanwise position and time. Uncertainty about circulation profiles parallels uncertainty about force coefficients in the blade-element theory.

There are strong similarities between Ellington's and Rayner's theories, and only two basic differences. Ellington's model is designed to study the effects of wake periodicity in isolation, and he assumes a uniform circulation profile. He invokes two approximations for the far wake structure to keep the mathematics simple and he assumes first that the ring area in the far wake is approximately half the initial area (as in the Rankine-Froude model), and second that the vortex rings are equally spaced along the far wake axis. Rayner's model considers a non-uniform profile with a rolled-up ring area, and avoids any approximations about the far wake structure by developing it from rest in a comprehensive numerical analysis. To estimate the induced power, Ellington uses the small periodicity approximations in the equations of a quasi-steady momentum jet, while Rayner derives it from the energy required to add a new ring to the chain of vortex rings. Ellington (1984e) compared his and Rayner's theories and found close agreements between them. Their models are briefly described below, but the reader is referred to their works for details.

8.5.1 Ellington's Hovering Model

Ellington (1984d-f) derived the induced velocity and power of the wake in stages, starting with the Rankine-Froude theory and applying a uniform, continuous pressure to the air. He then improved the model by taking into account variations in pressure and circulation over the disk area by considering a "modified" Froude disk that exerts a continuous and non-uniform pressure. With this approach he provided a correction factor, k, for the momentum-jet theory, which is a function of the vortex ring characteristics and includes a *spacing parameter* (Ellington 1984e). The induced power requirements was then be expressed as a function of the power obtained from the momentum jet theory, $P_{ind,RF}$, as

$$P_{ind} = k\, P_{ind,RF}. \tag{8.21}$$

The values of k for hovering insects with a horizontal stroke plane are estimated to 1.11–1.21 and with inclined stroke plane to 1.57–1.65 (Ellington 1984f), and these factors may become similar to those of birds and bats.

8.5.2 Rayner's Hovering Model

In Rayner's (1979a,c) theory, the wake is modelled by a chain of coaxial small-cored circular vortex rings stacked one upon another, and each member is generated by a single wingstroke. The theory involves lengthy numerical calculations, described in detail in Rayner (1979a). From these results, Rayner (1979c) derived formulae for the estimation of induced power for a hovering animal. Two dimensionless parameters were used to determine the wake geometry and power consumption, one depending on the initial vortex-ring radius (R′) and the other on morphology (feathering parameter, f). The resulting configuration of the vortex wake depends on the introduced *feathering* parameter f, which depends on the animal's morphology and which increases with the size of the animal. This non-dimensional parameter, which summarizes the relevant kinematic and morphological data, is the square root of the ratio between the mean induced velocity past the wings and the mean wingtip velocity, and is given by

$$f = (w_{RF}/u_t)^2 \tag{8.22}$$

This feathering parameter is proportional to Ellington's (1984e) spacing parameter s.

Rayner distinguished between *normal* hovering and *avian* (= asymmetrical) hovering, and his theory is applicable to both types. In avian hovering the upstroke is feathered and presumed to produce no useful forces. He introduced the wake period T_w as the period between successive powered strokes; in normal hovering $T_w = T/2$ and in avian hovering $T_w = T$, where T is the wingstroke period (= the inverse of the wingstroke frequency).

Using the induced velocity from the momentum jet theory, $(Mg/2\rho S_d)^{1/2}$ [Eq. 8.5], and the wingtip velocity $u_t = \pi b/2T_w$, where b is wingspan, Rayner defined the feathering parameter [Eq. 8.22] as

$$f = 2Mg\, T_w^2/\rho\pi^2 b^2 S_d = 8Mg\, T_w^2/\rho\pi^3 b^4. \tag{8.23}$$

The feathering parameter was calculated as 0.010 and 0.011 for two hummingbirds, and between 0.027 and 0.175 for some birds and a bat using asymmetrical hovering.

The other important parameter is the non-dimensional radius of each ring when released, R'. The initial radius is $2R'/b$, which is less than the wing semi-span $b/2$. Rayner derived the following values of R', by equating the momentum of the vortex ring to the momentum of the vortex sheet shed by the wings,

$$R' = (0.923\varnothing/\pi)^{1/2} \tag{8.24}$$

in normal hovering, and

$$R' = (0.808\varnothing/\pi)^{1/2}, \tag{8.25}$$

in asymmetrical hovering, where $\varnothing$ is the stroke amplitude, measured in radians [typically between $(2/3)\pi$ and π]. The induced power was calculated as a function of the power obtained with the momentum jet theory, $P_{ind,R,F}$, as

$$P_{ind} = P_{ind,RF}/R' \tag{8.26}$$

in normal hovering, and

$$P_{ind} = P_{ind,RF}\,(0.95/R' + 1.2f/R'^5) \tag{8.27}$$

in asymmetrical hovering. The inaccuracy of using the momentum jet theory becomes greater as f increases, and will give significant underestimation when f is larger than about 0.05 with large $\varnothing$, or for any f when $\varnothing$ is small. For small values of f the vortex rings are close together, and the situation is close to the momentum jet generated by an actuator disk of the radius $2R'/b$. Rayner concluded that the momentum jet may be a good approximation for the estimation of the induced power in symmetrical hovering, but not for asymmetrical hovering.

Rayner suggests that the optimum strategy for an animal to adopt hovering is to reduce disk loading (having a large wingspan), maximize R' by increasing the stroke amplitude (large $\varnothing$), and minimize f by increasing the wingstroke frequency (small T). Proportionally large flight muscles, as in the hummingbirds, would also be expected.

8.6 Animals with Sustained Hovering

Hummingbirds can hover for long periods, but many birds and bats using asymmetrical hovering hover only for short periods. Larger animals, and some small ones, often are unable to provide sufficient power output for hovering. Using the Rankine-Froude model for the induced power, Weis-Fogh (1977) made a dimensional analysis of hovering flight and concluded that, between specific aerodynamic powers of 50 to 200 W kg^{-1}, the absolute size of an animal capable of sustained hovering depends on the relative muscle mass m* (muscle mass divided by body mass). Many bats have a relative muscle mass of about 0.10, which would permit a 10-cm maximal wing length compatible with sustained hovering; the comparable values for hummingbirds are about 0.30 and 15 cm. The long-eared bat has a wing length of about 11.5 cm (Norberg 1970a), close to the estimated upper limit of the size of an animal capable of sustained hovering.

Rayner (1979c) noted that the wren, kestrel, mallard and pheasant have insufficient power to hover aerobically, but a pigeon with a large stroke amplitude could do so without incurring a significant oxygen debt. The pied flycatcher often hovers for short periods during insect captures, but this is probably not sustained hovering and results from anaerobic muscle work. Rayner suggested that pied flycatchers may be capable of sustained hovering with larger stroke amplitude over the 100° estimated by Norberg (1975). The long-eared bat, on the other hand, should be fully capable of sustained hovering.

8.7 Summary and for Ecologists and Others: Recipes for Power Calculation

This section will give some suggestions for calculating flight power in hovering animals and indicate which measurements are necessary for the different models. The choice of models takes into account the minimum amount of data required for a reasonable accuracy.

The doubly labelled water method (Sect. 3.6.3) estimates an animal's energy costs, and can be used to determine the cost for hovering, if the hovering sequences can be timed. This involves laborious work, and aerodynamic theory may be easier to use.

The power required for hovering is the sum of the induced power, profile power and the inertial power. Parasite power is negligible since forward flight is zero, and although the inertial power is very small (perhaps negligible), it could be included in thorough analyses.

Profile and inertial powers can be obtained with the blade-element theory, but the most accurate models for calculating the induced power are those of vortex theory which account for unsteadiness in air motion around the wings. The simplest model for calculating induced power is the Rankine-Froude momentum theory, which gives a *minimum* value of the induced power and may underestimate it by up to 20% in normal hovering and by 60% in asymmetrical hovering (Rayner 1979a; Ellington 1984e). Therefore this momentum theory should be used only in exceptional cases when kinematic data are unknown, and only the morphology of the animal is known. But if the Rankine-Froude value is multiplied by an estimated or assumed induced factor, the underestimation may be corrected for.

For calculating the different power components I have suggested different methods, but I prefer the first in each case. The power input P_i is then received from Eq. (2.20) which includes the muscles' mechanical efficiency η (0.20–0.25; Sect. 3.2), basal metabolism (e.g. Eqs. 3.3 and 3.4) and a circulation and ventilation factor (1.10).

8.7.1 Induced Power, Normal Hovering

1. *Model to be used:* The momentum jet induced power times a correction factor; Eqs. (8.6), (8.24) and (8.26) (Rayner 1979c).
   Data needed: Body mass M, wingspan b, wingstroke amplitude Ø.
2. *Model to be used:* The momentum jet induced power times a correction factor; Eqs. (8.6) and (8.21) and the mean values of the induced factor estimated for insects in normal hovering, $k = 1.16$ (Ellington 1984e,f).
   Data needed: Body mass M, wingspan b.
3. *Model to be used:* The momentum jet induced power times a correction factor; Eq. (8.6) times the induced factor $k = 1.2$ (Pennycuick 1975; Ellington 1984f).
   Data needed: Body mass M, wingspan b.

8.7.2 Induced Power, Asymmetrical Hovering

Model to be used: The momentum jet induced power times a correction factor; Eqs. (8.6), (8.23), (8.25), and (8.27) (Rayner 1979c).
   Data needed: Body mass M, wingspan b, wingstroke amplitude Ø, wingstroke period T (the inverse of stroke frequency).

8.7.3 Profile Power, Normal Hovering

Model to be used: Strip-analysis with the blade-element theory for power calculation, and the momentum-jet theory for calculation of the induced velocity. The local, instantaneous powers should be calculated and summed to get the average value for one wingstroke. Instead of a strip-analysis, the average instantaneous speeds found at $0.75 \times$ (wing length) could be used. Equations to be used: profile power Eq. (8.7), velocity equations (8.5) and (8.11), and profile drag coefficient $C_{D,pro} = 0.02$ (Rayner 1979b).
   Data needed: Wing length l_w, wingstroke amplitude Ø, wingstroke period T.

8.7.4 Profile Power, Asymmetrical Hovering

Model to be used: Eq. (8.8), muscle ratio $\mu = 1.1$ and profile drag coefficient $C_{D,pro}$ = 0.02 (Rayner 1979b).

Data needed: Wingspan b, wing area S, wingstroke amplitude Ø, downstroke ratio τ, wingstroke period T.

8.7.5 Inertial Power

Model to be used: Strip analysis for the calculation of moment of inertia (see Sect. 2.3); Eqs. (2.27), (2.28) and (8.9). The inertial power is regarded to be small and may be neglected.

Data needed: Wingstrip masses m_w' and radii r', wingstroke amplitude Ø, wingstroke period T.

Chapter 9

Forward Flight

9.1 Introduction

In slow forward flight unsteady aerodynamic effects are important but they decrease with increasing forward speed as the contribution of the flapping velocity to the resultant relative velocity decreases. In other words, as the value of the *reduced frequency* parameter (the ratio of flapping velocity to the forward speed) decreases, so do the unsteady aerodynamic effects. Therefore, slow forward flight is best understood with vortex theory, whereas fast flight may be analyzed accurately using conventional methods. Induced power decreases with increasing forward speed, whereas the profile power and, particularly, the parasite power progressively increase (cf. Fig. 2.9), so that fast-flying animals should have a very streamlined body and small wings to reduce the total power required for flight (see Sect. 2.3.2).

9.2 Wing Kinematics

Birds and bats move their wings in a similar way during the wingbeat cycle, but the birds usually move them with the downward path ahead of the upward path relative to the body, whereas bats do the reverse. Figure 9.1 shows the wing positions of a pigeon flying very slowly, with the body tilted at a large angle to the horizontal. As forward speed increases, the body is held more horizontally. During the downstroke the wings are extended and swung downwards and then forwards. In the upstroke the wings are more or less flexed at the elbow and wrist to reduce wing area and hence drag, and the primaries may separate as in hovering flight. The kinematics of bird flight have been described by many workers (e.g. Brown 1948, 1953, 1963; Bilo 1971, 1972; Oehme and Kitzler 1974; Nachtigall 1980; Oehme 1985, 1986).

To attain an optimal angle of attack along the entire wings during the wingstroke the animal must be able to twist the wings, which in turn requires adaptations in the skeletal and muscular systems (see Sect. 11.2 and 11.3). Twisting of the wings is particularly important during the downstroke and the later part of the upstroke when the wing extends again. The degree of wing twisting required is related to the flapping speed (Nachtigall 1980) and forward speed, which determine the direction of the relative airspeed, and hence the angle of attack. Bilo (1971) found that in the house sparrow in free flight, the angle of attack increased from $-3.9°$ at the wing base to $+20.1°$ at the tip in the

118

Fig. 9.1 Wing tracings from photographs taken at intervals of 0.01 s on a pigeon in slow flight: (*a*) downstroke, (*b*) upstroke. (After Brown 1963)

middle of the downstroke. Bat wings are twisted in a similar way (Norberg 1976a).

The kinematics of slow flight (2.4 m s^{-1}) have been thoroughly described in the long-eared bat *Plecotus auritus* (Norberg 1976a; Fig. 9.2). The body is held almost horizontally, the wingstroke angle is about 90° relative to the horizontal, and the stroke frequency is about 13.5 strokes s^{-1}. The stroke frequency decreases as speed increases and is 10 strokes s^{-1} at 3 m s^{-1}. The functional relationship between wingbeat frequency and flight speed is described in Chapter 10. The movement of the long wing-axis (the axis between the humero-scapular joint and the wingtip) is almost sinusoidal with respect to angular displacements about the mid-stroke position, which is also the case in birds in forward flight and in hovering birds and bats (see Sect. 8.3, Fig. 8.2).

In the long-eared bat, the downstroke starts with the wings extended at an angle of about 50° above the horizontal. The wings then sweep downwards and forwards fully extended and move essentially in one plane, tilted about 60° to the horizontal (Fig. 9.3a-c). During the first half of the downstroke the wings are twisted, with the middle and distal thirds being pronated (rotated nose-downwards). Just before they reach the horizontal level, the wings begin to supinate (rotate nose-upwards), and when they reach an angle of 20–25° below the horizontal the upstroke starts with a slight flexion at the elbow and wrist (Fig. 9.3d). The wrist begins to rise while the wingtip and trailing edge are still moving downwards and the whole wings are then brought upwards and slightly backwards relative to the still air. The wings remain twisted during the upstroke with the distal part most supinated. When extended during the latter half of the upstroke, the wings perform an upward and slightly backward flick (Figs. 9.3e and 9.4a).

At the level of the fifth digit the wing also is positively cambered during the upstroke. The feet are moved upwards during the upstroke, reducing the positive

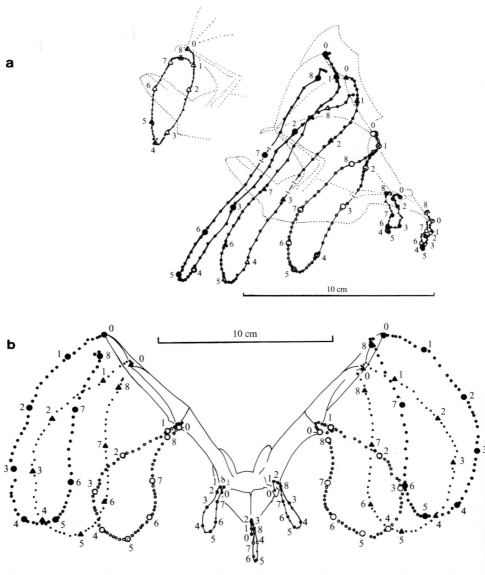

a

b

Fig. 9.2 *a* Lateral and *b* posterior projections of the tracks (traced relative to the body) of the tips of the thumb (only in *a*) and the third-fifth digits, feet, and tail over a complete wingstroke of the long-eared bat (*Plecotus auritus*) flying horizontally. The *numbers* indicate each 100th of a second starting from the uppermost position of the wings. In *a* the flight speed is 2.35 m s^{-1} and the stroke frequency is 11.9 Hz, and in *b* the speed is 2.4 m s^{-1} and the frequency 11.8 Hz. (Norberg 1976a, by courtesy of The Company of Biologists Ltd; from cine films run at 700–800 frames s^{-1})

Fig. 9.3a

Fig. 9.3 (pages 121–125). Long-eared bat (*Plecotus auritus*) in slow horizontal flight. Each plate contains three views of the same bat taken simultaneously with three cameras. *a* Beginning of the downstroke. The tail and feet are held straight backwards. *b* Middle of the downstroke. The twisting of the wings is clearly seen in the front view. *c* Latter part of the downstroke. The wings are sharply cambered and the tail membrane is fully lowered. *d* Beginning of the upstroke. The elbows and wrists are slightly flexed, the camber is still pronounced. *e* Latter part of the upstroke. The feet are raised, which reduces the camber at the wingbase, but the middle part is still cambered while the tips are momentarily inverted. Photographs by U.M. Norberg. (Norberg 1976a; by courtesy of The Company of Biologists Ltd)

Fig. 9.3b

Fig. 9.3c

123

Fig. 9.3d

Fig. 9.3e

125

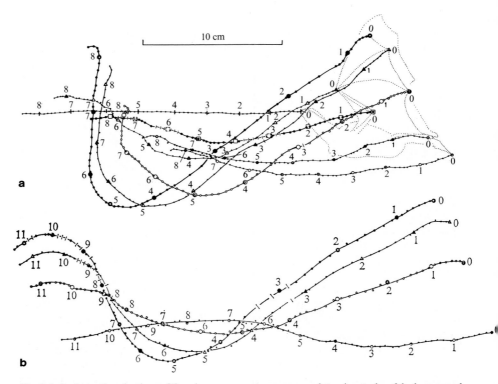

Fig. 9.4a,b. Lateral projections of the wing movements over a complete wingstroke of the long-eared bat (*Plecotus auritus*) with the tracks marked relative to the still air. Symbols as in Fig. 9.2. *a* Very slow flight at a speed of 2.35 m s⁻¹ and a stroke frequency of 11.9 Hz. The handwing is brought slightly backwards relative to the still air during the upstroke. This backward "flick" is more pronounced at still slower speeds. *b* Slow flight at speed 2.9 m s⁻¹ and stroke frequency 10.2 Hz. The wings are brought forward relative to the still air in the upstroke. (Norberg 1976a; by courtesy of The Company of Biologists Ltd)

chordwise camber of the proximal part of the wing. The wing membrane at the trailing edge bulges towards the morphological ventral side of the wing during the latter half of the upstroke (Fig. 9.3e, bottom). The movements of the legs cause the tail membrane (uropatagium) to move down and up in synchrony with the wings (Fig. 9.2).

In faster flight the wings are brought forwards relative to the still air throughout the entire upstroke (Fig. 9.4b). Both wingbeat frequency and stroke amplitude decrease with increasing forward flight speed, whereas the stroke plane angle increases. The wings flex only slightly during the upstroke in fast forward flight compared to slow flight. Aldridge (1986, 1987a) and Rayner and Aldridge (1985) found similar differences in the wingstroke of a number of bat species flying at different speeds, and Brown (1953, 1963) and Spedding (1987b) observed different flight kinematics at different speeds in birds. Rayner (1986) suggested that there is a distinct change in kinematic pattern with flight speed.

126

Aldridge (1986) found that in horizontal flight the greater horseshoe bats, *Rhinolophus ferrumequinum*, change gradually from one pattern to the other.

9.3 Relative Airspeeds and Forces

In flapping flight the aerodynamic forces are generated by actively moving wings. Since the airflow V_r meeting the wings in forward flight is the vector sum of the forward speed V of the animal, the flapping speed V_f of the wing, and the induced velocity w, the direction and magnitude of the resultant airflow, and hence of the local resultant airforce, will differ at each chordwise section along the wing length and during different phases of the wing stroke. If the resultant airforce is directed forward relative to the flying animal at any particular wing section and phase of the wingbeat cycle, it will produce a horizontal propulsive (thrust) component (T, Fig. 9.5a). When the local resultant force is directed backward, the horizontal component is a drag force (or *negative* thrust force, D_h, Fig. 9.5b). The total force experienced by the wings of a flying animal during the wingstroke is the vector sum of the forces on the different wing sections at the different stroke phases of the wingbeat cycle.

The wings do not produce constant lift and thrust throughout the wingstroke, for the local forces on the wings are different and vary during the stroke cycle. The basic aerodynamic requirements for sustained horizontal forward flight are that the wings generate enough lift so that the mean vertical force L_v balances body weight Mg, and the mean thrust T balances the overall drag D (Fig. 9.5c), while minimizing energy lost to drag. The total drag varies with the resultant airspeed and is least at the maximum range speed.

During flight with no vertical acceleration, when $Mg = L_v \propto L$, it follows from Eq. (2.8) that $Mg \propto V_r^2 S$, and $V_r \propto (Mg/S)^{1/2}$. Therefore, low wing loading enables an animal to fly at low speeds and still produce enough lift.

9.3.1 Downstroke Forces

In a flapping wing the flapping velocity V_f increases from about zero at the wing base to a maximum at the wingtip. For an animal flying horizontally at a constant forward speed V, the downwash velocity decreases with increasing forward speed and becomes very small at fast speeds. At the proximal part of the wing the resultant air velocity is determined primarily by the forward speed, and the resultant force will be directed slightly backwards. A lift and a drag force are provided (Fig. 9.6a) and both are proportional to V^2 [cf. Eqs. (2.7) and (2.8)].

At a section in the middle of the wing the flapping velocity is half that of the wingtip, and during the downstroke the resultant velocity meets the wings somewhat from below (Fig. 9.6b). The resultant force becomes directed more forward compared with that at the wing base, and the direction depends upon the relations between the forward and flapping velocities as well as on the lift:drag

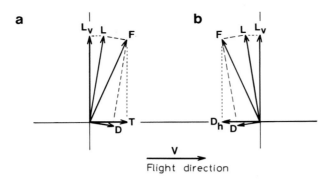

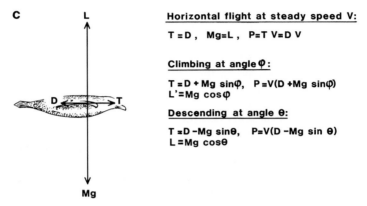

Horizontal flight at steady speed V:

$T = D$, $Mg = L$, $P = T\,V = D\,V$

Climbing at angle φ:

$T = D + Mg\,\sin\varphi$, $P = V(D + Mg\,\sin\varphi)$
$L' = Mg\,\cos\varphi$

Descending at angle θ:

$T = D - Mg\,\sin\theta$, $P = V(D - Mg\,\sin\theta)$
$L = Mg\,\cos\theta$

$L = (1/2)\rho V^2 S C_L$, $D = (1/2)\rho V^2 S C_L$

Forward flight: $D = D_{ind} + D_{pro} + D_{par}$

$$D_{ind} = \frac{k\,Mg^2}{2\rho V^2 S_d} = \frac{k\,Mg^2}{(1/2)\rho\pi b^2 V^2} \;,\quad D_{pro} = (1/2)\rho V^2 S C_{Dpro}\;,\quad D_{par} = (1/2)\rho V^2 S_b C_{Dpar}$$
$$= (1/2)\rho V^2 A_e$$

$$P = D\,V = \frac{2k\,Mg^2}{\rho\pi b^2 V} + (1/2)\rho V^3 S C_{Dpro} + (1/2)\rho V^3 A_e$$

Fig. 9.5a-c. Simplified diagrams for horizontal flapping flight and formulae for lift, drag and power according to classical aerodynamic theory. L is the average of the fluctuating lift force over an integral number of wingstrokes. The effective drag force D is a hypothetical average force the animal has to work against in powered flight. It includes three different components, of which the induced and profile drag components, D_{ind} and D_{pro}, have different direction and are dependent on different airspeed than the body (parasite) drag component, D_{par}. The induced drag factor k would be equal to 1 in the ideal case of an elliptical spanwise lift distribution; in airplane wings it is 1.1–1.2. S_d is wingdisk area (area of a full circle with span b as diameter, $= \pi b^2/4$), S_b is the projected frontal area of the body, and A_e is the equivalent flat plate area, that is, the area of a flat plate with $C_{Dpar} = 1$ giving the same drag as the body. *a* If the resultant airforce F is directed forwards relative to the flying animal, the horizontal component is a thrust force, T. *b* If the resultant airforce F is directed backwards relative to the flying animal, the horizontal component is a drag force, D. *c* Equilibrium diagram for horizontal flapping flight. (After Norberg 1985a)

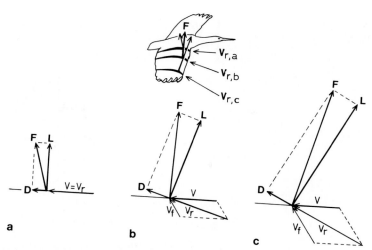

Fig. 9.6. Force and velocity diagrams for *a* the proximal part, *b* middle part, and *c* distal part of the wing in a downstroke in horizontal flight. *L* is lift force, *D* is drag force, *F* is resultant force, *V* is forward speed, V_f is flapping speed, and V_r is resultant speed

ratio of the wing. Therefore, the horizontal force can be a small drag, zero or even a small thrust force. The lift produced will be larger in the middle section than closer to the wingbase, since the resultant airspeed is larger (provided that the wing section area and lift coefficient remain unchanged). Near the wingtip the resultant airspeed is still larger and directed more from below, and the resultant force is directed forwards with large lift and thrust components (Fig. 9.6c).

During the downstroke the inner part of the wing produces lift and drag and the outer part lift and thrust. This can be seen in birds where the primary feathers are bent more or less in the direction of the resultant force and curve upward and forward (Fig. 9.7a). The net aerodynamic force for the whole wing in the downstroke is usually directed upwards and forwards, providing both lift and thrust (Fig. 9.8 middle).

9.3.2 Upstroke Forces

The major differences between various flight kinematics in birds and bats relate to the function of the upstroke. During the upstroke the flapping velocity is directed from above, so the resultant velocity reaches the wing more or less from above the horizontal at shallower, and sometimes negative angles. The wings are also partly flexed (and, in birds, flight feathers separated) to avoid large retarding effects, particularly in slow flight. Therefore, the local flapping velocity will not increase linearly with the distance from base to tip of the wing.

The wings may give negative lift, no lift at all, or small positive lift (Fig. 9.8 left), depending upon whether the resultant airstream meets the upper or lower surface of the wing. In birds the primaries can be rotated in the nose-up sense

Fig. 9.7 a,b. The pied flycatcher (*Ficedula hypoleuca*) in hovering flight. *a* Position of wings in the latter part of the downstroke. The tip feathers are bent upwards indicating an upward-forward directed resultant force. *b* Position of wings in the beginning of the upstroke. The wings are strongly flexed and the primaries are rotated in the nose-up sense letting air through. Photographs by the author. (*a* is from Norberg 1975)

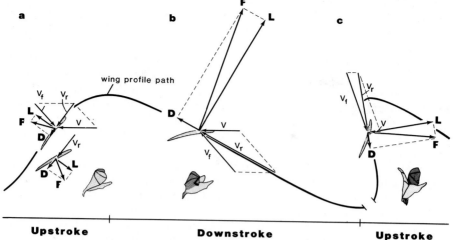

Upstroke **Downstroke** **Upstroke**

Fig. 9.8a-c. Force and velocity systems at the wings, averaged over the entire wing, in different phases of the wingstroke during horizontal flapping flight at speed *V. a* Upstroke in medium and fast forward flight. The resultant force *F* will become directed upwards-backwards, giving lift and drag, if the angle of incidence is positive, but directed downwards-forwards, giving negative lift and thrust if the angle is negative (the air meeting the wing from above). *b* Downstroke. The larger resultant airspeed V_r (resulting from a larger flapping velocity V_f) and larger wing area in the downstroke make the resultant force much larger than in the upstroke. A thrust force results from the forward inclination of *F. c* Late upstroke in slow forward flight. The air meets the wings from above (negative angle of incidence), resulting in a forward direction of the resultant force and, therefore, a thrust component. *L* is lift and *D* is drag. (Norberg 1985a; by courtesy of Harvard University Press)

during the first part of the upstroke to let air through (Fig. 9.7b), which reduces drag. Part of the upstroke may provide thrust, particularly in slow flight and hovering, when flapping velocity is large compared to the forward velocity. But to obtain a net thrust force during the upstroke in slow flight, the animal must move the wings backward relative to the still air (Fig. 9.8 right), as in the long-eared bat in slow flight (Norberg 1976a). This net thrust will then probably be rather slight because it is produced only during part of the upstroke and because the rest of the upstroke may give negative thrust.

The induced velocity may fluctuate during the wingbeat cycle. It is small in forward flight and will have only little effect on the magnitude and direction of the resultant velocity, so it has been ignored in Fig. 9.8.

9.4 Vorticity Action

The lift provided from the wings is related to the strength of the bound vortex. Interaction of the vortices in the sheet from the trailing edge forms the trailing vortices which have the same circulation as the bound vortex (see Sect. 2.4.1). If

131

the circulation varies in its direction, for example when there are differences in force action between downstroke and upstroke, a transverse vortex will be shed along the trailing edge, and it can be compared to the starting and stopping vortices.

Examination of the wake structure behind a flying animal may reveal which phases of the wingstroke are active and whether circulation is constant or discontinuous. If the upstroke produces no lift (negative or positive) the wake must consist of a series of vortex rings each associated with a single downstroke, because trailing vortices will be shed only during the downstrokes. Transverse vortices, connecting the two trailing vortices, will be formed at the beginning and end of the downstroke.

Wake flow visualization in a number of birds and bats has shown how the wake structure varies with flight speed. In hovering and slow flight the wake is generally composed of a stack of horizontal, coaxial, circular vortex rings (Fig. 9.9). During the (lift-producing) downstroke the travelling wing leaves behind a vortex sheet (Fig. 9.9a). During the upstroke the wings are more or less flexed, so they lack a bound vorticity, and contribute little or no lift (Fig. 9.9c). The flow is unsteady because of the wing's oscillations and the high value of the reduced frequency (the ratio of wing flapping velocity to induced velocity in hovering).

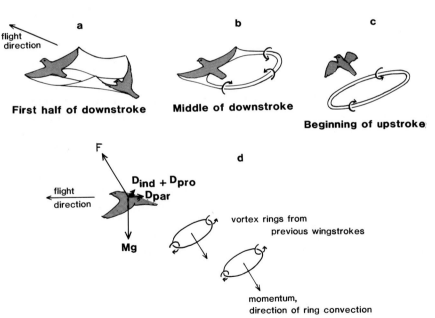

Fig. 9.9. The pattern of vorticity in one single downstroke (*a-c*) and during steady horizontal flight (*d*). A hypothetical picture of the vortex sheet formed during the first half of the downstroke is shown in *a*. A vortex ring is then formed (*b*) which convects downwards (*c*). The forces acting on the bird must be in equilibrium in steady horizontal flight (*d*). *Mg* is weight, D_{ind} is induced drag, D_{pro} is profile drag, D_{par} is parasite drag, and *F* is resultant force. (Norberg 1985a, by courtesy of Harvard University Press; after Rayner 1979c)

The rapid acceleration and deceleration of the wings at the beginning and end of the downstroke alter the circulation on the wing, and the shed starting and stopping vortices close the trailing vortex line into a loop, or torus, of concentrated vorticity (Fig. 9.9b). This pattern has been reported from small passerines (Kokshaysky 1979), for slow flight of the pigeon and jackdaw (Spedding et al. 1984; Spedding 1986), and for the long-eared and noctule bats (Rayner et al. 1986), which indicates that the upstroke gives only small or non-useful forces. Figure 9.10 shows Spedding's spectacular stereopair of the vortex wake behind a pigeon in slow flight, as it passed through a cloud of neutrally bouyant helium bubbles. As the animal continued forwards in slow flight, the successive downstrokes produced a stack of vortex rings behind the body, each inclined to the direction of flight and moving backwards and downwards. The rings are elliptical and elongated in the direction of flight.

The upward flick at the end of the upstroke, observed in many birds and bats in slow flight, is presumed to produce slight thrust (Norberg 1976a; Aldridge 1987a), to generate vorticity and to reduce the delay for building up maximum lift during the first part of the downstroke (Rayner 1980).

During flight at low wing-beat amplitude and frequency (e.g. in faster flights) the vortex wake consists of a pair of continuous undulating vortex tubes or line vortices behind the wingtips, with no concentration of vorticity across the wake (see Fig. 13.4), and with approximately constant circulation. This has been observed in kestrels in cruising flight (Spedding 1982, 1987b) and in noctule bats at medium and fast speeds (Rayner et al. 1986), indicating that the upstroke gives useful aerodynamic forces. During the upstroke in kestrels the wing flexes backward at the wrist so that the primary feathers are folded and aligned parallel to the incident air flow, while the secondary feathers remain extended. In this situation, the trailing vortices become less spaced than in the downstroke and appear to be shed from the end of the secondary feathers or the folded primaries (Spedding 1987b). This means that the arm wing performs aerodynamic work during the upstroke, while the hand wing is unloaded. During the downstroke the hand wing is extended and loaded again, and the vortices become more widely spaced.

9.5 Power Requirements for Horizontal Forward Flight

The power required to fly is equal to the total drag experienced by the animal times the resultant speed, $P = DV$. In hovering flight the power is needed to overcome induced, profile and inertial drag of the wings, while in forward flight the drag of the body adds to the total drag of the flying animal. The power required to overcome this drag is the parasite power.

As discussed above (Sect. 9.1), unsteady effects are small in fast flight and a quasi-steady assumption may give good approximations. For all speeds the profile, parasite and inertial powers can be calculated by conventional theory (see Sect. 2.3.2), but at slow speeds where unsteady flows are prominent, the vortex theory gives more accurate values for the induced power.

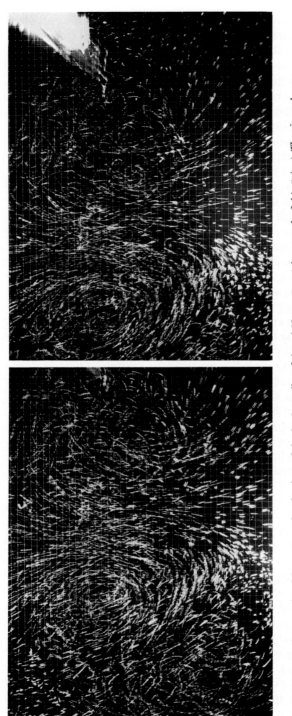

Fig. 9.10. Stereophotographs of the vortex wake of a pigeon flying horizontally at 2.4 m s⁻¹ (*from left to right* across the field of view). The pigeon leaves a vortex ring structure in the wake, as shown by multiple images of helium bubbles. Each bubble appears four times in the photographs, tracing out a double line as light from multiple flashes is reflected off the back and front faces of the bubble. The mean time between successive images is 5 ms. The background grid is 2 cm square. (Photographs by G.R. Spedding)

9.5.1 Induced Power

The induced power should be calculated with the vortex theory when slow flight is being considered. In fast forward flight, however, this power component is very small, and underestimations generated by using the momentum jet theory will be of minor importance.

In Rayner's (1979b) vortex model for horizontal level flight the upstroke is assumed to give no useful forces and the induced power is calculated from the kinetic energy increment in the wake in a single stroke, as in hovering. Although the vortex rings are elliptical and inclined to the horizontal, determining vortex ring size, strength and geometry is complicated and no simple formulation can be made. The kinetic energy increment has two components, the self-energy of the newly generated ring and the mutual energy of the new ring with each of the existing rings in the wake. While self-energy is a function of the force impulse, mutual energy dominates in hovering flight, falling rapidly with increasing speed. It is negligible at speeds above the minimum power speed. The induced power can be calculated as a function of forward speed from the total energy increment divided by the stroke period (see Rayner 1979b for details).

In medium and fast flights, with absent or weak transverse vorticity, the circulation remains constant or fluctuates slightly. In these conditions one can assume that the spanwise circulation differs little from the elliptical loading assumed in classical aerodynamics (Spedding 1987b), and the induced power can be expressed as

$$P_{ind} = (1/8)\pi\rho\Gamma_o^2 V \qquad (9.1)$$

[cf. Eq. (2.48); Milne-Thomson 1958], where Γ_o is the circulation at the centre line and equal to the circulation Γ of the vortices in the wake, and V is the freestream velocity. Spedding (1987b) solved circulation Γ in a momentum balance, assuming that the vertical components of the parasite and profile drags are small and could be neglected. An outline of his simple and elegant wake model for the kestrel in medium-speed flight is given in Section 9.9.5.

9.5.2 Profile Power

Profile, parasite and inertial powers are generally calculated with conventional methods (i.e. blade-element theory). Profile power, the most difficult to calculate and involving several uncertainties, is given by $P_{pro} = (1/2)\rho V^3 S C_{D,pro}$ [obtained from Eq. (2.15); Fig. 9.5c]. The profile drag coefficient $C_{D,pro}$ depends on the shape and smoothness of the wing surface, the boundary layer condition and Reynolds number. The incident velocity varies during the entire wingstroke and along the wing, and the wing area also takes different values in the upstroke when the wing flexes. In unsteady motion the skin friction component of the profile drag often is increased because the boundary layer is thinner and shear stresses are greater, but the pressure drag component may be lower if flow separation is reduced (Sects. 2.2.3 and 2.3.1). These two effects may cancel, and the profile drag may approach the quasi-steady value when complete separation and stalling are absent (Maresca et al. 1979). Pennycuick (1975), Rayner (1979b) and Ellington (1984d,f) have discussed the profile drag forces on flapping wings in some detail.

Rayner's (1979b) calculation of profile power is based on a method derived from that of Osborne (1951), Pennycuick (1968a) and Weis-Fogh (1972) as for the case of asymmetrical hovering (see Sect. 8.4.1). According to Rayner's model, the profile power is obtained from the total rate of working against profile drag on each wing section during a downstroke, while the upstroke is assumed to provide no forces. The angular velocity of a wing section was taken to be sinusoidal during the downstroke. Slightly simplified, Rayner's non-dimensionalized expression of the profile power is given by

$$P_{pro} = 0.0078 \mu \rho b^3 C_{D,pro} \emptyset \pi^2 S(r) V(r,t)^3 / \tau^2 T^3. \tag{9.2}$$

In Rayner's original expression, b refers to the halfspan but here to the entire wingspan, and I have reorganized his equation accordingly. The factor μ is defined as $(1 + m_s/m_p)$, where m_s is the mass of the muscles powering the upstroke and m_p the mass of the muscles powering the downstroke. $\emptyset$ is stroke amplitude, $S(r)$ and $V(r,t)$ the area and velocity at a strip at distance r from the fulcrum at time t. Although stroke amplitude and stroke frequency decrease with increasing forward speed, the wing's resultant speed will increase with the flight speed, albeit not linearly. Wing area can be supposed to be unchanged during the downstroke but it varies considerably during the upstroke. The downstroke ratio τ is the time the downstroke takes in proportion to the time T for the whole wingstroke. Rayner (1979b) used a constant, zero-lift, profile drag coefficient for the wings, $C_{D,pro} = 0.02$, and $\mu = 1.15$ for avian forward flight. With these values, with a constant wing area, and a mean resultant velocity, Eq. (9.2) can be written as

$$P_{pro} = (1.77 \times 10^{-3}) \rho b^3 \emptyset S V_r^3 / \tau^2 T^3. \tag{9.3}$$

This applies only for slow speeds when the upstroke forces may be small. In medium and fast flights the aerodynamic forces provided during the upstroke have to be taken into account. During the upstroke, the wing area changes and the various wingstrips take different values and velocities, which also must be accounted for. The profile drag coefficient may also vary.

Pennycuick (1975) assumed that profile power was independent of speed at medium flight speeds because as the forward speed increases, so does also the relative air speed at the wing, producing a decrease in the lift coefficient. The wing profile drag coefficient is a function of the lift coefficient and decreases with increasing velocity. The profile power is a function of both speed and the profile drag coefficient, which change in opposite directions and roughly compensate at medium flight speeds. Pennycuick defined the profile power in terms of a profile power ratio, meaning that profile power is a multiple of the absolute minimum power P_{am} (at minimum power speed) obtained for an ideal bird, one with dragless wings (profile power not included in the sum curve), $P_{pro} = X P_{am}$ (see Table 9.1). Based on experimental data, he suggested a profile ratio of $X = 1.2$. The minimum power P_{mp} would then be the sum of P_{am} and P_{pro} at the minimum power speed, $P_{mp} = P_{am} + 1.2 P_{am} = 2.2 P_{am}$.

136

Tucker (1973) proposed that profile power should be proportional to the sum of the induced and parasite powers at all speeds, giving a more vertically extended power curve compared to those of Rayner and Pennycuick. Pennycuick noted that there is no reason to expect this relationship between the power components because of the changes in relative airspeeds and profile drag coefficients at different speeds.

9.5.3 Parasite Power

Parasite power varies with speed as $P_{par} = (1/2)\rho V^3 S_b C_{D,par}$, where S_b is the frontal projected area of the body and $C_{D,par}$ the parasite drag coefficient. $S_b C_{D,par}$ can be replaced by A_e, which is the area of a flat plate, transverse to the air stream, with drag coefficient equal to 1 and which gives the same drag as the body [cf. Eq. (2.16)]. This "equivalent flat plate area" is the greatest frontal body area S_b multiplied by its drag coefficient, and was estimated by Pennycuick (1968a, 1975) to be

$$A_e = S_b C_{D,par} = (2.85 \times 10^{-3}) M^{2/3} \tag{9.4}$$

for an untilted body. A body drag coefficient of 0.43 is built into this formula because it was found in a pigeon and a vulture flying in wind tunnels, and Pennycuick (1968a) assumed that it remained the same for birds of any size, i.e. that it was independent of the Reynolds number. Later, regressions of drag measurements of frozen bodies in a wind tunnel led Pennycuick et al. (1988) to estimate the frontal body area in raptors and large waterfowl to be

$$S_b = (8.13 \times 10^{-3}) M^{2/3}. \tag{9.5}$$

The authors suggested a compromise estimate of $C_{D,par}$ for these birds of 0.40 at subcritical Re ($< 50,000$) and one of 0.25 at Re $> 200,000$, and at the transition region to vary with Reynolds number as

$$C_{D,par} = 1.57 - 0.108 \ln(Re). \tag{9.6}$$

These drag coefficients agree well with values in the literature for rounded bodies.

Tucker (1973) derived a slightly higher coefficient from seven different species,

$$A_e = (3.34 \times 10^{-3}) M^{2/3}. \tag{9.7}$$

Rayner (1979b) modelled the results of tilting the body as the drag of a tilted circular cylinder, and obtained the expression

$$A_e = [(X_1 \cos^3\beta + X_2 \sin^3\beta)(X_2 \sin\beta - X_1 \cos\beta) \sin\beta \cos\beta \times 10^{-3}] M^{2/3}, \tag{9.8}$$

where X_1 is equal to the constant in (9.4) or (9.7), X_2 was suggested to be 4.5×10^{-3} (based on morphological measurements on a pigeon), and β is the angle of tilt of the body to the horizontal. For $\beta = 0$, A_e may take the same value as in Eq. (9.4) or (9.7). For $\beta \neq 0$, the body drag and thus A_e become larger. For example, a slight tilt of $\beta = 10°$ gives $A_e = (3.81 \times 10^{-3})M^{2/3}$ with Pennycuick's (1968a, 1975) value and $A_e = (4.20 \times 10^{-3})M^{2/3}$ with Tucker's value. In natural conditions the body is held more or less horizontally in medium and fast flights, whereas in hovering and in slow flight the tilt is larger. However, this will not much affect the total power, since the parasite power is extremely low at slow speeds.

Streamlining of the body minimizes parasite drag and back-mounted radio transmitters, for example, can increase body drag and reduce flight range substantially (Obrecht et al. 1988).

9.5.4 Inertial Power

The inertial power can be calculated with Eq. (2.27). It is considered to be very low in medium and fast forward flight, since wing inertia is convertible into useful aerodynamic work at the bottom of the stroke (Pennycuick 1975). In slow flight inertial power was estimated to be just 2% of the aerodynamic power (Norberg 1976a), so it can be neglected in forward flight, although it may be of some importance in hovering and in very slow flight.

9.5.5 Flapping Flight with Constant Circulation

Spedding's (1987b) model is based on quantitative results from a three-dimensional velocity field calculated from a stereophotogrammetric analysis of medium-speed kestrel flight. The wake structure is consistent with a continuous undulating vortex pair with no transverse vorticity joining the two cores. This means that circulation can be assumed to be constant or fluctuate little during the two halfstrokes. The simple geometric description of the wake pattern with a constant value of the circulation around the trailing vortex line allows several simplifications in the aerodynamic analysis.

Assuming that the core is circular and contains all the vorticity, the circulation can be found by

$$\Gamma = 2\pi R w_t, \tag{9.9}$$

where R is the core radius and w_t is the tangential velocity at the core edge. Spedding found that in the kestrel, the downstroke circulation was $0.50\,\mathrm{m^2\,s^{-1}}$ and the upstroke circulation $0.60\,\mathrm{m^2\,s^{-1}}$. He noted that the difference between the two values is smaller than the uncertainty in their calculation, and that results from other wake cross sections were quite consistent with his result.

Figure 9.11 shows a simplified diagram of the vortex wake of a kestrel, which formed the basis for Spedding's general model of medium-speed flapping flight. The wings were fully extended on the downstroke, but flexed gradually during the upstroke and were then re-extended again at the end of the stroke. In the middle of the upstroke the effective wingspan is b_2. The wake separation is somewhat smaller than the wingspan, and theoretically a $\approx \pi b/4 = 0.785b$ for an elliptic load distribution [cf. Eq. (7.9)]. The dimensionless wake spacing, a' =

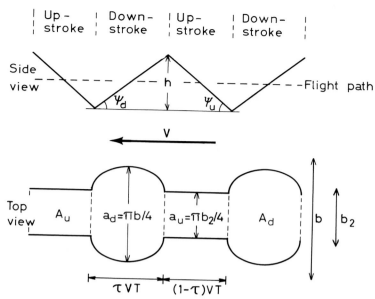

Fig. 9.11. Vortex-wake diagram of a kestrel (*Falco tinnunculus*). This diagram forms the basis for Spedding's (1987b) general model of medium-speed flapping flight. The wake is consistent with a continuous undulating vortex pair with no transverse vorticity. The *dashed lines* across the wake indicate the location where weak transverse vortices might appear at slightly lower flight speeds, or during acceleration periods. a_d and a_u are wake distances, A_d and A_u are wake areas, and ψ_d and ψ_u are wake inclinations for the downstroke and upstroke, respectively; h is wake amplitude, b is wingspan, b_2 is the effective wingspan during the upstroke, V is flight speed, T is stroke period, and τ is downstroke ratio. (After Spedding 1987b)

a/b, was 0.76 in the gliding kestrel (Spedding 1987a), which approaches the theoretical value, and 0.85–0.95 in the wake in flapping flight of the kestrel (Spedding 1987b).

The momentum normal to the plane of each wake segment is

$$Q_d = \rho \, \Gamma \, A_d$$

for the downstroke, and

$$Q_u = \rho \, \Gamma \, A_u \tag{9.10}$$

for the upstroke. The planar area for the downstroke, A_d, is the area of two half ellipses with a rectangle in between, whereas that for the upstroke, A_u, is a rectangle (Fig. 9.11).

For a momentum balance, the sum of the momenta represented by the horizontal projection of A_d and A_u must balance the weight Mg of the bird minus the vertical components of the profile and parasite drags, D_{pro} and D_{par} for one wingstroke period, T. Similarly, the rate of change of the horizontal component of momentum must balance the drag on the bird. Expressed into lift and drag force components (time averaged over T) normal and parallel to the direction of flight, this can be written as

$$L = \rho\Gamma \, (A_d\cos\psi_d + A_u\cos\psi_u)/T = (Mg - D_{pro}\sin\alpha_1 - D_{par}\sin\alpha_2) \qquad (9.11)$$

and

$$D = \rho\Gamma \, (A_d\sin\psi_d - A_u\sin\psi_u)/T, \qquad (9.12)$$

where D is the vector sum of the horizontal components of the profile and parasite drags,

$$D = D_{pro}\cos\alpha_1 + D_{par}\cos\alpha_2. \qquad (9.13)$$

Angles ψ_d and ψ_u are the wake inclinations for the downstroke and upstroke (Fig. 9.13). Angles α_1 and α_2 are the angles between D_{pro} and the horizontal line and between D_{par} and the horizontal line, respectively. These angles depend on the inclination on the body to the horizontal and on the stroke plane. In this mode of flight α_2 is assumed to be small ($\cos\alpha_2$ approaching 1). If T is known, α_2 is given by ψ_d and V. D_{pro} can be calculated with Eq. (2.15) and D_{par} with Eqs. (2.16) and (9.4). In Spedding's power model the induced power was calculated with Eq. (9.1), and Rayner (1979b) used a similar momentum force balance in his model.

9.5.6 A Method of Calculating Forces with Blade-Element Theory

For forward flight conventional aerodynamic theory can give results sufficient for power estimation if the lift coefficient is consistent with the quasi-steady assumptions (see Sect. 8.1). I used this approach for the long-eared bat (*Plecotus auritus*, body mass = 9 g) in slow flight, with a proof-by-contradiction (Norberg 1976a). If quasi-steady aerodynamics are adequate to explain horizontal forward flight, the lift coefficient, as averaged over the entire wing and the entire wingstroke, must not exceed the maximum coefficient of lift obtainable at the Reynolds number under which the wings operate. With this method the three aerodynamic power components are treated together in a force equilibrium model.

Both C_L and C_D were estimated by a method developed from Pennycuick's (1968a) and Weis-Fogh's (1972) expressions for estimating the average lift coefficient in flapping flight; the same approach as in my model for hovering flight (Sect. 8.4.3). My method differs from theirs in two ways. (1) the vertical components of the lift and drag forces are taken separately for the whole wing and wingstroke and added, to balance the body weight. Similarly, the horizontal components of the lift and drag forces, taken separately, were added to balance the body drag of the animal (cf. Fig. 8.4); (2) the upstroke was treated separately because it is so different from the downstroke. This means that the mean lift and drag coefficients and the L/D ratio can be directly solved from these relations. So, in contrast to other methods the actual lift:drag ratio can be calculated analytically.

The forces produced during a complete wingstroke can be found by dividing the wing into a number of chordwise strips and then adding together the calculated average forces for each strip. The movement (angular displacement) of the long wing-axis (the axis between the humero-scapular joint and the wingtip) is sinusoidal, as verified by analysis of high-speed cine films of the long-eared bat in free flight (cf. Sect. 8.4.1). The flapping velocity of a wing-element at distance r from the fulcrum at time t is thus $V_f(r,t) = r\pi\theta n \cos(2\pi nt)$ [Eq. (8.11)]. The induced velocity w was assumed to be constant along the wing during the entire wing stroke, although there probably are some differences between the down- and upstrokes. According to the momentum theorem for an ideal actuator disk, the minimum value of the induced velocity, at the level of the disk, can be obtained from the formula

$$w(V^2 + w^2)^{1/2} = Mg/2\rho S_d \qquad (9.14)$$

(Pennycuick 1968a). The instantaneous resultant velocity is the resultant of three speed components, the flapping speed and the projections of the induced velocity and the forward speed to a plane normal to the long wing axis. It is found by the cosine theorem (see Norberg 1976a for details).

The force coefficients and the lift:drag ratio are unknown, as are the directions of the resultant force F and the sign of its components. However, the directions of the lift L and drag D forces (but not their magnitudes) can be found, since the direction of the resultant airstream can be estimated. Therefore, the vertical and horizontal components of the lift and drag forces were used to estimate the vertical and horizontal forces, respectively. The sum of the vertical components of the lift and drag forces, taken separately, as integrated over the whole wingstroke and over the whole wings, must equal the weight of the animal as integrated over the same time. In the same way, the integral of the horizontal components of the lift and drag forces, taken separately, must equal the integral of the body drag of the animal. With this equation system the lift and drag coefficients and their ratio can be found.

This analysis of the forces generated in slow flight of the long-eared bat showed that weight support and thrust are provided by the downstroke, but a very small thrust force is also generated in the later half of the upstroke. Fig. 9.12 shows the vertical (a) and horizontal (b) forces divided by the lift coefficient elicited during one wingstroke. Since the lift and drag coefficients may be much smaller during the upstroke than during the downstroke, the positive horizontal (thrust) force during the upstroke may become very small. This is also confirmed by Rayner's et al. (1986) flow visualization of the long-eared bat in very slow flight (1.5 m s⁻¹), showing discrete vortex rings for the downstroke only, which indicates that no or only very small useful forces are generated during the upstroke. Any small thrust might have been obscured by the more dominant forces and vortices associated with the downstroke.

9.5.7 Comparison Between Different Power Models

In this section I will discuss and compare some of the methods of estimating the power required by an animal in forward horizontal flight. The blade-element and momentum jet theories frequently have been used (for example by Pennycuick 1968a,b, 1969, 1975; Tucker 1973, 1974; Greenewalt 1975; Norberg 1976a), while others (Ellington 1978, 1984e,f; Rayner 1979a,b; Spedding 1987b) used a combination of vortex and momentum jet theories. Vortex theory is used

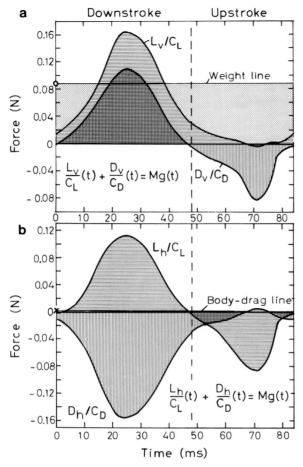

Fig. 9.12. a The virtual force indices, L_v/C_L and D_v/C_D, for the vertical components of the lift and drag forces, respectively, plotted against time for the long-eared bat (*Plecotus auritus*) in slow horizontal flight at 2.35 m s⁻¹. The area under the lift curve (vertical component of the varying lift force *horizontal hatching*) multiplied by the lift coefficient, C_L, plus the area under the drag curve (vertical component of the varying drag force *vertical hatching*) multiplied by the drag coefficient, C_D, must equal the body weight Mg of the bat as integrated over the whole wingstroke (*dotted area* between weight line and zero line). The weight is indicated by a *circle* on the ordinate ($Mg = 0.009$ kg $= 0.088$ N). *b* The virtual force indices, L_h/C_L and D_h/C_D, for the horizontal components of the lift and drag forces, respectively, plotted against time for the same flight as in *a*. In the same way as in *a* the area under the lift curve (horizontal component of the varying lift *horizontal hatching*) multiplied by the lift coefficient, C_L, plus the area under the drag curve (horizontal component of the varying lift drag force *vertical hatching*) multiplied by the drag coefficient, C_D, must equal the body drag D_b of the bat as integrated over the whole wingstroke cycle (the area between body drag line and zero line). The body drag is shown by a *cross* on the ordinate ($D_b = 0.00105$ N). By using the two pairs of equations from *a* and *b* above, the actual lift and drag coefficients can be calculated as well as the L/D ratio, given that quasi-steady aerodynamics is a reasonable approximation of the flow pattern. In this particular case C_L is 1.4 and C_D is 0.9. (After Norberg 1976a)

142

Table 9.1. Equations and proportionalities for various power components according to Pennycuick's (1969, 1972a, 1975) and Rayner's (1979c) models for forward flight

Power	Pennycuick	Rayner
Induced power	Momentum theory: $P_{ind} = k(Mg)^2/2\rho VS_d \propto b^{-2}V^{-1}$ where $k \approx 1.2$	Vortex theory: $P_{ind} \propto b^{-3/2}V^{-3/2}T^{-1/2}$
Profile power	Blade-element theory: $$P_{pro} = \frac{0.877Xk^{3/4}(Mg)^{3/2}A_e^{1/4}}{\rho^{1/2}S_d^{3/4}} \propto b^{-3/2}$$ where $X \approx 1.2$ at medium speeds and $k \approx 1.2$	Blade-element theory: $P_{pro} \propto b^5 T^{-3}$
Parasite power	$P_{par} = (1/2)\rho A_e V^3$	$P_{par} = (1/2)\rho A_e V^3$

Mg is body weight, A_e is equivalent flat plate area of the body (Sect. 9.5.3), S_d is wingdisk area ($\pi b^2/4$), b is wingspan, ρ is air density, X and k are constants, T is wingstroke period, V is speed.

only for the calculation of the induced power. Although the parasite, profile and inertial powers are calculated by conventional methods, the models differ for these power components (see Sects. 9.5.2–3). The energy requirements for flight have also been determined from various physiological measurements of flying animals (see Sect. 3.6). Using measurements of gas exchange, Tucker (1968b) demonstrated that the power-versus-speed curve is U-shaped in the budgerigar, as it is in pigeons, according to measurements in a wind-tunnel by Rothe et al. (1987).

Rayner's theoretical model on the shape of the wake structure behind a flying animal has gained empirical support from experimental flow visualization for flying finches (Kokshaysky 1979), pigeons (Spedding et al. 1984) and bats (Rayner et al. 1986). However, the calculations are rather tedious and in a simpler expression Rayner (in Norberg and Rayner 1987) related the minimum power and cost of transport as functions of the animal's body mass, wing area and wingspan (Sect. 10.3.5). Pennycuick's equations for the various power components (Table 9.1) and optimum powers are admirably simple, for they require total mass, wingspan, and flight speed to generate predictions of power consumption that may be correct to within 10% of measured values (Tucker 1973). A disadvantage of this model is that profile power is assumed to be constant over medium flight speeds ($P_{pro} = 1.2P_{am}$), which is not supported by theory, as Pennycuick emphasizes.

Greenewalt (1975) made a comprehensive dimensional analysis of bird flight and related size to the aerodynamics of flapping, and then predicted the energetic requirements as functions of airspeed and flight morphology. He used only two terms for the power calculation, namely the induced power and the profile power for the entire animal, and based his equations on theory for fixed-wing aircraft.

Pennycuick's (1975) and Rayner's (1979b) power curves for a 333-g pigeon are compared in Fig. 9.13. Induced power is lower with Pennycuick's modified

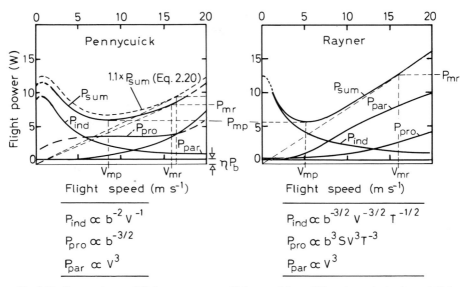

Fig. 9.13. Power curves of flight power versus flight speed for a 333-g pigeon in horizontal flight calculated from conventional aerodynamic theory according to Pennycuick (1975), and from the vortex theory for induced power and conventional theory for profile and parasite power according to Rayner (1979c). V_{mp} is minimum power speed, V_{mr} is maximum range speed, P_b is basal metabolic power, P_{mp} is total power required at V_{mp}, P_{mr} is total power required at V_{mr}, P_{ind} is induced power, P_{pro} is profile power, P_{par} is parasite power, P_{sum} is sum of the three aerodynamic power components, b is wingspan, η is mechanical efficiency, A_e is equivalent flat plate area, M is body mass, and T is stroke period

momentum theory (including the correction factor k) than with Rayner's vortex theory. The difference is largest for hovering flight. Profile power is calculated differently (discussed in Sect. 9.5.2), and parasite power is much higher with Rayner's model, since he assumed different degrees of tilt of the body. The sum curve in Pennycuick's model is more flattened than the one in Rayner's model, which agrees with physiological measurements. The sum flight power in Pennycuick's model is further multiplied by the ventilation and circulation factor 1.1 (see Eq. 2.20; Pennycuick 1975).

Torre-Bueno and LaRochelle (1978) measured oxygen consumption and carbon dioxide production during flight in unrestrained starlings and received a much more flattened power curved than predicted by theory. Tucker (1973) found some agreement between his predictions based on mechanical analyses and measurements of oxygen consumption. However, where the measured and predicted results differ, it is difficult to know whether the discrepancy is due to an error in prediction of one or more components because the physiological measurements do not allow different power components to be discriminated. Flying in wind tunnels somewhat changes the flight behaviour, which may influence on the metabolic rate. The drag and mass change due to head mask and tube further increases the metabolic rate by 12–13% or even more (Tucker 1972). So, even physiological measurements are marred by imperfection. The mechanical efficiency η, that is, the efficiency by which chemical power (as reflected in oxygen consumption) is converted by the animal into mechanical power, is not well known, nor do we know how it is affected by speed or power.

Rothe et al. (1987) compared power data of pigeon flight obtained by physiological measurements with those obtained by theoretical calculations. The physiological data, however, are obtained from pigeons of different masses and flying at different speeds, so they are not easy to compare with each other. The data on the metabolic rate of a 330-g pigeon in flight obtained by Rothe et al. (1987) may be the best for a comparison with Pennycuick's and Rayner's power models of the 333-g pigeon in Fig. 9.13. The power input in the experiments by Rothe et al. was about $P_i = 87$ W kg^{-1} = 28.7 W (inclusive of 12% correction for mask and tube) at $V_{mp} = 11$–13 m s^{-1}. The mechanical power for flight for a 0.333 kg pigeon flying at the theoretical $V_{mp} = 8.5$ m s^{-1} is about $P = 7$ W according to Pennycuick's model. Using the efficiency of $\eta = 0.25$ and assuming that the basal metabolic rate is $P_b = 1.7$ W (from Eq. 3.25), the power input becomes $P_i = (7/0.25) + 1.7 = 29.7$ W. The corresponding data for a pigeon flying at the theoretical $V_{mp} = 5$ m s^{-1} according to Rayner's model is about $P = 5.6$ W, giving $P_i = 24.1$ W. The two theoretical values both agree well with the value obtained by physiological measurements, keeping in mind that a small change in η changes the results. An efficiency of $\eta = 0.23$ (as suggested by Pennycuick 1975) would change Pennycuick's result to $P_i = 32$ W and Rayner's to $P_i = 26$ W, values lying on both side of that obtained by physiological measurement (28.7).

For bats, there is close agreements between the results obtained from physiological methods and the predicted values from my theoretical model (Norberg 1976a; Sect. 9.5.6), but the values obtained by Pennycuick's and Rayner's theoretical models differ a lot from these. Physiological measurements give that $P_i = 51.9$ M$^{0.71}$ (for five species; cf. Eq. 3.27), which means that a 9-g bat, such as Plecotus auritus, should need 1.83 W to fly. The mechanical power for flight for this bat at 2.35 m s^{-1} (slightly less than V_{mp}), as calculated with my theoretical model (Sect. 9.5.6), was $P = 0.36$ W, and for $\eta = 0.25$ and $P_b = 0.108$ (Eq. 3.25) the power input becomes $P_i = 0.36/0.25 + 0.11 = 1.55$ W. This means that the power needed to fly was 14 × BMR. Assuming a mechanical efficiency of 0.20 (η were found to vary between 0.12 and 0.40 in Phyllostomus hastatus: Thomas 1975), the power input becomes $P_i = 1.91$ and power needed to fly 17 × BMR. Using the doubly labelled water method, Racey and Speakman (1987) found that power input during free flight in Plecotus auritus was 21 × BMR. Rayner's theoretical bat model (Norberg and Rayner 1987) says that $P_{mp} = 10.9$ M$^{1.19}$ (Tab. 10.1), which means that the mechanical power for a 9-g bat flying at V_{mp} should be 0.041 W with corresponding power input $P_i = 0.27$ W, which is 2.5 × BMR (using Stahl's Eq. 3.25). Using a mechanical efficiency of 0.12 (cf. above) would change the picture to $P_i = 0.45$ W, which is 4 × BMR. Pennycuick's theoretical bird model gives $P_i = 0.47$ W for $\eta = 0.25$ and $P_i = 0.86$ W for $\eta = 0.12$, which is 4.3 × BMR and 7.8 × BMR, respectively.

The vortex ring model developed by Rayner (1979b,c) was used as a basis for the analysis of the wake structure of the pigeon by Spedding et al. (1984). They found, as predicted, that the wake consisted of a chain of discrete, small-cored vortex rings, but the measured ring radius was 23% smaller than predicted and the momentum in the wake was approximately half that required for weight support in unaccelerated level flight. Furthermore, the induced power calculated from the observed flow pattern was 47% lower than predicted by theory.

Spedding (1986) used the same techniques to investigate the wake of a jackdaw (*Corvus monedula*) in slow level flight, to find out if similar discrepancies prevailed in a species with different morphology and wingbeat kinematics. He found that the wake structure was similar to that in the pigeon and the momentum measured in the wake was only approximately 35% of that required for weight support, requiring an induced power of about one sixth of the predicted value. From these results Spedding concluded that a simple vortex model fails as an accurate theoretical description of the wake, and that the vorticity shed at the wingtips and trailing edges does not all roll up into a closed loop of concentrated vorticity. About half of the wake vorticity presumably lies outside the concentrated core regions, explaining the unreasonably low estimates of wake momentum. Wake vorticity distribution must therefore be more complicated than assumed in the simple vortex ring model. Nonetheless, Rayner's (1979c) vortex ring model seems to be the most appropriate representation available of the wake of a slow-flying animal. The vortex ring model predictions of the induced power requirements are, however, generally higher than those obtained with the momentum-jet theories.

With a simple vortex method Spedding (1987b; Sect. 9.5.5) estimated the induced power for a kestrel flying at medium speeds, where the wake consisted of a pair of continuous undulating trailing vortices, using the helium bubble method for calculating lift circulation. The estimated induced power agreed well with the value obtained by the momentum-jet theory (Pennycuick 1975).

9.6 Take-Off, Climbing and Landing

So far I have treated only hovering and level forward flight. But flight includes other components, such as take-off, climbing, landing and manoeuvring, all involving momentum changes in various directions. Grouse, pigeons and many raptors can take off almost vertically, while swans, geese, divers and others have to skitter a long way to lift off. Most species can land gently, but albatrosses bump clumsily onto the ground in calm weather. Wing form and muscle physiology are adapted to different behaviours and put some restrictions on different flight modes and abilities (see Chap. 11).

9.6.1 Take-Off and Climbing

During flight with no vertical acceleration the average, or effective, vertical force component equals the weight of the animal, $L_v = Mg$, and the average thrust equals the average drag, $T = D$, so these forces are in equilibrium (Fig. 9.14a).

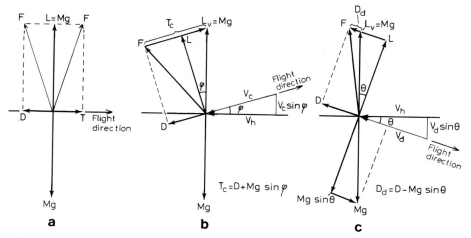

Fig. 9.14. Force diagrams for *a* equilibrium level flight, *b* climbing flight, and *c* descending flight. Mg is animal weight, L is lift, L_v is vertical lift, T is thrust in level flight, T_c is thrust in climbing flight, D is drag, D_d is drag in descending flight, F is resultant force, V_h is horizontal speed, V_c is speed in the climb path, V_d is speed in the descent path, φ is angle of climb, and θ is angle of descent

In ascending, flight work is done against gravity and when the animal is climbing at an angle φ to the horizontal, the thrust can still be supposed to act nearly in the direction of flight. If the motion is to be uniform, the forces must be in equilibrium so it follows from the diagram in Fig. 9.14b that thrust in a climb is

$$T_c = D + Mg \sin \varphi, \tag{9.15}$$

and that the lift force component should be replaced by

$$L = Mg \cos \varphi. \tag{9.16}$$

The power for climbing is

$$P_{climb} = VT_c = V(D + Mg \sin \varphi). \tag{9.17}$$

The vertical climbing speed is $V_c \sin \varphi$ and the climbing rate can be written as $dh/dt = V_c \sin \varphi$.

Drag is the sum of induced, profile and parasite drag. Using Eq. (2.17) for the total drag, the climbing power can be written as

$$P_{climb} = 2k(Mg)^2/\rho V \pi b^2 + (1/2) \rho V^3 (SC_{Dpro} + A_e) + MgV \sin \varphi. \tag{9.18}$$

Rayner (1985b) demonstrated in a hodograph the variation of mechanical power output with potential energy change in steady climbing flight at various horizontal speeds V_h ($= V_c \cos \varphi$), shown in Fig. 9.15. The climbing power is expressed as a function of maximum range power in level flight in the form P_{climb}/P_{mr}, and is plotted against $\tan \varphi$. $\tan \varphi$ is simply the vertical climbing speed divided by the horizontal speed, or height gain per unit horizontal distance flown (Fig. 9.14b). The intercepts with the line $\tan \varphi = 0$ correspond to points on the U-shaped power-velocity curve for level flight (Fig. 9.13), where minimum

147

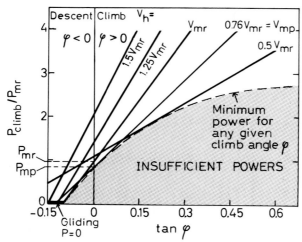

Fig. 9.15. Hodograph showing the mechanical power output for climbing (descending) as a function of climb (descent) angle φ at various horizontal speeds V_h. The mechanical power output is expressed as a function of the power required at maximum range speed in level flight, P_{mr}. The intercepts with the vertical line tan φ = 0 (representing horizontal flight) correspond to points on the level flight power-velocity curve (Fig. 9.13). Minimum power output, P_{mp}, in level flight is obtained at V_{mp}, and with lower powers the bird must descend. The minimum power for flight at any particular angle is shown by the *dashed curved line*. Minimum glide angle is obtained at the maximum range speed, V_{mr} = V_h. (After Rayner 1985b)

power output P_{mp} is obtained at V_{mp} and maximum range power P_{mr} at V_{mr}. The dashed line shows the minimum power for a given climb angle. Lower values are insufficient and unless potential energy is used to counter drag, the animal descends. The minimum gliding angle is obtained at $V_h = V_{mr}$. At stable gliding the flight power is nearly zero and energy is expended only for BMR and for isometric muscular contraction to keep the wings extended and down on the horizontal plane.

Birds with high wing loading, short wings and high aspect ratio include loons, mergansers, geese, swans, ducks and auks, species capable of fast and, for their speeds, rather economical flight, used for commuting and migration. Of these, mergansers and auks have the highest wing loadings, and the flight muscle mass of these birds and of the loons is small or rather small (7–19% of body mass; Magnan 1922, cited in Greenewalt 1962). These birds have to run (skitter) along the water to achieve the necessary relative air velocity as the feet contribute the extra vertical force needed to keep the bird out of water while acceleration is accomplished by feet and wings. The red-throated diver (*Gavia stellata*) skitters 15–40 m before taking off with its feet moving synchronously with the wings (2 foot-falls per complete wingstroke), and takes off at a speed of about 10 m s⁻¹ (R.Å. Norberg and U.M. Norberg 1971; Fig. 9.16). Since these divers breed in small tarns and are slow climbers, they have to circle around the tarn while climbing to tree-top level before setting off to the fishing lake. While circling, they spend a component of the potential vertical force for centripetal acceleration.

Fig. 9.16. Tracks in the water by the feet during skittering in the red-throated diver (*Gavia stellata*) at take-off. The splashes of the water indicate vigorous thrust by the feet. Photograph by the author. (R.Å. Norberg and U.M. Norberg 1971)

This necessitates generation of more lift than in horizontal flight, which puts a lower limit to the tree-free area for climbing turns.

Grouse also have very high wing loadings but, in addition, low aspect ratios and therefore very expensive flight. Their flight muscle mass is large (22–29%; Magnan 1922) and they can take off almost vertically and with high acceleration despite their short wings. But they can make only short flight bursts (see further Sect. 11.2.2); the very short wingspan in these species is an adaptation for flight within dense vegetation.

Birds and bats taking off from a vertical surface usually take to the air by falling, or gliding, a short distance while losing potential energy but gaining speed at low flight cost.

9.6.2 Landing

Some birds land on ground or water, others on tree branches or on vertical treetrunks. Bats usually land on branches or on vertical surfaces, either in a head-up position or head-down, after turning the body 180° about a dorso-ventral axis in the yaw plane. Before landing the animal has to perform various

149

manoeuvres or attain certain attitudes of the flight surfaces to keep the horizontal velocity low and to reduce the sinking velocity to avoid too hard impact on landing.

Landing is preceded by a gliding phase in many birds. In a descent at angle θ, which is shallower than that for equilibrium gliding, a component of the weight Mg along the direction of flight becomes less than the drag D, leaving a part of this drag to brake the speed. This component is $Mg \sin\theta$ (Fig. 9.14c), so the drag D_d braking the speed becomes

$$D_d = D - Mg \sin\theta. \tag{9.19}$$

The power for descending then is

$$P_{desc} = VD_d = V(D - Mg \sin\theta). \tag{9.20}$$

The descending speed is V_d and the vertical sinking speed $V_d \sin\theta$, and the sinking rate can be written as $dh/dt = V_d \sin\theta$.

There are various ways to decrease speed or the lift/drag ratio before landing. Flapping animals usually pitch up their wings to increase the incident angle above the stalling angle, which drastically reduces lift and increases drag. The turbulence this generates above the wings is visible as the raised and fluttering feathers in many landing birds (Fig. 9.17a). When landing on a vertical surface, bats and many small birds decrease speed by climbing in air before landing, to consume excess kinetic energy. Many bats land upside down and have to turn around first by performing a half roll while slowing to approach the perch.

The webbed feet of water birds and some vultures (of the genus *Gyps*) are used as air-brakes when lowered and spread (Pennycuick 1975). The feet of landbirds, when lowered into the airstream and with the toes spread, have a drag coefficient of 1.1–1.2 (Pennycuick 1968b, 1971b) so they effectively decrease the overall lift/drag ratio and steepen the glide angle.

Swans use their feet as water-brakes during landing by extending their legs forwards with the feet spread and angled in the toes-up direction. Boobies put down their landing-gear in a similar way before landing on the ground. Many water birds (such as loons) reduce their speed by dragging their feet behind them in the water on the landing approach (Fig. 9.17b).

Half rolls and sideslipping are effective ways of rapidly losing height before landing and are used by many bats (Norberg 1976c). Half-rolls are also used by birds, such as ducks and geese. Sideslipping is known to pilots as "fishtailing" and is an extremely efficient way of losing height rapidly. These manoeuvres are described in the next section.

Fig. 9.17a,b. Landing in the red-throated diver (*Gavia stellata*). *a* Landing approach with raised body and wings and with feet apart. Some upper wing coverts are lifted up, indicating a tendency to turbulence above the wing as a consequence of large angle of attack. The left foot makes a faint contact with the water. Photograph by the author. (R.Å. Norberg and U.M. Norberg 1971) *b* Later stage of landing with the body in a more horizontal position. The spread feet and tail are used as water-brakes. Photograph by R.Å. Norberg. (R.Å. Norberg and U.M. Norberg 1971)

151

9.7 Flight Manoeuvres

Flying animals perform aerial manoeuvres in different ways depending upon their purposes. For example, prey catching, avoidance of obstacles, and landing require different types of manoeuvres. Norberg and Rayner (1987) made a distinction between *manoeuvrability* and *agility* in their study of the ecological morphology of bats; they refer manoeuvrability to the minimum radius of turn the bat can attain without loss of speed or momentum, while agility is the maximum roll acceleration during the initiation of a turn and measures the ease or rapidity with which the flight path can be altered. It may, however, be convenient to include both types in the term manoeuvrability, as is frequently seen (Andersson and Norberg 1981).

9.7.1 Turning Ability

The theory of turning in powered flapping flight has been given in several papers (Pennycuick 1971a; R.Å. Norberg and U.M. Norberg 1971; see also Sect. 6.3.1). In a turn the animal loses some of the potential vertical force, because a component of it is necessary for centripetal acceleration to prevent sideslip. The animal must therefore bank its wings at some angle to the horizontal and develop more lift than in horizontal flight. At any particular angle of bank Φ the radius of turn, r, is given by $r = (Mg/S) (2/g\rho C_L \sin \Phi)$ [cf. Eq. (6.5) and Fig. 6.3]. At any given lift coefficient and angle of bank the wing loading (Mg/S) determines the turning radius. The ability to make tight turns is best in animals with low wing loading and with the ability to enhance the mean lift coefficient significantly above that in straight, level, flight. This is one reason why animals foraging within vegetation should have low wing loading and the ability to control camber of the wings (fore- and aft-curvature across the wing chord). Aldridge (1987b) found that there was a significant positive correlation between wing loading and turning radius in six species of bats, indicating that low wing loading improves manoeuvrability. Bats commonly gain height before entering a turn, which is a means of reducing speed to obtain a tighter turn, so kinetic energy is temporarily transferred to potential energy during a turn (Aldridge 1986, 1987b; Rayner and Aldridge 1985). When the lateral acceleration, V^2/r, is fixed (set by, for instance, the wing and flight muscle characteristics), r is least when V is least.

9.7.2 Maximum Roll Acceleration and the Initiation of a Turn

Flying birds and bats may detect insect targets at ranges of a few metres so they must make rapid manoeuvres to pursue and catch prey. To initiate a turn a net rolling moment must be produced and this can be done by differential twisting or flexing of the wings, or by unequal flapping of the two wings so that the aerodynamic roll moments of the two wings become asymmetrical. The aerodynamic rolling moment, or torque, τ_a, is the moment of the aerodynamic forces about the body's longitudinal axis (Fig. 5.4a, where τ_a is identical with M_r),

$$\tau_a = (1/2)\rho V^2 S b C_{rm}, \tag{9.21}$$

where C_{rm} is the coefficient of rolling moment. The torque is thus proportional to $V^2 S b$, so at a given speed, long broad wings will give large aerodynamic torque.

The fastest entry into a turn is achieved at the maximum angular acceleration available to the animal, which is

$$\alpha_{roll} = \tau/I_{b+w}, \tag{9.22}$$

where I_{b+w} is the total roll moment of inertia of the body (I_b) and wings (I_w). The moment of inertia I_{b+w} about the roll axis depends on body mass and wing mass and the radius of gyration r_{b+w} of body and wings and is

$$I_{b+w} \propto M r_{b+w}^2 \tag{9.23}$$

[cf. Eq. (2.23)]. When comparing manoeuvrability in animals of different mass at identical flight speeds (see Andersson and Norberg 1981), the angular roll acceleration would be related to body mass as

$$\alpha_{roll} \propto S b / M r_{b+w}^2 \propto M^{2/3} \times M^{1/3}/M \times M^{2/3} = M^{-2/3}. \tag{9.24}$$

If flight speed is assumed not to be constant but to vary as $V^2 \propto Mg/S$ (as expected for any characteristic speed on the power-versus-speed curve for geometrically similar animals), the angular acceleration would instead be related to body mass as

$$\alpha_{roll} \propto V^2 S b / M r_{b+w}^2 \propto M^{1/3} \times M^{2/3} \times M^{1/3}/M \times M^{2/3} = M^{-1/3}. \tag{9.25}$$

(Norberg and Rayner 1987). In both cases the angular roll acceleration increases with decreasing body mass.

Data on bats suggest that body and wing inertia are broadly comparable in magnitude, wing inertia being slightly larger than body inertia. In *Plecotus auritus* the measured wing inertia I_w about the humeral joint was 1.1×10^{-6} kg m^2 (Norberg 1976a) and the body inertia I_b was 0.84×10^{-6} kg m^2 (Aldridge 1987b). A decrease in either would enhance roll acceleration so that animals adapted to fast rolling should have thin bodies.

On dimensional grounds wing inertia is proportional to wing mass times wing length squared. Wing mass will rise with wingspan and wing area unless there is unusual thinning of the skeleton or relative reduction in muscle mass (which would be precluded for mechanical reasons). Norberg and Rayner (1987) suggested that wing mass will increase at least in proportion to wing area or to wingspan squared, and that the moment of inertia of the wings I_w will vary with wingspan and area at least as fast as Sb^2, and possibly as fast as Sb^3. To reduce wing inertia the wings should thus be short and light, especially in the distal region. The angular roll acceleration in Eq. (9.25) would then be proportional to $V^2 Sb/(M_b r_b^2 + Sb^2)$ or to $V^2 Sb/(M_b r_b^2 + Sb^3)$. Keeping speed V and body inertial moment $M_b r_b^2$ constant, α_{roll} would increase with decreasing wingspan but also slightly with increasing wing area.

Mass distribution decreases distally along the wing and is concentrated at the wingbase in birds (Fig. 2b in Weis-Fogh 1972) and with smaller peaks at the elbow and carpal joints in bats (see Fig. 2.5, Sect. 2.3.2). Wing inertia should therefore fall with rising tip length index T_1 (the ratio between handwing length and armwing length; Sect. 4.3). Small indices occur in hummingbirds and swifts and large values in many seabirds, such as albatrosses.

Summarizing, to enhance rapid manoeuvres a flying animal should have a thin body and short wings for high roll acceleration, and broad wings and wide wingtips for large aerodynamic rolling moment. High roll rates (up to 450 rad s^{-2}) have been recorded in horseshoe bats (*Rhinolophus ferrumequinum*; Aldridge 1986), that are slow-flying with rounded wingtips and average wingspan. But fast-flying bat species with pointed wingtips can also make rapid manoeuvres, and these species all have rather short wings.

9.7.3 Prey Catching and Landing Manoeuvres

The rolling movement was described in the noctule bat (*Nyctalus noctula*) during a manoeuvre to lose height rapidly during pursuit of an insect (Norberg 1976c). The rolling movement was preceded by a slight flexion of the wings at the elbows and wrists, reducing wing area. Figure 9.18 shows a pronation of the right wing (nose-down rotation of the wing about its long axis), which decreases angles of incidence and lift coefficients. A negative angle of incidence would cause the lift to be directed downwards. The left wing is supinated, increasing the angle of incidence and lift coefficient so the difference in pitching angle becomes large between the two wings. The difference in lift on the two wings causes a roll in the clockwise direction (right wing moving down) but the absence of rotation in the yaw plane indicates that the drag forces were about equal on the two wings. After half a roll (180°) the wings were extended almost symmetrically. The concave ventral side of the wings was then turned upwards with a positive angle of incidence, resulting in a downwardly directed lift force. This force, together with the force of gravity, produce a downward acceleration and hence a rapid descent or inverted turn downward. When the bat had descended in this upside-down position for about 0.08 s, the tail membrane was pouched by forward flexion of the hind legs and tail as the bat made its attack. The roll through 180° was performed in 0.08–0.10 s, corresponding to an average angular velocity of 39.2–31.4 radians s^{-1}.

When making turns in the looping plane, the centre of the turning curve is approximately in the direction of the lift force (that is, in the direction of the centripetal acceleration), and the turning radius is inversely proportional to the magnitude of the centripetal force. A turn downwards could also be obtained by pronation of both wings to obtain negative angles of attack and ventrally directed lift forces. The air would then meet the dorsal, convex surface of the wing and, in bats, the wing membrane would then tend to bulge ventrally. For structural reasons, however, the bat wing camber cannot be fully reversed, and the wing would give lower lift coefficients than at positive angles of attack with correspondingly larger turning radius. By rolling through 180°, however, the bat can make use of the high lift coefficients associated with the wing camber and positive angles of attack, resulting in the tightest downward turn possible.

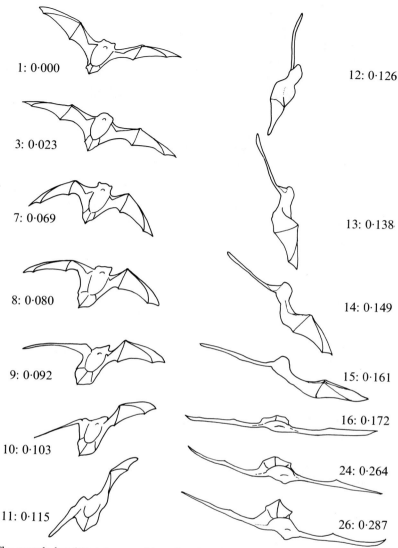

Fig. 9.18. The noctule bat (*Nyctalus noctula*) performing a roll through 180°. The bat is flying obliquely towards the camera. The *first number* indicates the frame number of the film and the *second number* the time in seconds. Filming rate is 87 frames s⁻¹. (Norberg 1976c; by courtesy of The Company of Biologists Ltd)

A more important reason for performing a roll before turning downwards is the relative sizes of the various flight muscles. If a turn downwards is brought about by wing pronation, the wing torque due to the inverted lift force would have to be counteracted by the wing elevator muscles which are very weak relative to the wing depressor muscles and not designed to carry heavy loads. In

155

the long-eared bat (*Plecotus auritus*) the mass of the elevator muscles is about 20% of that of the depressors (Betz 1958). By rolling through 180° before turning downwards, the bat can use its powerful wing depressors to oppose the torque of the centripetal lift force, and the load on the wings becomes oriented in the direction in which the wing skeleton and muscles are designed to accommodate. In aerobatic manoeuvres with aircraft, sharp turns downwards are preceded by a 180° roll to maintain the lift force of the wings in the direction in which they are designed to carry the heaviest load.

Half rolls often can be seen in ducks and geese during landing descent. This is obviously done to steepen the descent. Ravens use a similar manoeuvre in flight display, and Oehme (1968) described 180° rolls and "flight on the back" in the swift (*Apus apus*), while Pennycuick (1972b) observed tawny eagles (*Aquila rapax*) carry out complete rolls through 360°.

In bats, another landing manoeuvre for rapid descent is sideslipping (Fig. 9.19). The sideslip is usually preceded by a short gliding phase when the wings

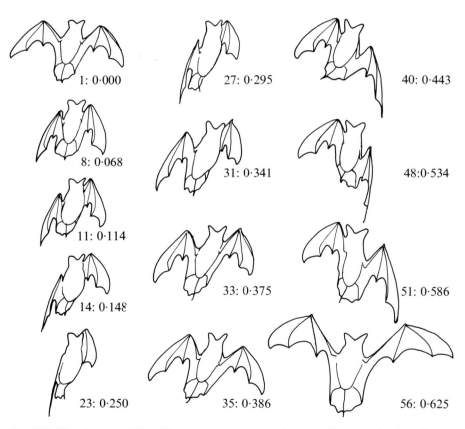

Fig. 9.19. The giant mastiff bat (*Otomops martiensseni*) performing sideslips to the right (frames *8–23*) and to the left (frames *35–48*). The bat is seen from obliquely below, facing the camera. The *first number* indicates the frame number of the film and the *second number* the time in seconds. Filming rate is 88 frames s^{-1}. (Norberg 1976c; by courtesy of The Company of Biologists Ltd)

are held in a strongly flexed position that increases wing loading and speed, and decreases lift. Pronation of the right wing causes a decrease of the lift and drag coefficients on the right wing, and supination of the left wing causes an increase of the coefficients of the left wing. The result is a clockwise body rotation in the roll plane and the animal begins to sideslip down to the right. The projection of the body on a plane normal to the incident air flow then becomes larger than the greatest frontal area of the body in level flight, and parasite drag will be extensively increased. The result is a large reduction in the total lift to drag ratio and a steepening of the equilibrium glide angle which causes the animal to more rapidly lose height. Sideslips are usually performed alternately to the left and right, and the duration of a single sideslip is usually very short (only ca. 0.24 s in the giant mastiff bat *Otomops martiensseni*; Norberg 1976c).

9.8 Energy-Saving Types of Flight

Since flapping flight requires very high energy output rates, there should be strong selection forces favouring adaptations for reduction of the flight costs. Migrating birds use soaring, gliding and formation flight to reduce power consumption, and several species of birds use *intermittent* flight during migration and foraging for this reason.

There are two types of intermittent flight strategies, *bounding* flight and *undulating* flight. *Bounding* flight consists of a few metres of flapping flight alternating with a few metres of passive flight with folded wings, making the flying bird rise and fall in an almost ballistic path. Smaller birds (such as tits, finches and wagtails) up to the size of the green woodpecker (*Picus viridis*) use this kind of flight. In *undulating* flight the active phase is a climb in a straight path, while the passive phase consists of gliding in a straight path on extended wings, and this is typical of birds larger than woodpeckers and of those that are good gliders (such as gulls and crows). Windhovering in falcons (Videler et al. 1983) is closely related to undulating. Other intermittent strategies include the climbing-gliding behaviour of certain forest birds (R.Å. Norberg 1981a, 1983), gliding reptiles and mammals (Rayner 1981; Scholey 1986), and probably of the ancient proto-fliers (Norberg 1985b,c, 1986a).

Birds and bats flying close to the ground or water make use of the *ground effect*, known in aircraft theory. By flying close to a surface the flying animal can reduce the induced power, which may be important in slow flights (see Sect. 9.8.3).

9.8.1 Bounding Flight

The energy cost of bounding flight can be estimated by a simple calculation (Lighthill 1977; Rayner 1977; Alexander 1982). During the passive phase the bird saves induced and profile drag by folding the wings, while during the flapping phase the wings must generate sufficient thrust and weight support to sustain the bird through both phases. Using the power Eq. (2.18) and neglecting the inertial power, the power required to fly can be written as

$$P = (A_b + A_w)V^3 + BL^2/V, \qquad (9.26)$$

where $A_b V^3$ is the parasite power [where $A_b = (1/2)\rho A_e$], $A_w V^3$ the profile power of the wings [where $A_w = (1/2)\rho S C_{Dpro}$], and e.g. BL^2/V the induced power (where $B = 2k/\rho\pi b^2$) (Alexander 1982). Assuming that the wings flap for a fraction a of the time (Fig. 9.20a), the lift L produced during the flapping phase must equal Mg/a and the power required to propel the bird is

$$P_{flap} = (A_b + A_w)V^3 + B(Mg)^2/a^2V. \qquad (9.27)$$

Assuming further that there is no lift during the period when wings are folded, that there is no induced drag, and that profile drag is minute, then the power required during the passive phase (profile power over the period (1−a)) is approximately

$$P_{fold} = A_b V^3. \qquad (9.28)$$

The mean power for bounding flight becomes

$$P_{bound} = a\, P_{flap} + (1-a)\, P_{fold} = A_b V^3 + a\, A_w V^3 + B(Mg)^2/aV. \qquad (9.29)$$

By differentiating this equation with respect to speed V and setting the derivative equal to zero the optimal value becomes

$$a_{opt} = (B/A_w)^{1/2}(Mg/V^2). \qquad (9.30)$$

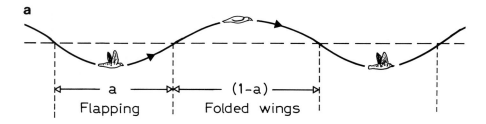

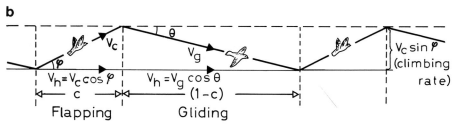

Fig. 9.20. Intermittent flight. *a* Bounding flight. The bird is flapping during fraction *a* of the time for one complete wingbeat cycle and has folded wings during fraction (*1−a*). *b* Undulating flight. The bird is flapping during fraction *c* of the time for one cycle and gliding during fraction (*1−c*). V_c is the climbing speed, V_g is the gliding speed, V_h is the horizontal speed, φ is the climb angle, and θ is the glide angle

The second derivative is positive, meaning that the extreme point is a minimum. Since for low speeds a_{opt} would be > 1, which is impossible, a slow-flying bird should flap its wings continuously ($a = 1$). For faster flights, however, a_{opt} would decrease as speed increases. But if a bird flew very fast at a low value of a, the muscle power needed during these short bursts would be very high. Since the power output of their wing muscles is limited (maximum 200 W kg^{-1}; Weis-Fogh and Alexander 1977), the optimum value of a may be larger than the value obtained by Eq. (9.30). Alexander (1982) suggested that birds should beat their wings continuously not only at low speeds but also at maximum speed, and that bounding flight is feasible and advantageous only at intermediate speeds. The wing muscles in larger birds can, however, produce little more power than is needed for continuously flapping flight, which precludes them from using bounding flight.

The maximum range speed is $V_{mr} = [B(Mg)^2/(A_b + A_w)]^{1/4}$, which is the speed at which drag is minimum (see Sect. 2.3.2). From this equation Mg can be expressed as a function of V_{mr}. By replacing Mg in Eq. (9.30) by this expression the optimum value of a can be written as

$$a_{opt} = (1 + A_b/A_w)^{1/2}(V_{mr}/V)^2. \tag{9.31}$$

According to the definition, bounding flight means that $a < 1$, and from Eq. (9.31) it follows that energy will be saved only at high flight speeds satisfying

$$V > V_{mr} (1 + A_b/A_w)^{1/4}. \tag{9.32}$$

The ratio A_b/A_w will likely lie between 0.5 and 2 (Rayner 1985b), and bounding should only reduce the flight power at air speeds greater than about $1.1V_{mr}$ (when A_b/A_w equals 2). Energy saving E_s is calculated from

$$E_s = 1 - (\text{Energy in bounding flight})/(\text{Energy in steady level flight at } V_{mr}) \tag{9.33}$$

(Rayner 1977). For example, a bird flying at 1.5 V_{mr} and with a $= 1/2$ would save about 17% of its used energy expenditure (Lighthill 1977), and the optimum saving (at the optimum value of a) for a chaffinch (*Fringilla coelebs*) flying at 1.3 V_{mr} would be 35% (Rayner 1977).

According to Eq. (9.8.6), the optimum value of a is independent of body size, but since bounding flight is confined to small birds, the aerodynamic explanation is inadequate. Bounding has been observed at slower flight speeds than V_{mr} (Csicsáky 1977; Rayner 1985b), and even in hovering (Zimmer 1943; Norberg 1975; Dalton 1982; Scholey 1983). Furthermore, even in fast flights, bounding flight does not reduce the cost of transport P/MgV below the minimum for steady flapping flight (Rayner 1985b), which means that migrants should not use bounding flight. Radar observations, however, show bounding flight in several migrating passeriform birds.

Rayner (1985b) modified the simple power model by assuming that the body produces a small weight support during the bound with completely folded wings and he neglected any induced drag associated with this force. The lift during the flapping phase became proportional to $Mg[1 - x(1 - a)]/a$, where x is a fraction of mean weight support, and with this value substituted in Eq. (9.27), bounding is advantageous at speed V if $x > (1/2)[1 - (1 + A_b/A_w)^{-1}(V/V_{mr})^4]$. In hovering and slow flight x will be very small with no useful savings, but at cruising flight speeds bounding would cost less if there is considerable weight support during the bound with folded wings. At the minimum power speed x must reach almost 0.5. Rayner noted that it is unlikely that weight support during the bound would contribute to power economy, but it might be advantageous at medium speeds.

As noted above, bounding flight cannot be explained solely as an adaptation to reduce the mechanical power needed to fly, but might be a compromise between conflicting demands of aerodynamic performance and flight muscle physiology (Goldspink 1977a; Rayner 1977). Rayner (1985b) considered in detail the hypothesis that bounding flight results from constraints on muscular efficiency. A brief summary is given here.

A bird can do one of three things to control power output (Goldspink 1977a): (1) vary contraction rate (strain rate or wingbeat frequency and amplitude), (2) recruit different proportion of the muscle fibres or motor units, or (3) introduce intermittent rest periods while the contraction rate remains almost constant and recruiting virtually all of the muscle. Wingbeat frequencies and amplitudes are rather uniform in birds using bounding flight as well as in other small birds eliminating the first way to control power output. Furthermore, slow fibres are rare in small birds, so they cannot glide. Bounding flight for intermittent rest periods seems to be the only way for smaller birds to keep muscle contraction dynamics close to optimum while obtaining a low mean power output.

Rayner (1977) showed that the flapping ratio a needed to balance cruise flight with peak power requirement was scale-dependent, rising approximately as $M^{1/6}$. Above a critical body mass (about 0.1 kg) there is not enough muscle capacity for bounding flight, or for sustained hovering or climbing. By changing a smaller birds can raise their body mass substantially before migration or breeding while keeping high flight ability, but without this ability extensive migrations would be impossible in smaller birds (Rayner 1985b).

Larger birds have a more heterogeneous muscle ultrastructure with fibres adapted to different flight behaviours. Slow fibres allow them to glide and to adopt undulating flight to reduce flight power. Bats use a number of independent muscles for the downstroke so that changes in muscle activity control power output (Norberg 1970b, 1972a; Hermanson and Altenbach 1983), making bounding flight unnecessary.

9.8.2 Undulating Flight

In undulating flight, power output can be reduced even at slow speeds. During the climb the animal must regain the potential energy lost during the glide, and weight is supported continuously and only thrust is intermittent. Rayner (1977, 1985b) derived a simple formulation of the energy optimization, and is related here. His 1985b paper was a reply to the model for undulating flight made by Ward-Smith (1984).

The animal is assumed to climb at an angle φ to the horizontal at an air speed V_c (Fig. 9.20b). The animal is then gliding at an angle θ to the horizontal with gliding speed V_d. For simplicity it is assumed that the horizontal speeds in climbing and gliding are equal, so that $V_c \cos \varphi = V_d \cos \theta = V_h$. The use of V_h as the horizontal speed component rather than true air speed along the actual flight path is convenient for the comparisons with flight ranges and speeds in horizontal flight.

Assuming that the bird is climbing over the period c and descending over the period $(1 - c)$, that the horizontal speed component is equal in both phases, and that the descent flight is equilibrium gliding, then the climbing ratio c can be defined as

$$c = c/[c + (1 - c)] = \tan \theta/(\tan \theta + \tan \varphi). \tag{9.34}$$

The climbing power can be calculated by Eq. (9.26) but with the addition of the power needed to provide the vertical force above that balancing the weight. Weight support in a climb must equal $L = Mg \cos \varphi$, and the power is $P_{climb} = V_c (D + Mg \sin \varphi)$ (see Sect. 9.6.2). Since gliding power is assumed to be zero, the mean power for undulating flight is

$$P_{und} = cP_{climb} . \tag{9.35}$$

If P_{und} is less than the power needed for horizontal flapping flight, then undulating flight reduces both power and cost of transport at constant speed V.

The optimum strategy depends on parameters such as climb ratio, climb angle, glide angle and glide performances, and so is difficult to evaluate. Rayner (1985b) demonstrated the power economy in a hodographic construction and concluded that undulating flight is energetically attractive at speeds below V_{mr}, but that it might result in small savings at speeds around V_{mr}. Differentiation of Eq. (9.35) gives an optimum climbing angle φ for a given horizontal speed. The power equation predicts only a modest power reduction (2.5% at V_{mp}) and the minimum cost of transport is not significantly lower than at steady flight. But the model assumes that A_b, A_w and B are constant. Since they rise with climbing angle, the power would be lower than predicted. Rayner noted that the model neglects the effects of overall changes in wingbeat kinematics, and he considered this in calculating the power for a starling (*Sturnus vulgaris*) in undulating flight (Fig. 9.21). Assuming that maximum power output for the flight muscle is 200 W kg^{-1} (Weis-Fogh and Alexander 1977), the instantaneous power output for the starling becomes 2.8 W, and maximum mean power output 2.8c W. Figure 9.21 shows that the flight muscles in the starling cannot achieve power output at c much below 0.25. At this value the power can be reduced by 14% at V_{mp} and by 10% at V_{mr} compared to steady flapping flight, and the minimum cost of transport can be reduced by about 11%.

Rayner also noted that the performance can be improved further by taking into account flexibility between gliding and flapping speeds. In the model, the horizontal speeds are assumed to be constant. The optimum strategy would be to glide at the minimum drag speed where rate of potential energy loss is least and to flap at the highest available aerobic power output. Power and cost savings are greatest when the ratio of maximum thrust (in the sense of maximum power output) to minimum drag is large (see, for instance, Gilbert and Parsons 1976), for the climbing period c can then be decreased. Power and cost savings depend

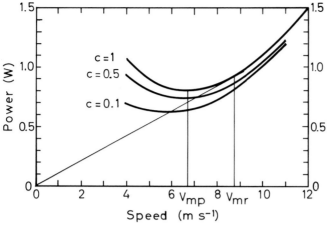

Fig. 9.21. Mechanical power in steady and undulating flight for a starling (*Sturnus vulgaris*) for various values of the flapping ratio c. c is unlikely to be much lower than 0.25 at low flight speeds, and should rise at higher speeds. Below 4 m s^{-1} the wing may stall in gliding, and undulating is impractical. Above 10–11 m s^{-1} the bird cannot generate sufficient thrust to climb during the flapping phase, but can fly horizontally. Undulating is possible between these limits, where it lowers the power curve. (After Rayner 1985b)

on wing morphology; animals with low wing loading and high aspect ratio have low flight costs and can save still more by undulating flight. This may explain why so many large-winged birds (such as crows and gulls) use undulating flight, while birds of low aspect ratios (passerines) or high wing loadings (ducks, auks) do not. Some bats occasionally use undulating flights. In general, bats have lower wing loadings than birds (U.M. Norberg 1981a).

Muscle physiology may not be as important a factor in undulating flight as it is in bounding. Undulating birds must have slow muscle fibres for gliding as well as faster fibres for flapping flight, and both fiber types appear in large birds.

9.8.3 Ground Effect

The influence of the ground or water surface on animals flying near to it can be investigated by the method of images. A wing at height h from the ground corresponds to a wing in a flow at height 2h above an extant similar wing image. The effect of the ground can be calculated in the same way as for the mutual interaction of two biplane wings, but the position of the image wing is that of the underwing of a biplane when reversed, so that the effect of the flow will be of opposite sign (see, for example, Betz 1963). The image aerofoil will cause a reduction in downwash angle and hence in induced drag and power of the real aerofoil. Characteristic speeds are also reduced, resulting, for example, in a longer glide path when gliding close to a surface (Blake 1983).

162

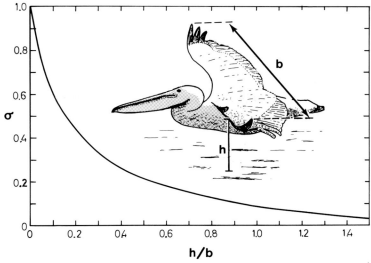

Fig. 9.22. Percentage reduction of induced drag (interference coefficient σ; where σ = 1 means 100% reduction) plotted against a non-dimensional index of the height above the ground (*h/b*, where *h* is the height of the wingstroke plane above the surface and *b* is wingspan). (Data from Reid 1932)

The extent to which ground effect reduces induced drag and speed depends upon the ratio between height h and the wingspan b, h/b, and is measured by an interference coefficient σ (giving the percentage reduction of induced drag; see, for example, von Mises 1959). Figure 9.22 shows values of σ as a function of h/b. If h > b the value of σ is negligible, but σ exponentially approaches 1 (= 100% saving) as h approaches 0. Useful savings are made when h is less than b/2 (Houghton and Brock 1960). Induced power with ground effect is

$$P_{ind,g} = (1 - \sigma) P_{ind}, \tag{9.36}$$

and the corresponding minimum power speed and maximum range speed are

$$V_{mp,g} = (1 - \sigma) V_{mp} \tag{9.37}$$

and

$$V_{mr,g} = (1 - \sigma) V_{mr}. \tag{9.38}$$

In their model Withers and Timko (1977) used the momentum and blade element theories to demonstrate the significance of ground effect to the flight energetics, foraging energetics, and daily energy budget of the black skimmer (*Rynchops nigra*). Induced power was calculated as

$$P_{ind,g} = (1 - \sigma)k(Mg)^2/2\rho VS_d \tag{9.39}$$

[cf. Eqs. (2.10) and (8.6)], with the use of k = 1.4 after Tucker (1973). By flying 7 cm above the surface (σ = 0.5) and somewhat faster than minimum power

163

speed, the black skimmer could reduce induced power by 50% and total mechanical power by 19%. Withers and Timko concluded that ground effect has profound influence upon foraging energetics and the daily energy balance of skimmers. They predicted that ground effect will be significant in other birds and in bats and insects flying close to a surface. They also noted that ground effect may be important for the take-off abilities of birds with high wing loadings and for hovering near ground or water. These factors may have been particularly important for large pterosaurs.

The actual flight speeds in the black skimmer should be compared with the minimum power and maximum range speeds in ground effect (which are both higher with ground effect), and not with those speeds calculated without ground effect (Blake 1985; Withers and Timko 1977). Observed flight velocities for the black skimmer during skimming are about 10 m s^{-1} (Withers and Timko 1977; Blake 1985), and estimated minimum power and maximum range speeds for σ = 0.5 are 5.8 and 7.6, respectively. The actual flight velocities are thus somewhat faster than maximum range speed, which might be expected for maximization of foraging efficiency (R.Å. Norberg 1981b).

9.9 For Ecologists and Others: Recipes for Power Calculation

The power for powered level flight is the sum of the induced, profile, parasite and inertial powers. The inertial power is minute in medium and fast flight and can be neglected. The induced power is also small at these speeds, so it is reasonable to use the minimum value obtained from the Rankine-Froude momentum theory, multiplied with an estimated correction factor. Profile and parasite powers are calculated with conventional aerodynamic theory. A strip-analysis is required for the calculation of the profile power.

Pennycuick (1989) gave a BASIC program for the calculations of bird flight performance. It has been designed primarily for those who wish to calculate the power requirements of foragers carrying food back to their nests and to estimate the range of a long-distance migrant.

9.9.1 Induced Power

Model to be used: Eq. (8.6) times the induced factor k = 1.2 (Pennycuick 1975).
  Data needed: Body mass M, wingspan b.

9.9.2 Profile Power

1. *Model to be used:* Eq. (2.15) times speed, and profile drag coefficient $C_{D,pro}$ = 0.02 (Rayner 1979b).
  Data needed: Body mass M, wingspan b, wing area S, resultant air speed V_r,

164

2. *Model to be used:* Eq. (9.3) (Rayner 1979b).

Data Needed: Wingspan b, wing area S, wingstrip areas $S_w(r)$, stroke amplitude $\emptyset$, wingstroke period T, downstroke ratio τ, resultant air speed V_r.

9.9.3 Parasite Power

Model to be used: Eq. (2.16) times speed, and Eq. (9.4) for equivalent flat plate area A_e, (Pennycuick 1968a, 1975). For birds > 100 g, the Eq. (9.4) for A_e should be replaced by Eq. (9.5) for frontal body area and by $C_{D,par} = 0.40$ at subcritical Re (< 50,000), by $C_{D,par} = 0.25$ at supercritical Re (> 200,000) and by Eq. (9.6) for $C_{D,par}$ at the transition region (Pennycuick et al. 1988).

Data needed: Body mass M, flight speed V and Re number.

9.9.4 Power Required for a Climb

Model to be used: Eq. (9.17) with the drag components used for forward flight (see Sects. 9.9.1–3 above) and Eq. (9.4).

Data needed: Body mass M, wingspan b, wing area S, resultant air speed V_r, equivalent flat plate area A_e (see Sect. 9.9.3 above), climb angle φ.

9.9.5 Power Required for a Descent

Model to be used: Eq. (9.20) with the drag components used for forward flight (see Sects. 9.9.1–3 above) and Eq. (9.4).

Data needed: Body mass M, wingspan b, wing area S, resultant air speed V_r, equivalent flat plate area A_e (see Sect. 9.9.3 above), descent angle θ.

9.9.6 Power Required for Bounding Flight

Model to be used: Eq. (9.29) with power components according to Sections 9.9.1–3 (above) and Eq. (9.4).

Data needed: Body mass M, wingspan b, wing area S, resultant air speed V_r, equivalent flat plate area A_e (see Sect. 9.9.3 above), period for flapping a, period for wing retraction (1-a).

9.9.7 Power Required for Undulating Flight

Model to be used: Eqs. (9.34), (9.35), (9.17) with the drag components used for forward flight (see Sects. 9.9.1–3 above) and Eq. (9.4).

Data needed: Body mass M, wingspan b, wing area S, resultant air speed V_r, equivalent flat plate area A_e (see Sect. 9.9.3 above), climbing ratio c, climb angle ψ, descent angle φ.

Chapter 10

Scaling

10.1 Introduction

study η growth

Scaling is widely used in studies of the biomechanics and energetics of animal flight, for with empirical allometric relationships one can explore how different mechanical, physiological and ecological constraints change in importance with size and wing morphology. Allometry can also be used to obtain a general norm for a morphological or physiological variable for a group of animals allowing the identification of deviations from the norm that may indicate adaptations to different biological niches and/or habitats. Chapter 3 discussed allometric relationships between various physiological characters and body mass, and here the theoretical basis, with empirical scaling data is given.

A character is usually scaled against the size of the animal, normally measured by body mass, and the relationships between a dimension or other character, y, and the body mass, M, is expressed by the power function

$$y = \alpha M^{\beta}, \tag{10.1}$$

where α is the y-intercept and β is the regression coefficient. When plotted in a double-logarithmic diagram (both axes of a coordinate system bearing logarithmic scales) a curve becomes a straight line with slope β (where $\tan^{-1} \beta$ is the angle between the line and the horizontal), and α is the y-intercept at the M-value 1, since log 1 = 0.

Several authors (in particular, Magnan 1922; Hartman 1961; Greenewalt 1962, 1975; Pennycuick 1969, 1975; U.M. Norberg 1981a; Norberg and Rayner 1987; Rayner 1988) have explored how the wing characters of birds and bats vary with body mass. The variation between species in wing dimensions for a given mass can be considerable, e.g. in a 1-kg bird wingspan and area can range from about 0.7 m and 0.052 m² in alcids (*Uria troille*) to about 1.6 m and 0.330 m² in ospreys (*Pandion haliaetus*). This means that compared to the alcid, the osprey has more than twice as long a wingspan with six times the wing area. Two of the shortest wingspans in a 10-g bird occur in Darwin's warbler finch (*Certhidea olivacea*, 0.165 m) and in the wren (*Troglodytes troglodytes*, 0.169 m). The 10-g hummingbird (*Colibri coruscans*) and *Parus* species weighing about 10-g have wingspans of 0.19–0.20 m.

10.2 Geometric Similarity

Different-sized animals of the same shape (= identical corresponding angles) are said to be geometrically similar. For geometric (isometric) similarity, any chosen volume M varies with the cube of any chosen length l, $M \propto l^3$, and any chosen area S with the square of any chosen length, $S \propto l^2$. So, for any characteristic speed, V (for example, minimum power speed V_{mp} or maximum range speed V_{mr}), and length, l, the following relationships (from Eqs. 2.29 and 2.30) apply to geometrically similar animals,

$$V \propto (Mg/S)^{1/2} \propto (l^3/l^2)^{1/2} = l^{1/2} \propto M^{1/6}. \qquad \text{i.e } C_L \text{ is same} \qquad (10.2)$$

Thus, if a bird is geometrically similar to another bird but has c times the wingspan, then it should fly $c^{1/2}$ times as fast; if it weighs d times as much it should fly $d^{1/6}$ as fast.

In steady horizontal flight the muscle power P_r required to fly at any speed can be represented as the product of the average, effective, drag D', acting backwards along the flight path, and the forward speed V as

$$P_r = D'V = Mg(D'/L')V \propto M \times M^0 \times M^{1/6} = M^{7/6} \propto l^{7/2} \qquad (10.3)$$

(Pennycuick 1975), where (L'/D') is the effective lift/drag ratio. If one bird weighs twice as much as another geometrically similar one, it will require $2^{7/6} = 2.24$ times as much power to fly at any (characteristic) speed under corresponding conditions. The cost of transport [(energy cost)/(weight · distance); e.g. Tucker 1970, 1975] should then vary as

$$C = P_r/(MgV) \propto M^{7/6}/(M \times M^{1/6}) = M^0. \qquad (10.4)$$

Hence, it is a non-dimensional expression that should be independent of body mass (see Sect. 3.7).

The power P_a available from the muscles may be represented as the product of the flight muscle mass m_m and the power available from a unit of mass of muscle. The latter is the specific work $Q_m{}^*$ done in each contraction times the maximum flapping frequency f_w [cf. Eq. (10.6) below]. The muscle mass is assumed to be a constant proportion of body mass, and the specific work independent of body mass (see Sect. 11.2.6). Thus, with geometric similarity, the maximum power available can be expressed as

$$P_a = m_m Q_m{}^* f_w \propto M \times M^0 \times M^{-1/3} = M^{2/3} \propto l^2 \qquad (10.5)$$

(Pennycuick 1975). The $M^{7/6}$ (10.3) and $M^{2/3}$ (10.5) scalings of power may be reasonable average estimates, but various constraints may differ for each species, and there are deviations from the rule. For insects, which have different constraints on the maximum wingbeat frequency, the slope is closer to one than to the two-thirds power of the body mass (Pennycuick 1975, 1986).

10.3 Estimated Relationships Between Flight Characteristics and Body Mass

All birds or bats are not geometrically similar, but deviations from isometry are small for most groups, so the relationships of speed, power and cost of transport to body mass shown in Eqs. (10.2)–(10.5) are good approximations. Important discrepancies between predicted exponents (on the basis of geometric similarity) and observed exponents may reflect mechanical and physiological constraints as well as adaptive factors.

Greenewalt (1962, 1975) calculated the relationships between various wing characters and body mass in several groups of birds ("passeriforms", "seabirds", ducks and hummingbirds) and Rayner (1988) did this for hummingbirds and other birds (the latter treated as a group), mainly using Greenewalt's data. Lawlor (1973), U.M. Norberg (1981a) and Norberg and Rayner (1987) calculated the relationships for bats. Greenewalt's passeriform group includes also species of other orders and families (such as Ardeidae, Apodidae, Laridae, Picidae, Falconiformes and Strigiformes), and his shorebird group includes, for example, Columbidae, Psittacidae, various waders, swans and geese.

U.M. Norberg (1981a) and Norberg and (Rayner 1987) gave separate regression equations for each family of bats, and the regression equations in their bat works (including 84 and 257 species, respectively) agree well with each other. Table 10.1 shows regression equations for wing morphology and aerodynamic characteristics for various birds, bats and pterosaurs, based on empirical data (birds and bats) or estimated data from reconstructions (fossil bats and pterosaurs).

The pterosaur data originate from Brower and Veinus (1981), who assumed that the wing's trailing edge attached to the pelvis instead of to the leg. I have assumed that the trailing edge instead attached to the legs, and I added 14% (based on my own reconstructions) to their estimated wing areas, which changes the y-intercepts but not the regression coefficients of their equations. The corrected values are given in Table 10.1 with their equations. However, because neither mass nor wingspan and wing area can be measured accurately on any pterosaur, and because there are no extant pterosaurs for comparison, the allometric data for them may not be reliable.

For the calculation of the regression equation, most authors (Greenewalt 1962, 1975; U.M. Norberg 1981a; Morgado et al. 1987; Brower and Veinus 1981) have used the least-squares method (LS), whereas others (Norberg and Rayner 1987; Rayner 1988) used the reduced major axes method (RMA) which is also used here for the fossil bats. The least-squares method compensates for errors in the y direction but not in the x direction, while RMA accounts for errors in both x and y directions. The regression equations derived by these two methods become about the same when variation about the regression line is small (the regression coefficient being close to one), since the RMA slope is equal to the LS slope divided by the regression coefficient (Sokal and Rohlf 1981). The slope can be markedly higher with the RMA method when variation is large.

Table 10.1. Power functions of wing dimensions and flight parameters against body mass (kg)

Animal group / Dimensions	Wingspan (m)	Wing area (m²)	Wing loading (Nm⁻¹)	Aspect ratio	Minimum power speed V_{mp} ms⁻¹	Maximum range speed V_{mr} ms⁻¹	Minimum power P_{mp} Watt	Minimum cost of transport C_{min}	Wingbeat frequency f_w Hz
Slope for isometry	0.33	0.67	0.33	0.00	0.17	0.17	1.17	0.00	-0.33
Birds, all[a,b]	–	–	–	–	$5.70M^{0.16}$	$15.4M^{1.10}$	$10.9M^{0.19}$	$0.21M^{-0.07}$	$3.87M^{-0.33}$
All birds except hummingbirds[a]	$1.17M^{0.39}$	$0.16M^{0.72}$	$62.2M^{0.28}$	$8.56M^{0.06}$					$3.98M^{-0.27}$
Hummingbirds[c,b]	$2.24M^{0.53}$	$0.69M^{1.04}$	$14.3M^{-0.04}$	$7.28M^{0.02}$					$1.32M^{-0.60}$
"Passeriforms"[c,b]	$1.13M^{0.42}$ to $1.65M^{0.42}$	$0.16M^{0.78}$ to $0.33M^{0.78}$	$60.2M^{0.22}$ to $29.4M^{0.22}$	$7.83M^{0.05}$ to $8.15M^{0.05}$	$5.04M^{0.13}$	$8.63M^{0.14}$	$12.9M^{1.05}$	$0.21M^{-0.08}$	$3.03M^{-0.36}$
"Shorebirds"[c,b]	$1.16M^{0.40}$	$0.13M^{0.71}$	$76.0M^{0.29}$	$10.4M^{0.10}$	$5.85M^{0.10}$	$10.5M^{0.06}$	$15.4M^{1.07}$	$0.21M^{-0.02}$	$4.00M^{-0.19}$
"Ducks"[c,b]	$0.90M^{0.41}$ to $0.93M^{0.41}$	$0.078M^{0.71}$ to $0.083M^{0.71}$	$12.6M^{0.29}$ to $18.0M^{0.29}$	$10.4M^{0.09}$	$6.78M^{0.22}$	$17.1M^{0.01}$	$20.8M^{1.14}$	$0.23M^{-0.08}$	$5.88M^{-0.24}$
Bats, all[d]	$1.20M^{0.32}$	$0.20M^{0.32}$	$67.2M^{0.44}$	$12.4M^{0.15}$	$8.96M^{0.21}$	$11.8M^{0.21}$	$10.9M^{1.19}$	$0.11M^{-0.036}$	–
Megachiroptera[d]	$1.23M^{0.35}$	$0.24M^{0.72}$	$45.4M^{0.33}$	$8.63M^{0.11}$	–	–	–	–	–
Microchiroptera[d]	$1.62M^{0.33}$	$0.18M^{0.61}$	$105M^{0.54}$	$17.1M^{0.21}$	–	–	–	–	–
Ancient Micro-chiroptera[e]	$0.90M^{0.28}$	$0.10M^{0.48}$	$108M^{0.53}$	$9.33M^{0.10}$	–	–	–	–	–
Pterosaurs[f]	$1.95M^{0.40}$	$0.29M^{0.79}$	$35.3M^{0.23}$	$16.0M^{0.10}$	–	–	–	–	–
Pterosaurs[g]	$1.95M^{0.40}$	$0.33M^{0.79}$	$31.2M^{0.23}$	$14.2M^{0.10}$	–	–	–	–	–

[a] Wing dimensions and flapping frequency from Rayner (1988). [b] Aerodynamic characteristics from Rayner (1979c). [c] Recalculated data from Greenewalt (1975). [d] Data from Norberg and Rayner (1987). [e] Calculated with data from Norberg (1989) (n = 9; r = 0.995 for wingspan, r = 0.971 for wing area, r = 0.976 for wing loading, r = 0.714 for aspect ratio). [f] Data from Brower and Veinus (1981). [g] Corrected values with 13% addition to the wing area, assuming that the membrane was attached to the legs.

10.3.1 Wingspan

Greenewalt (1962) found that wingspan in birds increased faster with increasing body mass than predicted by geometric similarity. Birds in his passeriform model increased in wingspan according to b $\propto$ M$^{0.42}$, birds in his shorebird model to b $\propto$ M$^{0.40}$, the ducks to b $\propto$ M$^{0.41}$, and the hummingbirds to b $\propto$ M$^{0.53}$ (Table 10.1). Morgado et al. (1987) found that the slope for 70 Chilean birds of different families was 0.39, also exceeding the slope for isometry. But a comparison between groups of animals should be at as low a taxonomic level as possible, because a slope of a regression line for a large bird group can differ from the slopes for the various families or genera in the same group. If family groups have similar slopes but different y-intercepts (for example, for some ecological reason) the main slope for the whole group will be quite different (Fig. 10.1), as discussed by Clutton-Brock and Harvey (1979).

For example, Morgado et al. (1987) reported that the regression line for wingspan in passeriform birds had the slope 0.27 (n = 21), charadriform birds 0.36 (n = 7), and falconiforms 0.29 (n = 8), which are all lower than the slope for all birds taken together (0.40). However, birds as a whole have a slope of around 0.40, and larger birds have proportionately longer wings than smaller birds. Samples representing lower taxonomic levels have shallower slopes and are more isometric to other birds in the same group. When comparing birds with bats or pterosaurs it is important to use taxonomic levels as similar as possible; this may be difficult, since taxonomy is based on different characters in the different animal classes.

For bats as a whole, wingspan varies with mass as b $\propto$ M$^{0.32}$ (Norberg and Rayner 1987) and the coefficient is close to that for geometric similarity. For families with larger numbers of species ($>$ 25) slopes were 0.35–0.37, but only one family (Vespertilionidae, slope = 0.37, n = 93) differs significantly from 0.33 (isometry). The fossil microchiropteran bats have shorter spans and a slope (0.27) below that for geometric similarity.

The wingspans of megachiropteran bats are about as long as those of members of the passeriform group (Fig. 10.2), but vespertilionid and molossid bats have proportionately longer wings than passeriforms of similar mass. The

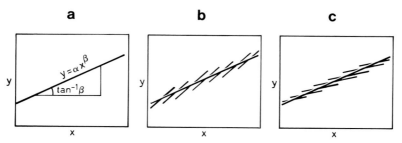

Fig. 10.1. a Regression line with slope β and y-intercept α in a double logarithmic diagram. *b* Groups of lower taxonomic levels with higher value of β and other values of α than that of the mean line for all individuals taken together. *c* Groups of lower taxonomic levels with lower value of β and other values of α than that of the mean line for all individuals taken together

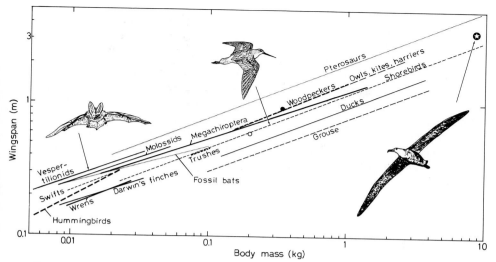

Fig. 10.2. Wingspan plotted on logarithmic coordinates against body mass for some animal groups. *Filled circle* shows the span of the fossil bat *Archaeopteropus transiens*, *open circle* that of the *Archaeopteryx lithographica*, and the *star* that of *Diomedea exulans*. (Data from Greenewalt 1975; Norberg and Rayner 1987, by courtesy of The Royal Society; Norberg 1989; U.M. Norberg unpublished)

lines for insectivorous bats lie nearest to hummingbirds, swifts (Apodidae), flycatchers (Muscicapidae), swallows (Hirundinidae) and the larger tyrannids of the passeriform group. The slope is steeper for hummingbirds than for bats and other birds, meaning that small hummingbirds have proportionately shorter spans than large ones. Ducks, wrens (Troglodytidae) and Darwin's finches (*Geospiza, Platyspiza, Camarhynchus, Certhidea*; U.M. Norberg and R.Å. Norberg 1989) have about the shortest wings among birds, and kites, harriers, swifts and several seabirds, the longest.

Pterosaur wingspan appears to have varied as b $\propto$ M$^{0.40}$ (Brower and Veinus 1981) and to have been longer (larger y-intercept, Table 10.1) than that in similar-sized birds and bats.

10.3.2 Wing Area

Wing area shows larger variation between groups of animals than wingspan (Table 10.1). Greenewalt's (1962) data give the following relationships between wing area and body mass: S $\propto$ M$^{0.78}$ for the passeriform group (and also for different families within this group treated separately), S $\propto$ M$^{0.71}$ for "shorebirds" and ducks, and S $\propto$ M$^{1.04}$ for hummingbirds. The latter value is much higher than the one predicted for isometry (0.67).

Recent bats are closer to isometry with the relationships S $\propto$ M$^{0.64}$ for all bats taken together, S $\propto$ M$^{0.72}$ for megachiropteran bats and S $\propto$ M$^{0.61}$ for micro-

171

chiropteran bats, and the slope for most bat families does not differ significantly from 0.67 (Norberg and Rayner 1987). Wing areas in bats include the tail membrane, which varies a lot between families and contributes to the large variation between the bat families. Fossil microchiropteran bats also had a well-developed tail membrane (Habersetzer and Storch 1987), but their wing area was somewhat smaller than that of recent bats. Larger ancient bats had a much smaller area in relation to body mass than the smaller species, in particular the ancient megachiropteran species *Archaeopteropus* (Norberg 1989).

Brower and Veinus (1981) estimated that pterosaur wing area scaled with body mass as $S \propto M^{0.79}$, so that large pterosaurs had proportionately larger wing area than smaller ones.

On average, members of the passeriform group (except for the wrens and Darwin's finches) have larger wing areas in relation to body mass than "shorebirds" and ducks (ducks have the smallest relative area). Hummingbirds have a much steeper slope of the regression line than the other birds and the bats. Small hummingbirds have an area as small as the wrens, but large ones have wings as large as most members of the passeriform group (e.g. tyrannids and parulids). Vespertilionid bats have much larger wing areas than any bird of similar mass, a consequence of incorporating the tail membrane in the total wing area. Small pterosaurs may have had wing areas similar to those of megachiropteran and molossid (free-tailed) bats, and many seabirds, kites and harriers.

10.3.3 Wing Loading

The regression lines for wing loading versus body mass for various animal groups (Fig. 10.3) mirror those for wing area. Wing loading actually increases less in pterosaurs and birds than predicted by isometry. The estimated values give $Mg/S \propto M^{0.23}$ for pterosaurs, $Mg/S \propto M^{0.22}$ for passeriforms, and $Mg/S \propto M^{0.29}$ for shorebirds and ducks, which approaches the prediction for geometric similarity. Hummingbirds have a constant wing loading that is independent of body mass, $Mg/S \propto M^{-0.04}$ (taken that $S \propto M^{1.04}$). Megachiroptera obey geometric similarity ($Mg/S \propto M^{0.33}$) but in recent and ancient Microchiroptera, wing loading is proportionately larger in large species than in small ones; the slope is 0.54 for recent bats and 0.53 for ancient bats.

The wing loading of animal gliders does not increase much with size (Rayner 1981), so small gliders have proportionately larger flight surfaces than large animals, which means that they should have similar gliding speeds.

10.3.4 Aspect Ratio

One might expect that an increase of size would have been accompanied by departures from geometric similarity that reduce the minimum specific power (power divided by mass) by increasing the lift/drag ratio L/D. The way to increase L/D as body size increases is to increase wingspan more rapidly than the square root of wing area, i.e. by increasing the aspect ratio b^2/S (Lighthill 1977)

172

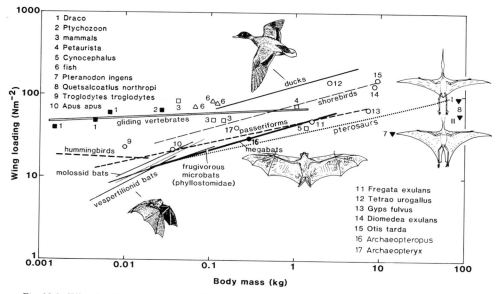

Fig. 10.3. Wing-loading plotted on logarithmic coordinates against body mass for some gliding and flying animals. (Data from animals numbered *1–6* are from Rayner 1981; after Norberg 1985a)

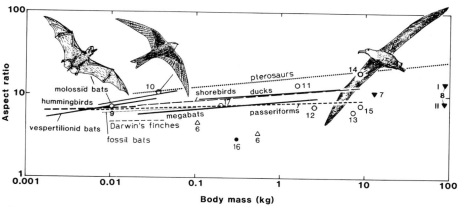

Fig. 10.4. Aspect ratio plotted on logarithmic coordinates against body mass for some animal groups. The *numbers* refer to the animals as identified in Fig. 10.3. (After Norberg 1985a)

because induced drag and power both decrease with larger span. This situation may explain why large birds have proportionately longer wings than small ones. For geometrically similar flying animals aspect ratio should not vary with size, but departures from constant aspect ratio (although very small) are statistically significant in birds and in microchiropteran bats (Fig. 10.4). In birds of the passeriform group, aspect ratio varies with body mass as $b^2/S \propto M^{0.05}$, while in ducks, shorebirds, megachiropteran bats, ancient microchiropteran bats and pterosaurs, it varies with body mass raised to about 0.1. In hummingbirds aspect ratio does not vary at all with body mass (exponent = 0). For all bats and for microchiropteran bats separately, departures from isometry are larger ($b^2/S \propto M^{0.15}$ and $b^2/S \propto M^{0.21}$, respectively), although the departures are less for several insectivorous bat families (U.M. Norberg 1981a; Norberg and Rayner 1987).

Greenewalt (1975, Appendix) noted uncertainty about whether or not wing area, which was measured by several investigators (and used by Greenewalt), included the projection of the body area between the two wings. Although Greenewalt decided that this area is relatively small and probably within the error of estimate, it accounts for about 20% of the combined area of the two wings in *Parus* species, and at least this in wrens and thrushes. The area is much smaller in harriers and kites, so the true regression lines for the different birds may lie somewhat higher for the wing area and lower for the wing loading and aspect ratio than they do in Figs. 10.2 and 10.3. Norberg and Rayner (1987) compensated for this in their calculations of regression equations for bats from their compiled data.

10.3.5 Flight Speed, Power and Cost of Transport

Rayner (1979c) calculated the relationships of flight speed and power to body mass for birds according to the vortex ring model, using empirical data from Greenewalt (1962). Using Rayner's (1986) model, Norberg and Rayner (1987) calculated the same relations with the mean morphology for each of 16 bat families, and their derived correlations correspond closely to the expected scalings for isometry, where speed should vary with body mass as $M^{1/6=0.17}$ and power with body mass as $M^{7/6=1.17}$. Table 10.1 summarizes the correlations with mass for minimum power speed V_{mp}, maximum range speed V_{mr}, minimum power P_{mp}, minimum cost of transport C_{mr}, and wingbeat frequency f_w. The minimum cost of transport is the power at V_{mr} divided by body weight and V_{mr}, $C_{mr} = P_{mr}/Mg\,V_{mr}$. The deviations from geometric similarity are rather small for birds but in "passeriforms" and "shorebirds" the flight speed for minimum power, and hence flight power, increases somewhat slower with increasing body mass than predicted for isometry. In ducks the speed increases more while the flight power increases less than predicted for isometry. Ducks in general have a higher optimum velocity than other birds following from their higher wing loadings.

In bats the regression equation for open-field flight speeds (assumed to be the maximum range speeds V_{mr}) against wing loading was $V_{mr} \propto (Mg/S)^{0.44}$ (n = 14), a slope not significantly different from that predicted for geometric similarity (0.50; Norberg 1987). But with additional data Norberg and Rayner (1987) obtained a higher slope (0.67, n = 26), which did not differ significantly from 0.50, because of variation. With the same data they estimated that speed varied with body mass as $V_{mr} \propto M^{0.32}$, a significant departure from the expected correlation [$V_{mr} \propto M^{0.17}$; cf. Eq. (10.2)]. Flight speed data for bats, however, are difficult to compare, because one cannot be sure if the bat was flying at the minimum power or maximum range speed or at some other speed.

174

10.3.6 Flight Muscle Masses

Different flight modes and kinematics put different demands on the size, arrangement and biochemistry of flight muscles. The main depressor of the bird wings is the pectoral muscle and the main elevator is the supracoracoideus muscle which is situated under the pectoral (Fig. 11.8, Sect. 11.2.9). Both muscles make up the main part of the thoracic musculature. The deltoid muscles may assist in the elevation of the wings to various degree in different bird species. In bats the wing movements are powered and controlled by several muscles (e.g. Vaughan 1959, 1970b; Norberg 1970b, 1972a; Strickler 1978).

Rayner (1985b, 1988) discussed the function of the supracoracoideus in birds at different flight modes and suggested that this muscle's contribution would not be necessary in flight with an aerodynamically active upstroke that generates a continuous vortex wake. Power for contraction of the supracoracoideus muscle would be used in slow flight and take-off in all birds and in fast flight in rounded-winged species when the upstroke is mainly a recovery stroke and the vortex wake consists of vortex-rings. Hummingbirds are an exception for they have aerodynamically active upstrokes (or backstrokes) with active supracoracoideus.

The relationships between flight muscle mass and body mass have been estimated in many birds of different families by Magnan (1922) and Hartman (1961). Using their data, Rayner (1988) estimated the regression equations for the weight of pectoralis m_p and supracoracoideus m_s muscles versus body mass to be $m_p = 0.15 \, M^{0.99}$ and $m_s = 0.016 \, M^{1.01}$, respectively. The pectoralis muscle makes up the largest part of the body mass in pigeons (Columbidae; $m_p/M = 0.240$), larks (Alaudidae; 0.226), sandpipers (Scolopacidae; 0.214), bustards (Otidiae; 0.209), Old World finches (Fringillidae; 0.206), trogons (Trogonidae; 0.203) and grouse (Tetraonidae; 0.202). This muscle is largest in relation to body mass in birds that make rapid bursts and those that take-off vertically upwards. In hummingbirds, m_p/M is 0.183. Small pectoral muscles occur in cuculids (*Piaya cayana*; $m_p/M = 0.076$), grebes (Podicipedidae; 0.091), rails (Rallidae; 0.097), divers (Gaviidae; 0.100), and pelicans (*Pelecanus occidentalis*; 0.102).

Judging from its size, the supracoracoideus appears to be of little importance in many birds. This muscle ranges in size from about 0.4% of the body mass ($m_s/M = 0.004$) in *Buteo* species to 12% (0.12) in some hummingbirds. Small supracoracoideus muscles ($m_s/M \approx 0.006$) occur in longer-winged species, such as Fregatidae, Accipitridae, Tytonidae and Strigidae, while the largest occur in hummingbirds (for which the mean of m_s/M is 0.089) and in species that make a rapid or steep take-off, such as Tinamidae (0.065), Phasianidae (0.063), Tetraonidae (0.058), and Columbidae (0.053). The greatest differences within a family in both muscles are found in larger species and hummingbirds, which have narrow power margins.

Unlike other New World finches ($m_p/M = 0.157$ and $m_s/M = 0170$), the Darwin's finches have small pectoral and supracoracoideus muscles and reduced flight performance. The muscle ratios are $m_p/M = 0.111$ and $m_s/M = 0.0102$ in the medium ground finch *Geospiza fortis* (unpublished data). There are at least three possible mainland ancestors to the Darwin's finches: *Melanospiza ri-*

chardssoni (Bond 1948; Bowman 1961), *Tiaris* spp. and *Volatinia jacarina* (Steadman 1982). The muscle ratios in the two latter species range from 0.139 to 0.156 (m_p/M) and from 0.0142 to 0.0185 (m_s/M). If the Galapagos finches are derived from any of these latter species, their pectoral and supracoracoideus muscles may have been reduced by 20–29% and 28–45%, respectively, assuming that the mainland species have not changed.

10.3.7 Wingbeat Frequency

The upper size limit of an animal is set by the strength of the muscles, tendons and bones. In geometrically similar animals the maximum wingbeat frequency varies with the minus one-third power of the body mass,

$$f_{w,max} \propto M^{-1/3} \qquad (10.6)$$

(Hill 1950), according to the following explanation (e.g. Pennycuick 1975). The force exerted by a muscle is proportional to the cross-sectional area of its attachment and hence to the square of the length, $F_m \propto l^2$, assuming that the stresses of the bones and muscles are constant. The moment J_m of the muscle about the centre of rotation of the proximal end of a limb equals the muscle force times the length of the limb, and is

$$J_m = F_m \cdot l \propto l^3. \qquad (10.7)$$

The moment of inertia of the limb about its fulcrum is the mass of the limb times the square of its length, and varies as

$$I = m \times l^2 \propto l^5, \qquad (10.8)$$

and the angular acceleration $\dot{\omega}$ that the muscle can impart to the limb varies with limb length as

$$\dot{\omega} = J_m/I \propto l^{-2} \qquad (10.9)$$

The time T taken for the stroke is inversely proportional to the square root of the angular acceleration $\dot{\omega}$, and becomes

$$T \propto \dot{\omega}^{-1/2} \propto l. \qquad (10.10)$$

The maximum wingbeat frequency is inversely proportional to T, so that $f_{w,max} \propto T^{-1} \propto l^{-1} \propto M^{-1/3}$ [Eq. (10.6);Fig. 10.5].

The lower limit of the wingbeat frequency is associated with the need to provide sufficient relative airflow over the wings to supply lift and thrust in hovering and in slow flight. For geometrically similar animals the minimum angular velocity, ω_{min}, needed to support the weight in hovering or in slow flight, is proportional to $(Mg/S^2)^{1/2}$ (Pennycuick 1975). Assuming a constant stroke angle, the angular velocity is proportional to wingbeat frequency and it should vary inversely with the square root of any representative length,

$$f_{w,min} \propto \omega_{min} \propto (Mg/S^2)^{1/2} \propto l^{-1/2} \propto M^{-1/6} \qquad (10.11)$$

(Pennycuick 1975; Fig. 10.5).

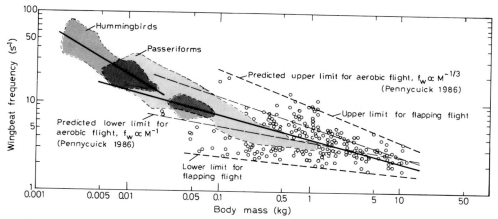

Fig. 10.5. Wingbeat frequency in birds. Body mass may not refer to the same individual for which the frequency was measured. (After Rayner 1988)

Lighthill (1977) and Weis-Fogh (1977) discussed a third constraint for hovering flight, namely one connected to the frictional and inertial power contributions; according to the "intraspecific rule" f_w should lie below a value $\propto l^{-2/3} \propto M^{-2/9}$.

There is also individual variation in wingbeat frequency according to the flight modes and speeds used, but it is usually higher at take-off and in accelerated flight, complicating comparisons between species. Morgado et al. (1987) found that wingbeat frequency varied according to $f_w \propto M^{-0.38}$ for 90 birds of different families. Rayner (1979c, 1988) compiled data (including those of Greenewalt 1962, 1975) for various birds (Table 10.1 and Fig. 10.5) and the resulting correlations vary among different groups. For example, hummingbirds, small passeriforms and galliforms have a proportionately higher flapping frequency than most other birds, whereas seabirds generally have low frequencies for species with long wings (gulls, frigates, albatrosses) or large wing areas (herons, raptors) generate sufficient thrust with low flapping frequencies.

10.4 Upper and Lower Size Limits

How big a bird can fly, and what restricts or permits an animal to use a certain flight mode? Pennycuick (1969, 1975, 1986) gave models for the power margin in birds and discussed them for various flight modes, and his theory is summarized here. The power margin (a power ratio) is defined as the ratio of the power available from the flight muscles P_a to that required to fly horizontally at the minimum power speed P_r, P_a/P_r. The power available depends upon the flapping frequency, which determines the upper and lower size of flying vertebrates.

Minimum and maximum wingbeat frequency lines converge as body mass increases (Fig. 10.6a, see also Fig. 10.5) until the point where flapping at

177

maximum frequency is the only way to achieve flight. An upper limit of 12 kg is based on actual sizes of the largest birds with powered flight (Pennycuick 1968a). Larger animals could not beat their wings fast enough to achieve the lift needed for horizontal flight, while small animals can use a wide range of frequencies. But animals of about 1 g are subjected to another limitation for the maximum wingbeat frequency, namely the time that vertebrate muscles require to reset the contractile mechanism after each contraction. Very small vertebrates cannot develop the maximum power needed for sprint manoeuvres, but insects do not have this limit for their fibrillar muscle can operate at much higher frequencies. The smallest hummingbirds weigh about 1.5 g, and the smallest bat (*Craseonycteris thonglongyai*) about 1.9 g.

Figure 10.6b shows the maximum power available for vertebrate muscle P_a and the power required to fly P_r. The continuous line for the power available represents aerobic muscle, used for sustained cruising locomotion, and the

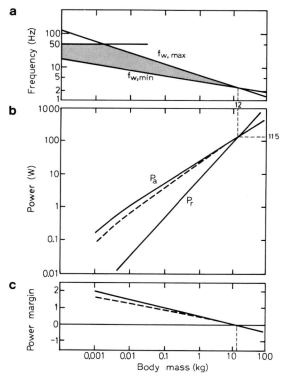

Fig. 10.6. a Maximum and minimum wingbeat frequencies converge as mass increases. The *stippled area* is available to vertebrate fliers. The *horizontal line* is approximate and reflects the minimum time taken to reset the muscle after each contraction. *b* The power required to fly, P_r, at the minimum power speed and the power available, P_a, from aerobic (*continuous line*) and anaerobic (*broken line*) flight muscles. *c* Power margin for vertebrate fliers. See text for further explanation. (After Pennycuick 1986)

178

broken line anaerobic muscle capable only of intermittent activity. Aerobic muscles produce less power for a given muscle mass because of the dilution with mitochondria, which are non-contractile, leaving a higher proportion of the muscle volume to (force-producing) muscle fibres. The difference between the muscles becomes larger for smaller animals.

Using the logarithm (to base 10) of the power margin ratio (Fig. 10.6c) an animal with a power margin of zero can produce exactly the amount of power to fly horizontally at minimum power speed, but cannot gain height. A power margin of + 1 means that ten times as much power is available, and –1 that only a tenth of the required power can be produced. Plotting of the power margin in the figure is based on the assumption that a 12-kg bird has a power margin of zero and requires about 115 W of mechanical power to fly (Pennycuick 1986; cf. Fig. 10.6b). For a 0.01-kg bird the power margin is 1.55, for a 0.1-kg bird it is 1.01, and for a 1-kg bird 0.55. The smaller the bird the easier for it to muster the power needed to fly horizontally, which is why expensive flight modes (such as hovering) are restricted to smaller animals. Birds up to about 2 kg can take off by jumping upwards, but heavier ones need a take-off run or they must take off downward from some elevation. Horizontal flight at the minimum power speed is possible in 12–15-kg birds (Pennycuick 1968a).

It is difficult to predict an absolute upper limit for flight because muscle efficiency varies among different-sized animals, and the structure and function of muscles and wings may differ among species. Other factors such as flock formation and the use of ground effect are most evident in larger birds with narrower energy power margins. Wingspans longer than predicted from isometry would reduce the power requirements below those predicted by theory ($M^{7/6}$), which is probably the reason why larger birds have proportionately longer wings than predicted from geometric similarity. Pennycuick (1969, 1986) noted that there is no discrete upper threshold for size of animals capable of powered flight, but the most strenuous forms of flight are restricted to the smaller animals. Larger birds typically spend more time gliding and soaring rather than using muscle-powered flight and this also may have been true for ancient fliers.

Bramwell and Whitfield (1974) estimated that the pterosaur *Pteranodon ingens* had a mass of about 17 kg and a wingspan of about 7 m. Lawson (1975) suggested a wingspan of 15 m for the largest known pterosaur *Quetzalcoatlus northropi*, but McMasters (1976) and Langston (1981) favoured Lawson's alternative estimate of 11–12 m, and Langston suggested that its probable mass was 86 kg. If these values are correct, these creatures must have flown mostly by gliding and soaring. The same would apply to the huge Pleistocene condor *Teratornis* with an estimated mass of 40 kg. The power margin for take-off and active flight must have posed immense problems for these incredibly large fliers.

The largest extant birds using flapping flight are near the assumed upper weight limit; e.g. the Kori bustard (*Ardeolis kori*), the white pelican (*Pelecanus onocrotalus*), the mute swan (*Cygnus olor*) and the Californian condor (*Gymnogyps californianus*). No bats weigh more than about 1.5 kg (*Pteropus giganteus*), a tenth of the weight of the largest birds, but we do not know what sets the upper size limit for bats. Physiological, behavioural, and ecological factors may act together in the evolution of their optimal size.

Chapter 11

Morphological Adaptations for Flight

11.1 Introduction

Because powered flight is very expensive (energy cost per unit time) even though the cost of transport over unit distance is fairly inexpensive as compared with running and swimming, it requires a high degree of morphological adaptation. For example, the wings have to be long to reduce the induced power, which is very high at low speeds. Wings have to be resistant to bending, but light to keep inertial forces within reasonable limits. Muscles must meet the different flight demands, and the muscular system must be arranged to transmit a great deal of power to the wings. Tendons play an important part because they transform the muscle force to the skeleton, and they must also store energy by elastic deformation. Aerodynamical and inertial properties of the wings and body determine the function and structure of the muscular and skeletal systems, and of feathers and membranes. Several arrangements in the wings increase the resistance to the aerodynamic forces or increase the aerodynamic performance.

11.2 Muscle System

The muscle tissue is highly diverse in structure and physiology among animals with different modes of locomotion. The biochemistry, mechanics, energetics and different requirements and constraints of muscles are described in several works (e.g. George and Berger 1966; Goldspink 1977a,b,c; White 1977; McMahon 1984; Goslow 1985; Taylor et al. 1985; Rosser and George 1986a). The requirements flight impose on muscles include the velocity of shortening, force production per unit volume or mass, force production over a range of different muscle lengths, and cost of isometric force and work production.

11.2.1 Muscle Fibre Structure and Function

Skeletal muscles are striated and composed of muscle fibres about 20–80 mμ in diameter. The structure and function of muscle fibres are described in various morphological textbooks (e.g. McMahon 1984), so I give only a brief account here. Each fibre is surrounded by a membrane (the sarcolemma), and each has a characteristic banded, or striated, appearance across the fibre seen under the light microscope. Each fibre is multinucleate and for oxidation contains many mitochondria among the threadlike contractile elements (myofibrils). Fibres shorten in length to develop force. The myofibrils are about 1 mμ in diameter and arranged in thin and thick filaments. The thin filaments are held together by transverse Z disks, and the structures between two disks are called sarcomeres. A sarcomere is the basic unit and consists of two sets of thin filaments and one set of thick filaments. The thick filaments have myosin crossbridges which interact with the thin filaments when activated, and which generate force. Each crossbridge acts independently in a cyclic way, and the energy required for the force generation is supplied by ATP.

The fibres in a vertebrate muscle are arranged in groups, or motor units, each of which is activated by a nerve fibre (neuron). A muscle can be composed of motor units of different types which need not contract simultaneously. The force generated per unit cross-sectional area of a muscle depends on the number of filaments in parallel, so muscles with a large fibre cross-sectional area and with high myofibril density are well suited to produce large forces, usually by glycolytic (anaerobic) processes. Narrow fibres usually have a large volume fraction of mitochondria (Fig. 11.1), and are mainly aerobic. In most vertebrate muscles the maximum force is of the order 100–400 kN m^{-2} of fibre cross-sectional area.

Flapping flight involves mainly *isotonic* contractions, meaning that the muscle fibres exert a constant force while shortening. In gliding flight the wings are held down near the horizontal in an outstretched position and this involves mainly *isometric* contraction; i.e. the muscle fibres develop force (tension) without appreciable change in their length. Muscle fibres may, under certain circumstances, be forcibly extended while they are still generating force, as when they are shortening. From a mechanical point of view, the muscle is doing *negative work* on its environment, but the metabolic cost is still positive, albeit less

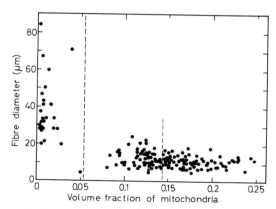

Fig. 11.1. Muscle fibre diameter versus volume fraction of mithochondria for the quail pectoralis, showing the distinction between anaerobic (left) and aerobic fibres. The *long vertical dashed* line marks the limit between the two fibre types. The *short dashed line* marks the mean volume fraction of mitochondria for the aerobic fibres (14.3%). (Pennycuick and Rezende 1984; by courtesy of The Company of Biologists Ltd)

than when positive work is done by muscle shortening. This is the case, for example, during the decelerating phase at the end of the upstroke of the wings.

The rate at which a muscle fibre shortens and develops force is determined by the rate at which individual crossbridges work (the *intrinsic rate of shortening*) and by the number of sarcomeres in series. For rapid contractions there must be many sarcomeres and the intrinsic rate of shortening must be high. The intrinsic rate of shortening can be expressed as either the rate of shortening in muscle lengths per second or as the rate of shortening per sarcomere. The rate of shortening differs considerably among animals, but the fibres in flight muscles all have about the same intrinsic rate of shortening and so form a one-geared system. Two-geared systems, however, have been found in bats (Hermanson and Foehring 1988). Pennycuick and Rezende (1984) reported that vertebrate striated muscles work over ranges of 15% of the maximum length, a value confirmed by Cutts (1986), who found that the minimum sarcomere lengths of the wing muscles of pigeons were 1.7–1.9μm and the maximum values 2.1–2.3 μm.

Carbohydrates, lipids and proteins may provide energy for muscular contraction, while glycogen acts as short-term energy supply used for short bursts of intense activity and at the onset of any activity. Lipids are the main energy source for sustained activities, but proteins may not be used to provide energy for muscular activity except when all other sources are exhausted.

In vertebrates the locomotory movements are produced by striated muscle which shows diverse structure, physiology and biochemistry. The muscle fibres can be described by their mATPase activities after acidic and alkaline preincubations, and by their metabolic capacity as indicated by an oxidative enzyme such as succinate dehydrogenase (SDH) (Rosser and George 1984). The concentrations and the forms of the enzymes are important. Dawson et al. (1964) showed that lactic dehydrogenase (LDH) is composed of subunits of the H (heart) and M (muscle) molecules (named after their source tissue), where the H form is adapted for aerobic metabolism and the M form for anaerobic metabolism.

There are two main types of striated muscle fibres, *tonic* fibres and *twitch* (phasic) fibres. Tonic fibres are slow-contracting while twitch fibres are either slow- or fast-contracting. Tonic fibres are multiply innervated (en grappe motor end-plates) and occur in the appendicular muscles of all vertebrates except mammals. A twitch fibre responds to a single propagated nerve action potential as a single propagated muscle action potential with a relatively rapid force development ("twitch" response). Twitch fibres are found in the somatic muscles of all vertebrates.

The difference between dark and light meat has been known for a long time and the terms "red" and "white" and "slow twitch" and "fast twitch" have been used to describe the fibres in light and dark meat. But, as Rosser and George (1986a) noted, "the nomenclature of avian skeletal muscle fibre types is in a near chaotic state". The red colour relates to the myoglobin and mitochondrial concentration within the muscle, so it reflects the muscle's capacity for aerobic contraction and endurance. The early associations of "slow" muscle with "red" colour often implied that slow twitch muscle was resistant to fatigue while fast twitch ones were fatiguable. This generalization is not correct for many twitch fibres in vertebrates are "fast" and "red" (resistant to fatigue).

Between the extreme fibre types there are usually several intermediate variants. The characteristics of the main types of muscle fibre occurring in avian and mammalian flight muscles have been described elsewhere (e.g. George and Berger 1966; Goldspink 1977a; Johnston 1985; Rosser and George 1986a), and they can be classified according to the following simplified scheme (Table 11.1).

Fast-twitch glycolytic fibres (white fibres, W, FG) are broad, fast contracting, fast fatiguing fibres with high power output, low isometric economy and fairly low isotonic efficiency. The intrinsic speed of shortening is high, and these fibres are used for rapid isotonic contractions. White fibres contain few mitochondria and they cannot supply ATP as fast as they use it, so they fatigue rapidly. The metabolism during contraction is mainly glycolytic (anaerobic) and white fibres are adapted for intense, brief bursts of activity such as rapid ascent, and occur in both mammals and birds.

Fast-twitch oxidative-glycolytic fibres (red, R, and intermediate, I, fibres, FOG) and *fast-twitch oxidative fibres* (red, R, FO) are narrow, fast contracting, fatigue-resistant fibres. They are similar to the white fibres, but have a high concentration of mitochondria and oxidative enzymes, particularly the red fibres. Red and intermediate fibres are adapted to fast repetitive movements and can recover fairly quickly so they are suitable for sustained flight, and occur frequently in bird and bat flight muscles. The intermediate fibres are intermediary between red and white fibres in many respects and are therefore useful for animals engaging in several different activities. They are common in soaring birds and indicate a certain degree of flexibility in contractile capacities

Table 11.1 Main properties of fibre types in vertebrate flight muscles

Fibre type	Twitch fibres			Tonic fibres
	W, FG (white, fast-glycolytic)	R, I, FOG, FO (red, intermediate, fast-oxidative-glycolytic or fast-oxidative)	SO (slow oxidative)	ST (slow tonic)
Myofibril diameter	Broad[a]	Narrow[b]-intermediate	Intermediate?	Broad
Contraction rate	Fast	Intermediate-fast	Slow	Very slow[c]
Oxidative capacity	Low	High-intermediate	High	Low-moderate
Glycolytic activity	High	Low-intermediate	Low	Low
Myoglobin content	Low	High	High	High
Glycogen content	High	High-low	Low	Low?
Lipase and fat content	Low	High	High?	High?
Resistance to fatigue	Low	High-intermediate	Very high	Very high
Isotonic efficiency	Reasonably high	High?	High	High
Isometric economy	Low	High?	Very high	Very high

[a] Pigeon: 69 μ, [b] Pigeon: 30 μ (George and Nail 1958a). [c] Contraction speed: ≤ 1.0 muscle lengths s^{-1}.

(Johnston 1985). They have moderate oxidative capacity, typical of slow fibres (high efficiency and low rate of energy consumption).

Slow-twitch fibres (red fibers, SO) and *slow-tonic fibres* (pale fibres?, ST) are slow contracting, slow fatiguing fibres with high isotonic efficiency and isometric economy. The *slow-twitch fibres* have high levels of oxidative enzymes and mitochondria but the oxidative capacity of the slow tonic fibres is low or moderate. Both fibre types are very fatigue-resistant and they are responsible for maintaining posture and for carrying out slow repetitive movements, and are therefore suitable for sustained flight. *Slow-tonic fibres* can develop and maintain isometric tension (Goldspink et al. 1970; Matsumoto et al. 1973), making them suitable for gliding flight. Slow-tonic fibres occur in birds but not in mammals, which have slow-twitch fibres that are absent in birds.

When present, the slow muscle fibres are probably always recruited first followed by the fast oxidative (red and intermediate); the fast glycolytic (white) fibres are recruited only when high acceleration or speed (= strong force) is required. Slow muscles usually occur deeper than the other fibre types and the white fibres are concentrated at more peripheral portions of the muscles.

During gliding the wings are outstretched and held forward primarily by the triceps and deltoid muscle groups while the wings are maintained in the horizontal position by the pectoral muscles in birds. In gliding, fewer groups of muscle fibres are recruited than in flapping flight, and the muscle activity is primarily isometric contraction. The energy turnover is lower in isometric contraction than in isotonic contraction (Hill 1938). Because slow muscles can produce isometric tension more economically than fast muscle fibres, the former should be involved when there is an isometric component to the flight, such as gliding and soaring flight.

11.2.2 The Structure of Bird Muscle Fibres

In both birds and bats there are apparent relationships between fibre structure, the biochemical properties of the flight muscles, and flight habits. The pectoral muscle of carinate birds (birds with a sternal ridge and most of which can fly) consists mainly of fast-twitch fibres (Kiessling 1977; Talesara and Goldspink 1978; Rosser and George 1986b), while slow-tonic fibres constitute large proportions of the muscle mass and are widely distributed in the flightless ratites (birds without a sternal ridge; McGowan 1982; Rosser and George 1986b). In flying birds, the slow fibres are limited in number and restricted to one small region of the muscle (Suzuki 1978; Rosser and George 1986a).

George and Berger (1966) classified the pectoral muscle of birds into six groups (the types of fibres are given within parenthesis in the respective order of predominance): (1) Fowl type (WIR), (2) Duck type (RWI), (3) Pigeon type (RW), (4) Kite type (mainly I), (5) Starling type (RI), and (6) Sparrow type (R) (including the more specialized hummingbird pectoral muscle). This classification was intended as a convenient generalization open to modification.

The pectoral muscle of soaring and gliding birds consists mainly of intermediate (1) fibre types (George and Berger 1966). There is an accessory deep belly of the pectoral muscle consisting entirely of slow tonic fibres, in the turkey vulture (*Cathartes aura*: Rosser and George 1986b). Slow fibres are most effective in stabilizing the wing during gliding flight. They comprise a small proportion (8%) of the pectoral muscles of herring gulls (*Larus argentatus*), but the location of these fibres was not specified (Talesara and Goldspink 1978). Slow fibres make the deep belly lighter in colour than the superficial belly consisting of faster fibres. Pale deep bellies, which may consist of slow fibres, have also been found in other soaring birds, such as the magnificent frigatebird (*Fregata magnificens*), black kite (*Milvus migrans*) and Indian white-backed vulture (*Gyps bengalensis*) (George and Berger 1966; Kuroda 1961).

In addition to dark (fast-twitch) fibres, pale fibres also occur in the wing-elevating supracoracoideus muscles of passerine birds that perform bounding flight, while their pectoral muscles contain only red fast-twitch fibres (Salt 1963). These pale fibres may be slow tonic like those of the pectoral deep belly of soaring birds, and bounding flight is a series of rapid wingbeats followed by a phase when the wings are folded towards the body (see Sect. 9.8.1). Salt suggested that the dark fibres in the supracoracoideus muscles produce the powered upstrokes, while the pale fibres hold the wings towards the body in the passive stage.

The pectoral muscles of the pigeon consist of two distinct fibre types, one red, narrow and fat-loaded, the other white, broad with a sparse distribution of fat globules, and instead glycogen-loaded (George and Naik 1958a,b). The large numbers of the narrow fat-loaded fibres suggest that fat is the principal source of energy for prolonged flight. In pigeons, the white fibres (the first to contract) are used for short bursts of flight, such as steep take-offs. White fibres are found in several other birds that often perform power bursts, such as herons, geese, ducks, mergansers, and grouse (Chandra-Bose and George 1965a,b; Rosser and George 1986a). For example, in the spruce grouse (*Canachites canadensis*) and the ring-necked pheasant (*Phasianus colchicus torquatus*), the white fibres make up more than 80% of the fibres in the pectoral muscle. These species have poor abilities for sustained flight correlated with limited numbers of oxidative fibres (Kaiser and George 1973).

Red and intermediate (fast-twitch) fibres are the major types in the pectoral muscle in birds flying within vegetation by slow manoeuvrable flight, or hovering, such as owls, small passerines and hummingbirds. The red fibres are closely packed with mitochondria and lipid bodies and have higher oxidative capacity than the intermediate fibres. They are highly suitable for sustained flight and repetitive movements. The hummingbird pectoral muscle is similar to that of the smaller passeriforms, but has a much denser network of blood capillaries, and each fibre is studded with numerous mitochondria, giving it a very high metabolic efficiency.

Unlike passeriforms, the supracoracoideus muscles of hummingbirds consist solely of red fibres. This muscle acts normally as a wing elevator in birds, but in hummingbirds the "upstroke" is a backstroke providing lift and is not a recovery stroke.

The metabolic demands may vary considerably within a single individual, and the needs for sustained flight and short flight manoeuvres may be different at different times. For example, during migratory flights there are great demands for aerobic catabolism of fatty acids and increase in oxidative capacity, but during the breeding season, foraging often involves short bursts of flight, meaning demands of high glycolytic and high anaerobic capacity. Lundgren and Kiessling (1985, 1986) and Lundgren (1988) found seasonal variations in catabolic enzyme activities in the pectoral muscles of some migratory passerines. One pattern was high oxidative capacity and low glycolytic capacity during migration, and the converse during the breeding season. Fat oxidation capacity increased as migration approached, and this pattern occurred in nocturnal migrants that deposit large amount of fat prior to migration. Examples include the long-distance migratory reed warbler (*Acrocephalus scirpaceus*), and the short-distance migratory robin (*Erithacus rubecula*) and blackbird (*Turdus merula*). A second pattern occurred in the short-distance migratory great tit (*Parus major*) and reed bunting (*Emberiza schoeniclus*), which do not deposit large stores of fat. In the second pattern there was an increase in the oxidative capacity before migration, but no significant change in the fat oxidation capacity. A third pattern was found in the sedge warbler (*A. schoenobaenus*), which does not increase its aerobic capacity during migration, even though it is a long-distance migrator. Bibby and Green (1981) suggested that this difference between the reed warbler and the sedge warbler could be due to differences in migration and feeding strategies. Among the aforementioned species, only the blackbird has pectoral muscles with red and white fibres; the other species have only red fibres (Lundgren and Kiessling 1985, 1988).

11.2.3 The Structure of Bat Muscle Fibres

The fibre structure of the pectoral muscles of bats have been analysed in a few species. Using the amount of mitochondria, Ohtsu and Uchida (1979a,b) identified three fibre types in bats: (1) Mitochondria-more rich fibres (narrow, high-oxidative, high SDH), (2) mitochondria-less rich fibres (intermediate in size and biochemistry), and (3) mitochondria-moderate fibres (broad, low-oxidative, low SDH). Judging from observations on the biochemical structure of the fibres, the authors concluded that the first type corresponds to fast-twitch R fibres and the second type to fast-twitch I fibres, while the third type corresponds to slow-twitch SO fibres. They then divided the bats into three categories: Type 1 (R, SO, I): horseshoe bats, *Rhinolophus ferrumequinum* and *R. cornutus* (Rhinolophidae); Type 2 (R, I): common bent-winged bats (*Miniopterus schreibersi*; Vespertilionidae); Type 3 (R): the vespertilionids *Pipistrellus abramus, Myotis macrodactylus,* and *Vespertilio superans.*

George and Naik (1957) found that the pectoral muscle of Schneider's leaf-nosed bat (*Hipposideros speoris*; Hipposideridae) consisted of a white, broad, glycogen-loaded variety with few mitochondria, and a red, fat-loaded, narrow variety with larger mitochondrial content corresponding to the pigeon type. The pectoral muscle of the megachiropteran *Pteropus* contained R and I, while that of the megachiropteran *Rousettus leschenaulti* (both Pteropodidae) and the microchiropteran *Pipistrellus* (Vespertilionidae) consisted only of narrow R fibres with much fat (George and Jyoti 1955; George 1965).

Valdivieso et al. (1968) and Muller and Baldwin (1978) used the distribution and functional properties of LDH isoenzymes and biochemical data on the activities of glycolytic and Krebs cycle enzymes to assess the importance of aerobic and anaerobic pathways for energy production in the pectoral muscles of a number of bats with different flight behaviour. Bats with low dependence on anaerobic metabolism are the Australian little brown bat (*Eptesicus pumilus*; Vespertilionidae), white-striped bat (*Tadarida australis*), little flat bat (*T. planiceps*), and *Molossus fortis* (Molossidae), which are all fast and continuous fliers. Species that are intermediate in this respect were Gould's wattled bat *Chalinolobus guoldii* and the lesser long-eared bat *Nyctophilus geoffroyi* (Vespertilionidae), which have a more variable flight with the need of a heterogeneous distribution of fibre types for a more complex motor control programme. Bats with high dependence on anaerobic metabolism include the eastern horseshoe bat (*Rhinolophus megaphyllus*; Rhinolophidae), common bent-winged bat (*Miniopterus schreibersi*), and hoary bat (*Lasiurus cinereus*; Vespertilionidae), which are all insectivorous, and the Jamaican fruit-eating bat *Artibeus jamaicensis* and buffy flower bat *Erophylla bombifrons* (Phyllostomidae) which are frugivorous and nectarivorous, respectively. Both *M. schreibersi* and *L. cinereus* are migratory species and would benefit from aerobic metabolism, but they also have a rapid, manoeuvrable and powerful flight and may depend more on anaerobic glycolysis than, for instance, the fast-flying molossids. But these data conflict with the findings of Ohtsu and Uchida (1979a,b), who found that the pectoral muscle of *M. schreibersi* consisted solely of high-oxidative red and intermediate fibres, while the flight muscles of *A. jamaicensis* (the pectoral muscle in particular) consisted of two types of fast-twitch high-oxidative fibres (Foehring and Hermanson unpublished, cited in Altenbach and Hermanson 1987). However, as in birds, the metabolism might change from anaerobic during the breeding season to aerobic prior to migration.

The twitch contraction time for bat muscles has been measured for the short head of biceps brachii of the Mexican free-tailed bat (*Tadarida brasiliensis*). Four specimens were studied and they showed contraction times of 15–31 ms (mean 21.25 ms; Altenbach and Hermanson 1987), suggesting fast-twitch contraction times (Burke 1978). Histochemical studies showed that both the pectoral, short head of biceps brachii and serratus ventralis (downstroke adductors), and subscapularis (adductor and extensor of the humerus) muscles probably contained only one type of fast-twitch high-oxidative fibres (FO fibres; Foehring and Hermanson 1984).

Accessory flight muscles of bats usually contain more than one fibre type. The acromiotrapezius and acromiodeltoideus (abductors initiating the upstroke) of the little brown bat (*Myotis lucifugus*) contained three types of fast-twitch fibres with different oxidative enzyme activity (Armstrong et al. 1977). The long head of the biceps brachii (adductor during the downstroke) and the triceps brachii muscles (abductor during the upstroke) of *T. brasiliensis* contained a mixture of fast and slow twitch fibres (Foehring and Hermanson 1984); the slow fibres were supposed to play a postural role and stabilize the shoulder joint during wingbeat movements.

11.2.4 Contraction Rate and Wingbeat Frequency

The pectoral muscles of birds comprise a large part of the body mass (up to 24%, in the pigeon, Sect. 10.3.6). It has been suggested that diversity in the fibre composition would add to the mass, and explain why small birds have mainly one fibre type, while larger birds have three types. A more reasonable interpretation may be that larger birds have a more heterogeneous flight repertoire, requiring more than one fibre type.

Pennycuick and Rezende (1984) noted that the wingbeat frequency is determined by the aerodynamic and inertial properties of the wings and body, and that the properties of the muscles then have to match this frequency. They proposed that the adaptation consists of adjusting the maximum strain rate (intrinsic speed) of which the myofibrils are capable, so as a general rule there is little intraspecific variation in the rate at which the muscle fibres of the pectoral muscle contract, and the wingbeat frequency remains more or less the same. To increase flight speed, the birds change their wingbeat kinematics.

Compared to small birds, the muscle fibres in large birds are very long with lots of sarcomeres, meaning that the intrinsic speed of shortening must be lower to withstand the strain rates that otherwise would be too high. If large birds had the same intrinsic speeds of shortening as smaller ones, the muscle fibres would shorten (i.e. develop strain) at a higher rate than the muscles and tendons could withstand. Therefore, there is an optimum rate of shortening for producing maximum power and so each bird species should show a definite optimum wingbeat frequency. Since larger animals have slower muscles, only large birds use gliding to save energy, and only the largest bats have been observed gliding under some circumstances. When more rapid movements, such as take-offs, are required, fast contracting fibres must be used and, therefore, many flight muscles contain more than one type of fibre.

11.2.5 Body Size, Wing Shape and Flight Muscle Fibres — a Summary

Birds. Figure 11.2a shows the muscle fibre types in birds with different wing shape, measured by aspect ratio (b^2/S) and relative wing loading ($Mg/SM^{1/3}$). The pectoral muscle of very small birds (small passerines and hummingbirds) usually have only one fibre type — red and fast-twitch (FOG) — that are suitable for sustained, slow and fast, manoeuvrable flight. A lack of slow (ST) or intermediate fast-twitch (FOG) fibres in the pectoral muscle makes these birds poorly adapted for gliding flight, when the wings must withstand the aerodynamic force by isometric contraction. To save energy during flight, several small species use bounding flight that includes half-passive phases, when the wings are held near the body between the flapping phases. Salt (1963) suggested that the wing-elevating supracoracoideus keeps the wings near the body during these phases, by the isometric action of the (slow?) fibres of this muscle. Most of the birds with predominantly or solely red (FOG) fibres in the pectoral muscle have wings of low aspect ratio, so their flight is expensive. Their wing loading is mainly intermediate and they fly at intermediate speeds, and energy budget constraints

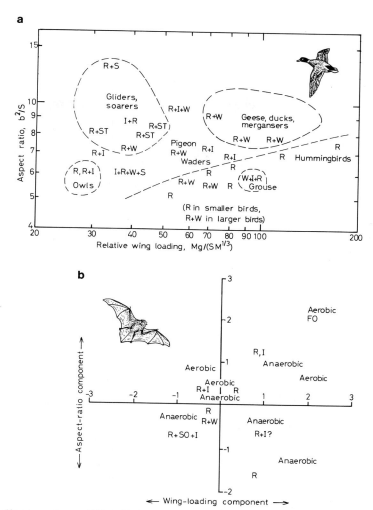

Fig. 11.2. a Muscle fibre types (see p. 183) in birds related to their wing shapes, measured by the non-dimensional aspect ratio (b^2/S) and relative wing loading ($Mg/SM^{1/3}$). (Data on aspect ratio and relative wing loading from U.M. Norberg and R.Å. Norberg 1989) *b* Muscle fibre types in bats related to their wing shapes, measured by the non-dimensional aspect ratio and wing loading in a PCA diagram from Norberg and Rayner (1987). The fibre type data are from the different sources mentioned in the text. Low total power is obtained with high aspect ratio, particularly in combination with low body mass and low wing loading; low wing loading permits slow and manoeuvrable flight (see Chap. 12)

mean that most of them cannot afford to fly continuously, so they must perch between foraging bouts. Nevertheless, many of them migrate long distances, but before migration the biochemistry of the muscles may change to achieve a higher aerobic capacity. Hummingbirds have high wing loading (fast flight) and average or somewhat below average aspect ratio.

Medium-sized birds cannot effectively use gliding flight to save energy, because they do not have any slow muscle fibres in their pectoral muscle and their FOG fibres are not slow enough. But they can use undulating flight which includes short gliding phases. Birds using undulating flight usually have both red and intermediate (FOG) fibres in their pectoral muscle, just like most soaring birds, and they are about average in both wing loading and aspect ratio.

The larger the bird, the longer the sarcomeres, and the slower the intrinsic speed of shortening. Therefore larger birds can use gliding and soaring flight that requires isometric contraction of the pectoral muscle. The main part of the pectoral muscle in larger birds is made up of red and intermediate (FOG) fibres, such as in owls. Owls have very slow (low wing loading) and expensive flight (low aspect ratio). Many large soaring birds have a deep belly of the pectoral muscle containing slow tonic (ST) fibres for isometric economy. Like owls, soaring birds are usually slow fliers (low wing loading) but they have wings of high or average aspect ratio so they can fly more or less continuously during foraging. They also are usually long-distance migrators (see further Sect. 12.3).

The pectoral muscles of birds that need rapid flight bursts, such as steep take-offs, include the fast glycolytic white fibres (FG) in addition to the red and sometimes intermediate (FOG) ones. The majority of these birds have a rather high wing loading (fast flight) and rather low aspect ratio. White fibres are predominant in the pectoral muscle of gallinaceous birds, which together with their expensive flight (because of the high wing loading and low aspect ratio) make them unable to fly longer distances.

Bats. Figure 11.2b shows the distribution of fibre types within bats with different wing shapes. The smallest bats investigated have mainly red (FO and FOG) fibres in their pectoral muscles, but a megachiropteran (*Rousettus*) also has just red fibres. The rather small molossid bats have one type of FO fibres. Red fibres may occur in bats of different sizes with different aspect ratios and wing loadings.

Rhinolophid bats have very low aspect ratio wings and red, intermediate (FOG) as well as slow-twitch (SO) fibres. Although they have slow fibres, they do not glide. Their small size may make even the slow fibres too short and thus not slow enough for isometric economy. The only bats which use gliding in slope lifts belong to the genus *Pteropus*, which includes the largest bats; the one *Pteropus* investigated has red and intermediate (FOG) fibres, like many soaring birds.

The dependence on aerobic metabolism has been investigated in a few bats and those with high dependence on aerobic metabolism have high aspect ratio and high wing loading and have thus fast and fairly inexpensive flight. Most of the bats with low dependence on aerobic metabolism have low aspect ratio but differing wing loading. These species, and in particular those with a high wing loading, have to perch more between the foraging bouts.

11.2.6 Flight Muscle Power

Power Output. The power output of a muscle is the rate at which it does mechanical work. If a muscle shortens through a distance ΔL exerting a tension force F, then the work done becomes $Q = \Delta LF$ (Fig. 11.3). If the contraction rate

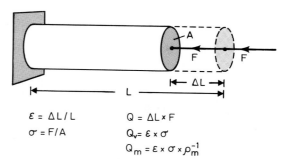

$$\varepsilon = \Delta L / L \qquad Q = \Delta L \times F$$
$$\sigma = F / A \qquad Q_v = \varepsilon \times \sigma$$
$$Q_m = \varepsilon \times \sigma \times \rho_m^{-1}$$

Fig. 11.3. A schematic muscle attached to a skeleton at left. A is muscle cross-sectional area, L its extended length, ΔL its shortening length, F is force, ε is active strain, σ is stress, Q is work, Q_m is mass-specific work, Q_v is volume-specific work, and ρ_m is muscle density. (After Pennycuick and Rezende 1984; by courtesy of The Company of Biologists Ltd)

is n, then the power output (rate of doing work) is $P = n\Delta LF$. It is often convenient to use the specific power output which can be expressed as volume-specific and mass-specific power output, which are, respectively, the rates of doing work per unit volume of muscle and per unit mass. The specific power output depends strongly on the contraction frequency (Hill 1950).

Strain and Stress. If the extended (relaxed) length of the muscle is L (Fig. 11.3), then the "active strain" ε of the muscle is defined as

$$\varepsilon = \Delta L / L, \tag{11.1}$$

and the "strain rate" (intrinsic speed) as

$$\dot{\varepsilon} = \varepsilon / t_s, \tag{11.2}$$

where t_s is the inverse of the wingbeat frequency n, $t_s = 1/n$. The muscles are adapted to work at a particular contraction frequency, the "operating frequency", which also defines the strain rate and the time it takes the muscle to shorten.

The strain rate determines the stress σ which the myofibrils can exert. The muscle stress is defined as

$$\sigma = F / A, \tag{11.3}$$

where A is the cross-sectional area of the muscle. The work done by the muscle can then be expressed as

$$Q = \Delta LF = L\varepsilon \times A\sigma, \tag{11.4}$$

and the power output as

$$P = nQ. \tag{11.5}$$

The product of L and A roughly equals the volume of the extended muscle, and the "volume-specific work" Q_v^* becomes

$$Q_v^* = \varepsilon \times \sigma. \tag{11.6}$$

191

The mass-specific work Q_m^* can be found by dividing by the density ρ_m of the muscle,

$$Q_m^* = \sigma\varepsilon/\rho_m. \tag{11.7}$$

Hill (1938) discovered the relationship between stress and strain rate in the slightly different form of a relationship between force F and velocity V of shortening of the muscle. This relationship can be expressed by the hyperbolic function

$$(F + c)V = d(F_0 - F), \tag{11.8}$$

where F_0 is the maximum force which the muscle can exert in an isometric contraction, and c and d are constants where c has the dimensions of force and d those of velocity. The instantaneous power developed during shortening equals force times velocity, that is

$$P = FV = dF(F_0 - F)/(F + c). \tag{11.9}$$

Differentiating this with respect to F shows that the maximum power is obtained when

$$F = (c^2 + cF_0)^{1/2} - c. \tag{11.10}$$

The force constant c is related, but not equal, to the muscle's mechanical efficiency. Hill found that, in frog skeletal muscle, c was about $F_0/4$, which can be taken as representative for vertebrate skeletal muscle (Pennycuick and Rezende 1984). With this value, Eq. (11.10) shows that maximum power is obtained at a force equal to $0.31\,F_0$. Zero force or zero speed would give zero power. Hill (1950) re-expressed his force-velocity relationship in the form of strain and stress. Figure 11.4 shows the curves of strain rate (instead of velocity) and specific power output as functions of stress (instead of force) calculated from Hill's equation. Maximum isometric stress (at the righthand end of the scale) is obtained when the strain rate is zero. The maximum isometric stress that can be exerted by the myofibrils of vertebrate skeletal muscles was estimated by Weis-Fogh and Alexander (1977) as between 250 and 400 kN m^{-2}. Pennycuick and Rezende (1984) used a value of 300 kN m^{-2} to define the scale of Fig. 11.4, which compare two hypothetical muscles, one "fast" and one "slow". The maximum strain rate, at which each muscle shortens against zero stress (that is, the intrinsic speed), and the volume-specific power (stress × strain rate) differ greatly between the two muscles. The power curves show rather flat peaks at a stress about 31% of the maximum, as predicted by Hill's equation (11.10).

Power Density of Mitochondria. Weis-Fogh and Alexander (1977) investigated the limits of specific power output and the requirements for converting and supplying energy to the contractile proteins by mitochondria. In a simple model, Pennycuick and Rezende (1984) showed that the volume ratio of mitochondria to myofibrils should depend on the power density of mitochondria and on the operating frequency, but not on the mechanical properties of the myofibrils. This would mean that the specific power output could be determined by examination of electron micrographs.

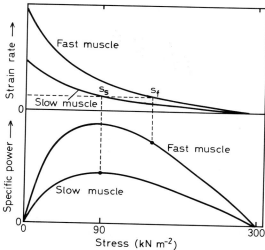

Fig. 11.4. Strain rate (intrinsic speed) and instantaneous specific power output plotted against stress for two hypothetical muscles, calculated from Hill's equation. The *horizontal dashed line* represents the strain rate fixed by factors external to the muscle. This particular value of strain rate results in the slower muscle extering the stress σ_s which corresponds to its maximum power. At the same strain rate, the faster muscle exerts a higher stress σ_f, and therefore produces more power than the slow muscle, even though the corresponding specific power is displaced well to the right of the peak power. (Pennycuick and Rezende 1984; by courtesy of The Company of Biologists Ltd)

Pennycuick and Rezende considered a muscle with a volume v_c of myofibrils and a volume v_t of mitochondria. Neglecting other components in the fibre, such as nuclei and sarcoplasmatic reticulum, the volume v of the whole muscle then is $v = v_c + v_t$. Assuming that the volume of mitochondria required is directly proportional to the mechanical power output P, then

$$v_t = kP, \tag{11.11}$$

where k is the volume of mitochondria required to sustain unit mechanical power output ("inverse power density of mitochondria") and a constant with the dimensions volume/power. If σ is the stress exerted across the myofibrils (not across the whole muscle), then, from Eq. (11.5)

$$P = \sigma \varepsilon n v_c. \tag{11.12}$$

Inserting this in Eq. (11.11), we obtain

$$v_t = k\sigma \varepsilon n v_c. \tag{11.13}$$

The volume-specific power output for the whole muscle then is

$$P_v^* = P/(v_c + v_t) = \sigma \varepsilon n v_c/(v_c + k\sigma \varepsilon n v_c) = \sigma \varepsilon n/(1 + k\sigma \varepsilon n), \tag{11.14}$$

and the mass-specific power output becomes

$$P_m^* = \sigma \varepsilon n/\rho(1 + k\sigma \varepsilon n). \tag{11.15}$$

At very low operating frequencies (= wingbeat frequencies n; below about 10 Hz), the factor $(1 + k\sigma\varepsilon n)$ is negligibly greater than 1, but at very high frequencies $k\sigma\varepsilon n$) is much greater than 1, and for high frequencies Eq. (11.15) can be approximated to

$$P_m^* = 1/\rho k. \qquad (11.16)$$

Knowing the ratio of myofibrils to mitochondria from an electron micrograph would make it possible to estimate the specific power output from this model. If the ratio of v_t to v_c is defined as q, then the Eqs. (11.13) and (11.15) can be written

$$q = k\sigma\varepsilon n \qquad (11.17)$$
and
$$P_m^* = q/\rho k(1 + q). \qquad (11.18)$$

As a starting hypothesis, k is assumed to have a constant value for all muscles working at a particular operating temperature. Pennycuick and Rezende suggested that the quantity of mitochondria in the flight muscles of a migratory bird may be subject to short-term variations, or that the volume fraction of mitochondria may be genetically fixed. In the Japanese quail (*Coturnix coturnix*), reared in a laboratory, the area fraction occupied by mitochondria was 14.3% and in the sparrow (*Passer domesticus*) 29.9%, corresponding to mean ratios of mitochondria to myofibrils of 0.18 and 0.45, respectively. Table 11.2 shows the estimates of k for these species and for a hummingbird, a locust and a fly. The fly has fibrillar muscles while those of the locust and the birds are non-fibrillar. In all species but the Japanese quail the estimates of k are almost 10^{-6} m^3 W^{-1}, meaning that about 1 ml of mitochondria is required to sustain 1 W of mechanical power output. The Japanese quail has less than half of this, and does not have enough mitochondria to sustain aerobic flight. Qualitatively, the Japanese quail's pectoralis resembles the pigeon type (see Sect. 11.2.2), which is adapted for sustained cruising flight and rapid bursts. Pennycuick and Rezende suggested that these characteristics must have been retained from migratory ancestors through many generations of captivity. They noted that if there is any short-term variation in volume fraction of mitochondria in the muscles of an individual bird, a laboratory quail could be trained to perform aerobic flight, like the wild, migratory, form of the same species.

Table 11.2. Estimates of k for three birds (a quail, a sparrow and a hummingbird), a locust and a fly. n is wingbeat frequency and q is the ratio between volume of mitochondria and volume of myofibrils. The stress is assumed to be 150 kN m^{-2} and the strain to be 0.15. (Pennycuick and Rezende 1984)

Species	n Hz	q	k (m^3 W^{-1})
Coturnix coturnix	18	0.18	4.4×10^{-7}
Passer domesticus	20	0.45	1.0×10^{-6}
Amazilia fimbriata	35	1.0	1.3×10^{-6}
Schistocerca gregaria	19	0.43	1.0×10^{-6}
Phormia regina	164	0.73	1.1×10^{-6}

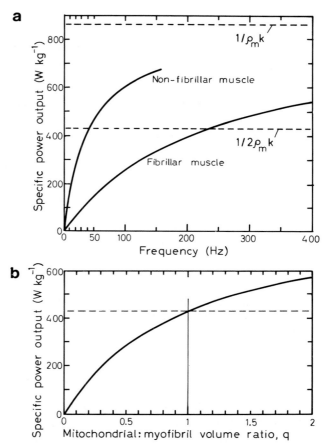

Fig. 11.5. a Estimated mass-specific power output plotted as a function of operating frequency for non-fibrillar, vertebrate striated, muscle and for insect fibrillar muscle. The respective stress and strain values assumed during shortening are $150\,kN\,m^{-2}$ and 0.15 for non-fibrillar muscle and $200\,kN$ m^{-2} and 0.02 for fibrillar muscle. The *upper dashed line* is the asymptote for both curves, and the *lower one* corresponds to $q = 1$ (volume mitochondria equals volume myofibrils). ρ_m is muscle density. *b* Estimated mass-specific power output as a function of the mitochondria:myofibril volume ratio, q. This curve is independent of the mechanical properties of the myofibrils. (Pennycuick and Rezende 1984; by courtesy of The Company of Biologists Ltd)

Figure 11.5a shows the mass-specific power output as a function of the operating frequency for vertebrate striated (non-fibrillar) and insect fibrillar muscles, calculated from Eq. (11.15) for $k = 1.1 \times 10^{-6}\,m^3\,W^{-1}$. At about 40 Hz the vertebrate muscle reaches the level where the volume of mitochondria equals that of myofibrils (giving $P^*_m = 1/2\,\rho k$), as they do in hummingbirds and large insects. The corresponding value for insect fibrillar muscle is 230 Hz. At these values the muscle would be able to deliver a specific power of about $429\,W\,kg^{-1}$ muscle mass. At very high frequencies both muscles approach an asymptote at

195

about 860 W kg^{-1}. The muscles would, however, consist mainly of mitochondria long before the limiting value is reached.

Figure 11.5b shows the mass-specific power output as a function of the ratio of mitochondria to myofibrils (q), calculated from Eq. (11.18) and for k = 1.1 × 10^{-6} m^3 W^{-1}. The specific power is 429 W kg^{-1} when the volume of mitochondria equals that of myofibrils (q = 1) for both non-fibrillar and fibrillar muscles.

Less mitochondria are required to sustain a given mechanical power output at higher temperatures than at lower temperatures. Pennycuick and Rezende suggested that this would be the main reason why it is advantageous for flying animals to maintain a high body temperature.

Structural Limitations. Pennycuick and Parker (1966) estimated the structural limitations on the power output of the pigeon's flight muscles. The upper limit to the power which could be transmitted to the humerus by the pectoralis and the supracoracoideus muscles was estimated by measuring the breaking tension of the insertion of each muscle, its amplitude of movement, and the maximum wingbeat frequency at take-off. The maximum amounts of work that could be done by each muscle in contraction were 2.48 J for the pectoralis and 0.21 J for the supracoracoideus, and the maximum possible work per cycle was thus 2.48 + 0.21 = 2.69 J for each side and 5.38 for the muscles on both sides together. At a wingbeat frequency of 8.9 strokes s^{-1}, the maximum possible average power output would be P_{max} = 5.38 × 8.9 = 48 W. The maximum possible power outputs per kg muscle (specific power output) are P^*_{max} = (2.48 × 8.9)/0.038 = 580 W kg^{-1} for the pectoralis and (0.21 × 8.9)/0.0067 = 280 W kg^{-1} for the supracoracoideus, together 860 W kg^{-1}.

The lower limit was obtained from observed climbing performance. A pigeon weighing W kg climbing at a vertical speed V m s^{-1} is doing work against gravity at the rate WV Watts. This can be regarded as the minimum power exerted by the flight muscles. If the wings were mechanically and aerodynamically perfectly efficient the flight muscles cannot produce less power than WV Watts. With an average body mass of 0.373 kg (or body weight of 0.373 × 9.81 N) and a rate of climb of 2.5 m s^{-1}, the minimum power becomes at least P_{min} = 0.373 × 9.81 × 2.5 = 9.1 W. The weight of both flight muscles on the two sides is 0.0896 kg, so the minimum possible average specific power output would be P^*_{min} = 9.1/0.09 = 101 W kg^{-1} (Pennycuick and Parker 1966).

11.2.7 Flight Muscle Structure

The individual fibres in a muscle can be arranged either parallel to the muscle's long axis (called *parallel-fibred* muscles) or at an angle to it (called *pennate* muscles) (Fig. 11.6) and both types are represented among flight muscles. Pennate muscles can be either unipennate, bipennate, or multipennate. Unipennate muscles have the tendon running along one side and the fibres attach on its side. In bipennate muscles the tendon passes up the centre of the muscle.

Contracting parallel-fibred muscles shorten through a longer distance than pennate muscles of similar shape and exert a relatively small force. Pennate muscles are generally more powerful than parallel-fibred muscles of the same

196

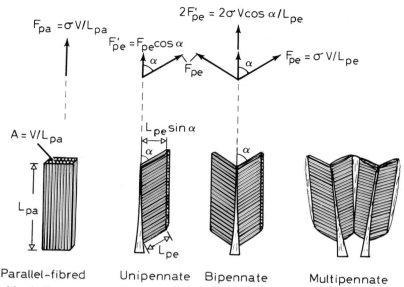

$$F_{pa} = \sigma V/L_{pa}$$

$$2F'_{pe} = 2\sigma V \cos \alpha / L_{pe}$$

$$F'_{pe} = F_{pe} \cos \alpha$$

$$F_{pe} = \sigma V/L_{pe}$$

$$A = V/L_{pa}$$

$$L_{pe} \sin \alpha$$

Parallel–fibred   Unipennate   Bipennate   Multipennate

Fig. 11.6. Muscle fibre arrangements in skeletal muscle. See text for explanations

mass, because they allow a larger number of fibres to work in parallel. Assume that two muscles, one parallel-fibred muscle with fibre length L_{pa} and one bipennate muscle with fibre length L_{pe}, have equal volume V and fibres that exert a stress σ (force per unit cross-sectional area, $\sigma = F/A$). The cross-sectional area A is then V/L_{pa} and V/L_{pe}, respectively, for the two muscles, and the corresponding forces along the fibres become $F_{pa} = \sigma V/L_{pa}$ and $F_{pe} = \sigma V/L_{pe}$ [cf. Eq. (11.3)]. Since L_{pe} is much shorter than L_{pa} the pennate muscle can exert much more force than the parallel-fibred muscle. The fibres of the pennate muscle run at an angle α to the direction of tension, so the force component along the tendon is $F'_{pe} = (\sigma V/L_{pe})\cos\alpha$, and this is the force that the unipennate muscle exerts. If α is small the force is approximately equal to $F'_{pe} = (\sigma V/L_{pe})$. A bipennate muscle of volume 2V exerts $2F'_{pe}$ along the tendon (Fig. 11.6).

The force component along the tendon also can be expressed as a proportion of the area A_{pe} of each face of the part of the tendon to which the muscle fibres attach (Alexander 1983). The breadth of the bipennate muscle is $2L_{pe}\sin\alpha$ (Fig. 11.6) and the volume is 2V, so the area of attachment becomes $A_{pe} = 2V/(2L_{pe}\sin\alpha)$. The force $2F'_{pe} = (2\sigma V/L_{pe})\cos\alpha$ can now be written as $2\sigma A_{pe}\sin\alpha \cos\alpha = 4\sigma A_{pe}\sin2\alpha$.

11.2.8 Muscle Arrangements

The effectiveness of muscles can be increased in different ways. For example, insertions of muscles on ridges and tuberosities of the humerus better permit rotational movements and give larger attaching surfaces, while attachments on tubercles, extending from the main axis of the bone, give better mechanical outcome.

Force lever systems in the wings of both birds and bats increase the mechanical outcome of muscle contraction, a situation exemplified by two muscle arrangements in bats. The first involves the extensor carpi radialis longus muscle, which pulls the second (leading edge) digit forwards. The muscle passes along the anterior side of the forearm, and its tendon in front of the carpal bones at a distance from the fulcrum of the second metacarpal. The trapezium bone of the carpus projects dorsally, preventing the tendon of the muscle from sliding posteriorly when the muscle contracts. The force of the muscle constitutes the applied force F of a lever system (Fig. 11.7a) operating about the fulcrum, which is the second metacarpal. The force lever arm l is the distance from the fulcrum to the tendon. Since

$$\text{the moment of action} = l \times F, \tag{11.19}$$

its increases proportionally to l. Figure 11.7b shows a similar force lever system where contraction of the abductor digiti quinti muscle pulls the fifth metacarpal ventrally, maintaining the chordwise camber of the wing during the downstroke. F is the applied force of the force lever system in the direction of the tendon of the muscle at the origin, and F' is the force of the muscle in the direction of the tendon at the insertion. $F = F'$, l is the force lever arm on which F acts, and the moment of action is l F. The longer l is the larger the resultant moment becomes.

Similar muscular arrangements exist at several places in the wing. They increase the efficiency of the muscles and so reduce the demands for heavy muscle bellies and contribute to low weight of the wings. Low weight is also obtained by skeletal arrangements, which will be treated in the next section.

11.2.9 The Main Flight Muscles

In birds, bats and pterosaurs, the pectoral muscle is the main depressor (adductor) and usually a pronator (nose-down rotator) of the wing. The pectoral muscle is strongly developed, and varies from about 8% to about 24% of the animal's body mass, with the larger values occurring in animals with high wing loadings that need to be compensated for by high flight speeds and/or high flapping frequencies.

Figure 11.8 shows the direction of pull of the main muscles acting during the wingbeat cycle in birds (a), pterosaurs (c), and bats (e). The size and structure of the various flight muscles vary somewhat between species with different flight modes, but in general, the same muscles are involved in particular movements of the wing.

Birds. The arrangements of the muscles and skeletons of the bird locomotor system have been well studied by, for example, Sy (1936), Swinebroad (1954), George and Berger (1966), Herzog (1968), Shyestakova (1971), and Raikow (1985). The downstroke is performed mainly by the pectoralis and the coracobrachialis muscles (Fig. 11.8a,b). The pectoral muscle originates on the sternum, clavicle, ribs and coracoid, passes directly onto the humerus and inserts ante-

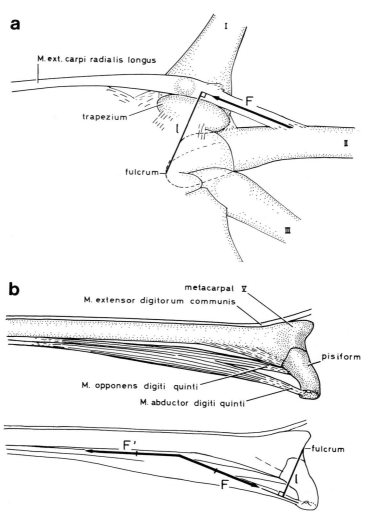

Fig. 11.7a,b. Skeleto-muscular force lever systems. *a* Bat wrist, where an extensor muscle pulls the second metacarpal forwards. *b* Fifth digit of a bat, where contraction of an abductor muscle pulls the fifth metacarpal ventrally. *F* is the applied force (= force *F'*) and *l* is the force lever arm on which *F* acts. The moment of action = *l* × *F*, and hence proportional to *l*. (Norberg 1970b)

riorly on the pectoral crest, thereby rotating the humerus in the nose-down sense while pulling it downwards. The pronating moment is needed to balance the opposite moment caused by the aerodynamic force on the wing.

Although the supracoracoideus muscle originates deep to the pectoral muscle on the sternum and coracoid, it acts as a wing elevator. The muscle passes through the foramen triosseum, between the proximal ends of the scapula, coracoid, and clavicle, and inserts posterodorsally on the humerus; it thereby

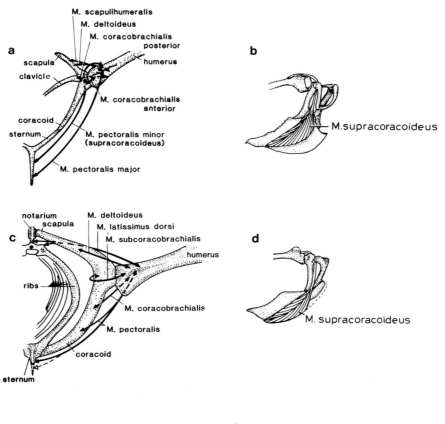

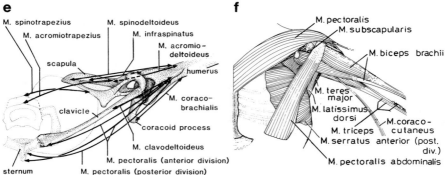

Fig. 11.8. Muscle arrangements in birds (*a,b*), pterosaurs (*c,d*) and bats (*e,f*). The anterior views, *a*, *c* and *e*, show the directions of pull of the main flight muscles, while the lateral views, *b* and *d*, show the paths of the tendon of the supracoracoideus muscle in a bird and a pterosaur, respectively. *f* shows the ventral view of the left shoulder region of *Plecotus auritus* with the posterior division of the pectoralis muscle removed. *a, c, e* are from Norberg 1985a, by courtesy of The Company of Biologists Ltd; *b* and *d* are modified from Padian 1983; *f* is from Norberg 1970b)

exerts an upward pull on the humerus (Fig. 11.8b). The deltoideus muscles act as additional wing elevators, while the latissimus dorsi and scapulihumeralis muscles retract the wing and also are involved in the upstroke.

Rayner (1985a, 1988) suggested that the supracoracoideus may not be active in fast flight with unpowered upstroke but only at slow flight and take-off (cf. Sect. 10.3.6). In slow flight and in hovering, the upstroke gives some lift and/or thrust, whereas in fast flight it may be more or less passive, so the supracoracoideus should be most developed in hovering and slow-flying birds. Greenewalt (1962) found that the pectoral muscle of his passeriform group has about 12 times the mass of the supracoracoideus. Hummingbirds are exceptional because of the aerodynamic power produced in the upstroke (= backstroke), and their pectoral muscles have only twice the mass of the supracoracoideus.

High-speed film and electromyographic studies of free-flying pigeons in slow level flight (at 12 m s^{-1}) reveal that the supracoracoideus muscle is strongly activated during wing elevation (Dial et al. 1988; Goslow et al. 1989; Fig. 11.9a). The activity of the pectoralis begins in late upstroke and continues into the downstroke, while the supracoracoideus exhibits a biphasic pattern. The major burst begins in late downstroke and continues into the upstroke, and a second, and more variable, burst occurs during the middle of the downstroke.

The arrangement and morphology of various flight muscles differ between species with different flight modes. For example, the two main extensors (metacarpi radialis and ulnaris) and one flexor (carpi ulnaris) of the handwing have two bellies in the kite (*Milvus migrans*), a soaring bird, and only one belly in the parakite (*Psittacula krameri*), a flapping bird (Nair 1954). Nair suggested that the splitting of the muscles in the kite is a means of facilitating soaring by preventing over-fatigue in the muscles.

Pterosaurs. In pterosaurs the pectoralis and coracobrachialis muscles may have been the main depressors of the wing (Fig. 11.8c). The humerus in pterosaurs differs from those of birds and bats for the head is saddle-shaped rather than convexly rounded and has a very large anterior, hammer-like, process called the "deltoid crest" by Bramwell and Whitfield (1974) and the "deltopectoral process" by Wellnhofer (1975a). These names imply that it served as insertion for the deltoid and pectoral muscles, but the pectoral muscle may not have inserted on the crest, but rather on its anterior side. Otherwise the pectoralis would have exerted too large a rotating moment of the wing in the nose-down sense, because of the large moment arm (Pennycuick 1988). Pennycuick suggested that the crest served mainly as an origin for a wing-finger protractor muscle. This muscle is hypothetical, but a large groove at the end of the metacarpus (illustrated by Wellnhofer 1975a, 1978) suggests the existence of a massive tendon that inserted on the first phalanx of the wing finger. The anterior process also may have been the origin for the leading edge tendon (see below).

The subcoracoscapularis, deltoideus, and latissimus dorsi muscles may have been involved in the upstroke in pterosaurs (Bramwell and Whitfield 1974). Wellnhofer (1975a) suggested that a supracoracoideus muscle could also have elevated the wing (see also Sect. 11.3.2). Pterosaurs lacked a foramen triosseum because they lacked clavicles, but they had a groove on the medial face of the coracoid where the tendon of a supracoracoideus muscle may have passed (Fig.

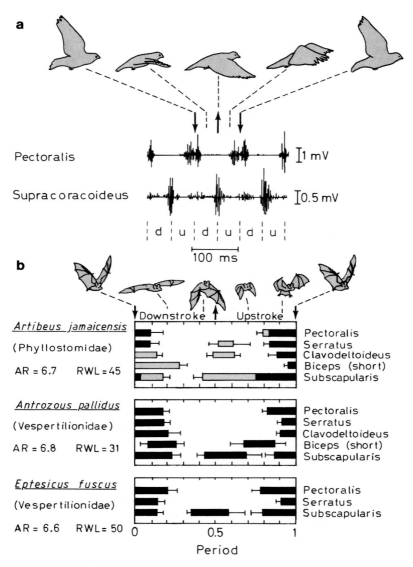

Fig. 11.9. *a* Electromyographic activity patterns for the major downstroke and upstroke muscles of the pigeon in slow, flapping flight. *Downward arrow* designates the initiation of downstroke (*d*) and *upward arrow* that of the upstroke (*u*). The pectoralis muscle begins activity in late upstroke and continues into the downstroke. The supracoracoideus muscle exhibits a biphasic pattern, where the major burst begins in late downstroke and continues into the upstroke. The second, more variable burst, occurs in the middle of the downstroke. (After Goslow et al. 1989). *b* Electromyographic activity patterns for selected shoulder muscles in three bats during slow, accelerating flight. *Horizontal lines* indicate one standard deviation from the mean times of onset and termination. *Closed bars* represent maximal amplitude and open bars half maximal amplitude. *AR* is aspect ratio, *RWL* is relative wing loading. (After Altenbach and Hermanson 1987)

202

11.8d), suggesting that such a muscle functioned as a wing elevator in pterosaurs as in birds.

Bats. The flight muscles in bats (Figures 11.8e,f) have been described in a number of species (e.g. Vaughan 1959, 1966, 1970a; Norberg 1970b, 1972a; Kovtun 1970, 1981, 1984; Strickler 1978; Altenbach 1979 and Hermanson 1979). Kovtun and Moroz (1974), Altenbach (1979), and Hermanson and Altenbach (1981, 1983, 1985) provided electromyographical analyses of the flight musculature during locomotion. The main downstroke muscles are the pectoralis, serratus anterior, subscapularis, clavodeltoideus and latissimus dorsi, while the pectoralis and the serratus anterior are the main adductors of the wings, the clavodeltoideus mainly extends the humerus whereas latissimus dorsi pronates it. The upstroke is controlled mainly by the spinodeltoideus and acromiodeltoideus muscles, but also by the trapezius group and the long and lateral head of the triceps brachii. The trapezius muscles move the scapula towards the vertebral column and act indirectly as wing elevators. Several muscles are bifunctional, controlling humeral orientation and stabilizing the shoulder joint. The complex downstroke muscular system in bats, unlike the system in birds, reflects the use of the forelimbs in terrestrial locomotion and climbing.

With electromyograms, Hermanson and Altenbach (1981, 1983, 1985) showed that the downstroke muscles already begin to contract in the later part of the upstroke to initiate wing pronation and adduction (Fig. 11.9b). In the same way, the upstroke muscles begin to contract in the later part of the downstroke to initiate the upstroke. Only a few wing muscles are active in the middle periods of the downstroke and upstroke.

11.3 Skeleton System

11.3.1 Trunk Skeleton

Vertebrate flyers have short, streamlined, and stiff trunks. The vertebrae have fusions in different regions, and when not fused they are shaped to limit or prevent motion. In bats, fusions are common between adjacent trunk vertebrae, while pterosaurs had fusions (the notarium) in the thoracic region, and birds have 12–20 vertebrae fused in the synsacrum. Some birds (such as pheasants and falcons) have additional fusions in the thoracic region.

The ribs are usually flattened and provide an area for muscle origins. The uncinate process of bird ribs provides additional area and bracing, and most of the ribs are fused to the sternum in all three animal groups.

The size of the sternum has often been associated with the size of the pectoralis muscle and with the ability of flight. Bats have three units in the sternum, while birds and pterosaurs have only one. Birds (Fig. 11.10a,b) have a large sternum with a well-developed sternal ridge (cristospine), which provides a large surface for the origins of the pectoral muscles. Pterosaurs (Fig. 11.10c,d) also had a large sternum, but with a cristospine that was not as large as in most

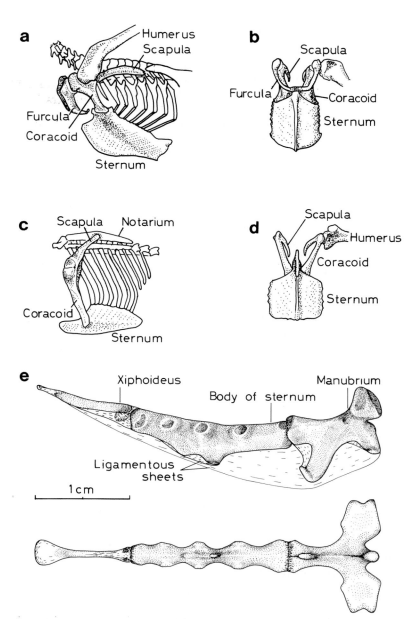

Fig. 11.10. Sternum and shoulder girdle in birds (*a,b*) and pterosaurs (*c,d*) in lateral (*a,c*) and ventral (*b,d*) views. The figures are not to scale, and lateral and ventral views are not from the same species. (*a* is modified from Herzog 1968; *b* and *c* are from Padian 1983). *e* Bat (*Rousettus aegyptiacus*) sternum in lateral (*top*) and ventral (*bottom*) views. (Norberg 1972b; by courtesy of Springer-Verlag)

birds, and only prominent anteromedially, leading to the suggestion that pterosaurs could not produce as powerful a downstroke as birds. Some bats have a well-developed sternal ridge, whereas others have scarcely any, but the anteriormost part of the manubrium of the sternum usually has a ventrally projecting tubercle from which a median ligamentous sheet passes ventrally (Fig. 11.10e). Together with the sternum, this sheet forms an attaching surface, on either side, for the pectoral muscle. This ligamentous sheet and the sternal ridge are together analogous to the sternal ridge in birds as an area of origin for the pectoralis muscle. The ventrally projected enlargement of the sternum need not be rigid, because the pectoral muscles on both sides pull in opposite directions balancing each other's forces. Not even the strong sternal ridge in birds could withstand the muscle forces if this force balance did not prevail. Since ligamentous sheets do not fossilize, pterosaurs might have had similar ligamentous sheets along the low keel, allowing them better developed flight than had been proposed (Norberg 1970b). The rugose ventral edge of this sternal keel and the elliptical bony prominence anteromedially on the sternum imply a further cartilaginous or ligamentous extension (Padian 1983).

The sternal ridge in birds has an additional function for pectoral muscles contain air cavities (for cooling purposes) connected to the interclavicular air sacs. Pennycuick (1986, Fig. 15) pointed out that the cristospine prevents these cavities from collapsing when the muscles contract. The lack of a bony ridge in pterosaurs suggests that they lacked such air cavities in their pectoral muscles and bats do not have air cavities in their muscles, and so do not need a bony ridge. The sternal ridge may have little or nothing to do with the possibilities for flight muscles to develop strong forces, and its absence in pterosaurs may not indicate anything about their flight ability.

11.3.2 Pectoral Girdle

The shoulder girdle in flying animals is strong, large and firmly attached to the sternum by the anterior coracoid (birds, pterosaurs) or the clavicle (bats). In birds and pterosaurs the scapulae and coracoids are long, narrow and fused together where they meet to form the glenoid cavity, which receives the proximal head of the humerus (Fig. 11.10a-d). A prominent acrocoracoid process near the glenoid cavity is present in both groups. In birds the clavicle articulates with the coracoid and often also with the scapula, while in birds and in bats the scapula has no firm anchorage, but is suspended by muscles and ligaments. In pterosaurs the scapula articulated against the notarium of the fused trunk vertebrae, and these animals lacked clavicles, and a foramen triosseum, which in birds is a passage for the supracoracoideus muscle, that acts as an elevator of the wing. In pterosaurs the presence of the acrocoracoid process suggests a passage of a tendon of the supracoracoideus muscle, so that it also acted as an elevator in this group (Wellnhofer 1975a; Padian 1983).

The two clavicles in birds are usually fused together to form the familiar "wishbone", or furcula, which provides an increased site of origin for the pectoral muscle and it was assumed to have evolved for this purpose (von Stegmann 1964;

Feduccia and Tordoff 1979). Birds also have a triangular fascial sheet stretched between the furcula, coracoid and the anterior part of the sternum, that serves as an additional area of origin for flight muscles. Jenkins et al. (1988) and Goslow et al. (1989) examined skeletal movements during flight by high-speed X-ray movies of the European starling (*Sturnus vulgaris*) flying in a wind tunnel (Fig. 11.11a-c) and found that the furcula expands its width by almost 50% in each wingbeat cycle because its two legs bend laterally during the downstroke and recoil during the upstroke. The spreading of the clavicles begins as the distal end of the humerus is depressed and continues throughout the downstroke. The dorsal ends of the coracoids and the cranial ends of the scapulae are displaced laterally, and the posterior ends of the scapulae translate medially as the shoulder spread. The sternum exhibits elliptical movements in the median, sagittal plane. The furcular spreading seems to be performed primarily by the coracoid displacement (by the sternocoracoideus muscle) and by the extended wings via the pectoral muscles during the downstroke. Goslow et al. (1989) proposed that the main function of the furcular movements may be to help with the bird's breathing, for during the downstroke the bending furcula would cause expansion of the first group of air sacs, while the sternum compresses the second group of sacs by pressing against them. Then, as the wings are raised, the first set of air sacs are compressed and the second can expand. It has been suggested (Schaefer 1975, cited in Norberg 1985a) that a function of the furcula may be to store elastic energy during the wingstroke (see Sect. 11.5.5) and the kinematic analysis of the starling provides indirect support for this hypothesis, showing how elastic strain energy is stored during the wing's downstroke and recovered as elastic recoil during the upstroke.

The bat pectoral girdle is highly movable and composed of a well-developed and faceted scapula and a clavicle that is fused to the sternum. The coracoid is reduced to the acromion process on the scapula (Fig. 11.11d) and the facets of the scapula provide large attaching surfaces for several flight muscles.

Hermanson (1981) and Altenbach and Hermanson (1987) used single-frame X-ray exposures to correlate skeletal movements with the wingbeat cycle and to

---

Fig. 11.11. a-e Cineradiographic analysis of a flying European starling (*Sturnus vulgaris*). *a* Dorsal view of some phases in the wingbeat cycle. The *dashed lines* indicate the positions of the joint between furcula and coracoid and show the spreading of the furcula during the downstroke. *b* Resting positions of furcula and coracoid (*solid lines*) and their positions at the end of the downstroke (*dashed lines*). l_r is the furcular distance at rest and l_d that at the end of the downstroke. *c* Sternal excursion determined by means of implanted markers (*dots*). The *arrows* indicate typical movements from the position at the end of the upstroke, *u*, to the position at the end of the downstroke, *d*, and back. The position of the *stippled* sternum relative to the markers at *u* shows one of the largest excursion observed. (*a-c* are modified from Jenkins et al. 1988 and Goslow et al. 1989). *d* Bat (*Plecotus auritus*) shoulder girdle; dorsal exploded view of the right shoulder (*top*) and lateral view of the right scapula (*bottom*). (Norberg 1970b). *e* Model of the scapulo-humeral locking mechanism in bats viewed from in front. *Left* generalized pectoral girdle. *Right* the pectoral girdle at initiation of downstroke. *F* is the force vector applied by pectoralis contraction, f_1 is the fulcrum of humeral adduction without scapulo-humeral lock (= scapular fulcrum) and f_2 that with dorsal scapulo-humeral lock (= clavicular fulcrum), l_1 is the moment arm of pectoralis with scapular fulcrum and l_2 that with clavicular fulcrum. (After Altenbach and Hermanson 1987)

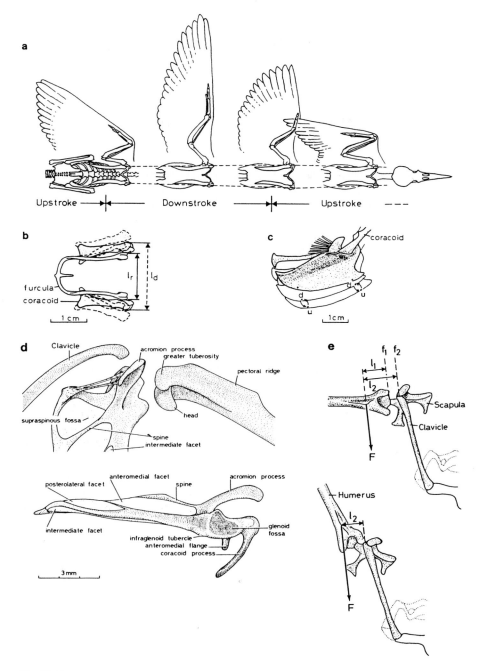

a

Upstroke ──→├←── Downstroke ──→├←── Upstroke ── ─ ─

b

furcula
coracoid

l_r l_d

1 cm

c

coracoid

d u
d u

1 cm

d

Clavicle
acromion process
greater tuberosity
pectoral ridge
supraspinous fossa
head
spine
intermediate facet

anteromedial facet
posterolateral facet
spine
acromion process
intermediate facet
infraglenoid tubercle
anteromedial flange
coracoid process
glenoid fossa

3 mm

e

f_1 f_2
l_1
l_2
Scapula
Clavicle
F

Humerus
l_2
F

Fig. 11.11 a-e

207

describe the role of individual muscles in bats. In the three bats investigated (*Artibeus*—Phyllostomidae, and *Antrozous* and *Eptesicus*—Vespertilionidae), the clavicle is adducted synchronously with downstroke movements of the humerus. The authors suggested that the outward rotation (adduction) of the clavicle would result in (1) dorsoventral flattening of the thoracic profile and lower parasite drag, (2) reduced excursion of the pectoralis muscle during contraction, which would enable the muscle to operate close to its maximum value of a length-tension range, and (3) passive lengthening of several flight muscles.

Scapulo-Humeral Lock. In bats the scapula and humerus interlock as the humerus is abducted during the upstroke by scapular muscles (e.g. Vaughan 1959). Pennycuick (1982) found that the albatrosses *Diomedea* and *Phoebetria* have a similar lock on the humerus that works with a tendonous sheet that is in parallel with the pectoral muscle. The lock prevents the wing from elevating above the horizontal when the humerus is fully protracted, although the arm can be raised when retracted. Pennycuick (1982) also found that giant petrels (*Macronectes*) are similar functionally, but that their anatomy is a little different; a lock was not found in small procellariform species. A locking arrangement reduces the energy cost of gliding flight by relieving wing-depressing muscles from the work of keeping the wings down on the horizontal plane. A functionally similar locking mechanism occurred also in the large pterosaur *Pteranodon*, so that as the humerus moved upwards there was one position where its articular head locked in the glenoid cavity (Bramwell and Whitfield 1974). The locking was due to the peculiar shape of the cavity, which had its upper and lower surfaces set at an angle to each other (Fig. 11.10c,d). The humerus was calculated to have been directed upward at 20° and swept backward at 19° in the locked position, presumably the natural position for gliding (Bramwell and Whitfield 1974).

Vaughan (1959) was the first to suggest that in bats the dorsal interlocking of the greater tuberosity of the humerus and the dorsal articular facet of the scapula form a wing-adducting mechanism. The movable clavicle is a component of this mechanism and the clavicle (instead of the scapula) acts as a fulcrum, so that depression of the lateral scapular border by the serratus muscles would rotate the scapula about its long axis and begin adduction of the interlocked humerus (Altenbach and Hermanson 1987). The adduction would then be completed by the pectoralis, subscapularis and coracoid (short) head of biceps brachii, using the scapula as a fulcrum. Anatomical data (Norberg 1970b, 1972a; Strickler 1978; Altenbach 1979) and electromyographic recordings in *Desmodus* (Altenbach 1979) support this hypothesis. Figure 11.11e illustrates the scapulo-humeral lock as the humerus is abducted during the upstroke by the scapular musculature. The interlock is maintained until the midpoint of the downstroke by contraction of the pectoralis. After the humerus has locked into the scapular facet, the scapula rotates about its articulation with the clavicle, and the scapulo-clavicular articulation takes over the function as the fulcrum for the humerus. This situation lengthens the moment arm for the pectoralis, as well as the mechanical advantage for adduction.

Padian (1983) compared the skeletons of pterosaurs, birds, and bats and concluded that pterosaurs shared skeletal characters important for flight with both birds (such as reinforced pectoral girdle braced by the sternum and with an acrocoracoid process) and bats (such as the cristospine, which is analogous to the manubrium of the bat sternum). Pterosaurs had also pneumatic foramina and insulatory covering, suggesting increased body temperature and, taken together, these characters indicate that pterosaurs were active fliers.

11.3.3 Wing Skeletal and Membrane Arrangements

The force of the airstream subjects the wings of flying animals to great strains during flight. The wings of all vertebrate flyers, whether feathered or membraneous, are supported by a framework of skeletal elements controlled by muscles. Special wing arrangements reduce the demand for powerful muscles and large cross-sectional areas of the wing bones reducing the mass of the wings, and so their inertial loads, thus reducing the total mass of the animal.

Pneumatic bones that reduce weight characterize pterosaurs and most birds but not bats. A given strength of a bone element can be achieved with less bone material (less mass) if it is made tube-shaped than if solid, but the solid skeletal elements of bats, especially in the wings, are very slender, contributing to low weight. The leading edge of the wings are subjected to especially great strains during flight and the wing skeleton in pterosaurs, birds and bats forms stay systems in the armwing, and in birds and bats also in the handwing. These systems provide good support for the anterior parts of the wings (Fig. 11.12). In all three groups the humerus makes an angle less than 180° with the ulna and radius, permitting a suitable wing profile (Herzog 1968). Because the elbow joint is more elevated than the shoulder and wrist joints when the wing is outstrecthed laterally (in birds and bats and probably also in pterosaurs), and because the muscle along the leading edge of the armwing tightens the patagium (membrane) anterior to the arm, the chordwise profile of the armwing becomes very convex, which promotes lift production. The muscle and the membrane of the leading edge prevent the angle between the humerus and the ulna and radius from opening excessively.

Birds. In birds the armwing unit is formed by the humerus, the ulna and radius and the tensor patagii longus muscle (Fig. 11.12a) which arises from the shoulder (mainly from the apex of the clavicle) and in many birds inserts on the extensor process of the carpometacarpus and usually also into the fascia of the manus (George and Berger 1966). The ulna and radius are separated from each other and form a slightly convex unit, an arrangement that greatly increases resistance to bending forces in the plane along which they lie. The bending rigidity is proportional to the distance between the bones. The bending strength in the feathers resides in the feather shafts, which are robustly supported at their proximal ends, and the bending loads are transferred to the rigid arm skeleton (Pennycuick 1986). The radius and ulna move in a fashion similar to a pair of "drawing parallels", causing automatic extension and flexion of the manus when the elbow extends and flexes (described by Coues 1872 and analyzed by Fisher 1957; Fig. 11.13). As a consequence, the radius is mainly subjected to tension and the ulna mainly to compression during flight when the wing is extended and

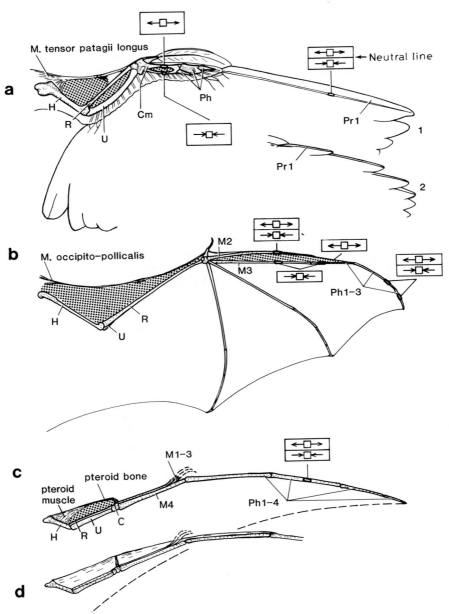

Fig. 11.12. The wings of a bird (*a*), a bat (*b*) and a pterosaur (*c*). The *stippled areas* show the parts of the leading edge supported by skeletal elements forming stay systems. *H* is humerus, *R* is radius, *U* is ulna, *C* is carpus, *Cm* is carpometacarpal, *M* is metacarpal, *Ph* is phalanx, and *Pr* is primary feather. The *boxes with arrows* are schematic representations of lengthwise sections of various supporting elements with forces indicated; the *arrows* show compression (*inwardly directed arrows*) and tension (*outwardly directed arrows*) forces in skeleton, rachis and ligament. *d* Shows the forward position of the pterosaur's pteroid bone. (After Norberg 1985a)

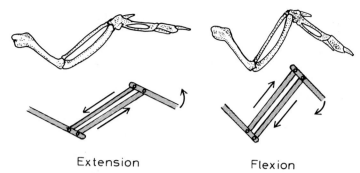

Extension   Flexion

Fig. 11.13. Ring wing skeleton of a bird. The radius and ulna function as "drawing parallels" (*bottom figures*) and cause automatic extension and flexion of the handwing when the elbow opens and flexes

subjected to bending forces. This mechanism largely reduces the need for large muscles located within the wing to move (extend and flex) the manus and this work is done instead by the arm-extending and -flexing muscles, which need to be stronger. This arrangement adds mass to the proximal part of the wing, which is more advantageous since it adds little to the wing's moment of inertia. In bats the ulna is rudimentary, and extension and flexion of the manus are caused by extensor and flexor muscles of the arm and manus.

The proximal part of the leading edge of the handwing in birds owes its rigidity to the first, second and third digits and in particular to the carpome-tacarpus and phalanges of the second digit. The distal part of the leading edge is formed by the anteriormost primary (or primaries when graded; Fig. 11.12a). The two longer metacarpals form a slightly convex unit that encloses an inter-metacarpal space resembling the one in the forearm between the ulna and radius. This unit increases the resistance to bending forces in the plane along which the fused metacarpals lie. The phalanges of the second digit are thick and dor-soventrally flattened, and the first primary feather attaches to the tip of the outermost phalanx. In many birds the first primary is shorter than the next, so the bending forces caused by the aerodynamic force are spread over more than one primary feather. The air pressure on the feathers behind those of the leading edge cannot be transformed into backward pull on the leading edge of the handwing because the feathers take up their own bending forces more or less inde-pendently.

Bats. In bats the triangular unit in the armwing is formed by the humerus, the radius (and the very rudimentary ulna), and the occipito-pollicalis muscle (Fig. 11.12b) which generally arises from the neck, is joined to the clavodel-toides, passes ventral to the thumb and inserts in the wing membrane near, and anterior to, the second metacarpal.

The aerodynamic forces during flight bulge the wing membrane, and the tension set up in the membrane pulls at the lines of attachment in the leg, arm, and digits. The tension occurs in the tangent planes of the membrane at these lines. Although the resultant force of the wings is directed nearly normal to the wing chord, this force is transformed into tension in the membrane. The strain

211

is especially great on the skeletal elements of the leading edge (particularly of the handwing), which stretches out the membrane and leads the wing movements. The profile shape during flight is controlled by the plagiopatagialis muscles which run across the membrane area at the proximal part of the wing (plagiopatagium; Fig. 11.14a), and these muscles are especially powerful in the large megachiropteran bats (Norberg 1972a). The thin wing membrane further contains a fibrous network with elastin and collagen and individual bundles of the net are similar to elastic ligaments (Holbrook and Odland 1978) and tense the membrane during flight.

Rigidity of the leading edge of the handwing is obtained by a special arrangement of the second and third digits (Norberg 1969; Fig. 11.15a). Determining factors are (1) a ligamentous connection from the distal end of the second digit to the anterior base of the second phalanx of the third digit, and (2) the bending of the third metacarpophalangeal joint so that the joint angle is somewhat less than 180° in the membrane plane anterior to the third digit. Because of the convexity, this arrangement constitutes a rigid unit between the thumb and the joint between the first and second phalanges of the third digit in the plane of the dactylopatagium minus (membrane between the second and third digit). This patagium lacks elastic strands and is kept very taut. During the wingstroke the second digit (metacarpal 2 and phalanx 1) and the second and third phalanges of the third digit are subjected to bending forces (tension in the anterior parts, compression in the posterior parts of the elements). The ligament connecting the tip of the second digit and the base of the second phalanx of the third digit is subjected only to tension, and the metacarpal and first phalanx of the third digit mainly to longitudinal compression. This rigid unit of the leading edge is very broad in large and broad-winged bats (Figs. 11.14a,b) and there is a similar rigid unit between the third and fourth digits in the large and broad-winged (particularly megachiropteran) bats, and this releases the wingtip from large tension forces and acts to keep the joints of the fourth digit steady without involving large muscular forces (Norberg 1977a,b; Figs. 11.14b, and 11.15b).

The digits of bats are shaped so that their greatest cross-sectional diameters are in these planes where the bending forces are largest. This keeps their mass low while still maintaining rigidity in the important directions (Norberg 1970b, 1972a; Fig. 11.16) and is generally more pronounced in small bats than in larger bats. Reduced curvature of the membrane means a tauter membrane and larger tension forces at the lines of attachment. Since the wing is divided into sections by the digits, each patagium can be regarded as a unit for the acting forces. The

---

Fig. 11.14a-c. The megachiropteran bat *Rousettus aegyptiacus. a* The plagiopatagiales muscles run across the membrane behind the arm (plagiopatagium) and tighten it during flight. The second to fourth digits are angled so that the membrane around the proximal parts of these digits is kept taut without any need of large muscular forces. *b* The large wrinkles in the membrane parts (dactylopatagia) between the second and third digits and between the third and fourth digits indicate the location and direction of the largest tension forces between these digits. *c* Diagrammatic representation of the distribution of the largest tension forces between the digits. Photographs by the author. (Norberg 1972a; by courtesy of Springer-Verlag)

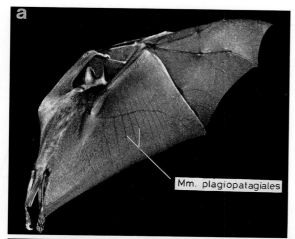

a

Mm. plagiopatagiales

b

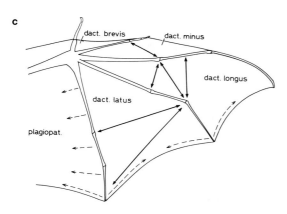

c

dact. brevis   dact. minus

dact. longus

dact. latus

plagiopat.

213

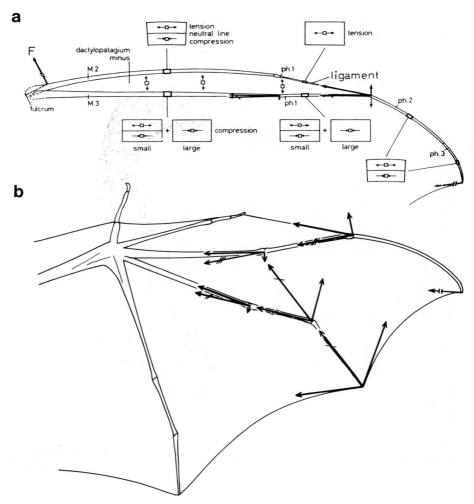

Fig. 11.15. a Simplified diagram showing the arrangement of the second and third digits in a static condition of the bat wing, occurring in both Mega- and Microchiroptera. Because of the convexity, this arrangement makes a leading edge that is rigid in the plane of the membrane. The relative magnitudes of the forces acting on the wing are arbitrarily indicated, but the relative magnitudes of forces within and between the two vector diagrams are correct. *F* is the force of Musculus extensor carpi radialis longus, acting on the second metacarpal (actually, the force component acting perpendicular to a line connecting the fulcrum of the second metacarpal and the point of insertion of the tendon on the second metacarpal). The *backward broken arrow* at the first interphalangeal joint of the third digit indicates the backward component of the pull of the entire wing membrane at this joint. This force cancels out the forward-directed component of the pull from the ligament. A *broken arrow* also indicates the pull of the membrane on the tip of the third digit. The *arrows at the squares* indicate compression (*inwardly directed arrows*) or tension (*outwardly directed arrows*) occurring within the various elements. *M* is metacarpal, *ph* is phalanx. (After Norberg 1969, 1970b). *b* Diagrammatic representation of forces acting on the third and fourth digits in a megachiropteran bat. The fourth metacarpophalangeal joint is angled backwards and the fourth interphalangeal joint forwards; this arrangement tightens the proximal part of the membrane between these digits and keeps the digital joints very steady without any need of large muscular forces. (Norberg 1972a; by courtesy of Springer-Verlag)

214

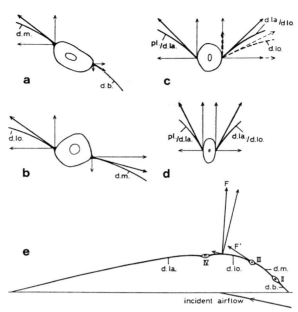

Fig. 11.16a-e. Diagrams showing the tension forces of the membrane acting on the digits (seen in cross section) during the downstroke. The digits are flattened in those planes in which the largest bending forces occur. The *dotted lines* indicate the condition in megachiropteran bats and those microchiropteran species which have very taut dactylopatagium longus. The *bottom figure* shows a cross section of the right wing seen from the distal end with the second, third and fourth digits in place. The aerodynamic resultant force *F* of the wing is transformed into tension *F'* in the wing membrane on the third digit. *d.b.*, *d.m.*, *d.lo.*, and *d.la.* are dactylopatagium brevis, minus, longus, and latus, respectively, *pl.* is plagiopatagium. (Norberg 1972a; by courtesy of Springer-Verlag)

second digit is exposed to bending mostly in the membrane plane, and flattening of the second metacarpal in this plane makes it resistant to these bending forces. The end phalanx of the third digit is exposed to pull only from behind, and when the membrane curves, the phalanx rotates so that the main axis of the cross section lies mainly in the plane of the membrane. Dorsoventral flattening of this phalanx makes it most resistant to bending forces in this plane. In equilibrium the tensions transmitted to digits four and five from the membranes are equal on each side, unless lateral muscle forces are involved, so these digits are subjected mostly to dorsoventral bending and their phalanges are laterally compressed.

Pterosaurs. In the flying reptiles, the triangular unit in the armwing may have been formed by the humerus, the ulna and radius, the pteroid muscle and the long and slender pteroid bone (Fig. 11.12c). The pteroid muscle may have attached to the pteroid bone and to the anterior process of the head of the humerus (Short 1914; Pennycuick 1988), but it may also have originated on the head or body as in bats. Frey and Riess (1981) suggested that the pteroid bone pointed forwards in the extended wing during flight so it would have acted as a compressive element because the leading edge membrane would have attached to it at acute angles. In this position the membrane could easily rupture at its attachment to the pteroid bone, unless the tip of the bone was cartilaginous (resilient) and/or bifurcated, as are the fifth digit and tail in some bats. A functionally more appropriate arrangement in this sense would have seen the pteroid bone passing inwards in the direction of the tendon as in Fig. 11.12c. Frey and Riess noted that in this position the pteroid would have pressed too much against the proximal part of the carpal bone, against which it attaches, but this moment could have been counteracted by a ligament between the anterodistal part of this carpal bone and the metacarpals. Pennycuick (1988) found it difficult to accept that a bony element would have been required in a position subjected only to tension and he found that the pteroid bone of the medium-sized Cretaceous pterodactyl *Santanadactylus spixi* could be articulated in two different positions against the same articular surface on the carpus. The pteroid could have operated with an automatic snap action if it was controlled by a leading-edge tendon running to the first interphalangeal joint. Protraction of the fourth digit deployed a drooped leading edge with the pteroid pointing forward-downward (Fig. 11.12d) and retraction caused the bone to point inward towards the humerus and furled the propatagium (membrane anterior to the arm; Fig. 11.12c). Pennycuick suggested that both facets were functional articulations, and that the pteroid could snap from one position to the other. The "furled" position along the leading edge would be used in fast flight, and the "extended" position would increase wing camber and be used in slow flight (cf. Sect. 11.5.1).

Based on imprintings of the rhamphorhynchs *Sordes pilosus* (Sharov 1971) and *Pterodactylus kochi*, Frey and Riess (1981) suggested that the membrane in front of the armwing continued through the metacarpals 1–3 and along the first two phalanges of the fourth digit. But these skin flaps found along the finger may have been folds of the membrane behind the fourth digit as the membrane was flaccid in the folded wing.

The ulna was much thicker than the radius which passed along it. Bramwell and Whitfield (1974) suggested that these bones may have caused slight automatic extension and flexion of the manus. The enormously elongated fourth digit, with its four long phalanges, formed the leading edge of the handwing together with the much reduced first to third, fused, metacarpals. The metacarpals were short in the long-tailed pterosaurs (Rhamphorhynchoidea) and longer in the short-tailed ones (Pterodactyloidea).

The common opinion has been that the pterosaur wing was made up by a thin membrane transversed by elastic fibres, much like that of bats, and that the trailing edge of the membrane was reinforced by an elastic strand and attached to the legs near the feet. This arrangement means that the fourth digit alone had to resist the bending forces caused by air resistance and by the pull of the membrane of the handwing. As predicted because of the strong bending forces on the leading edge, the fourth digit was remarkably thick. The feet were supposed to have been held backwards and rotated round dorsally, as in bats, so that the fifth toe faced medially. Rhamphorhynchs had a very long fifth toe; the first of its two phalanges was angled slightly outwards and could be rotated into various positions against the metatarsal, while the second phalanx articulated with the first to form a right-angle bend. If the pterosaurs held their feet backwards like bats, the fifth toes pointed towards the tail and could have supported a tail membrane (uropatagium) or small flaps along the legs like the ones in pteropodid and some phyllostomid bats. This position of the fifth toes occurs in undisturbed, fully articulated rhamphorhynch specimens. The fifth toe was short in pterodactyls and had only one phalanx.

A different reconstruction was made by Brower and Veinus (1981) and Padian (1983), who suggested that the trailing edge of the pterosaur wing membrane was attached to the side of the body or to the pelvis instead of to the legs, giving a very narrow, pointed wing shape, but this notion is based on negative evidence. There is no sign in any specimen that the posterior edge of the wing membrane attached to the body for the proximal part of the membrane is indistinct in most pterosaur fossils and it is extremely difficult to see if it joins to the leg or to the body or to the tail. But in *Pterodactylus kochi* the membrane seems to attach to the knee and there is a clear impression of something resembling a tail membrane (Fig. 11.17a). Padian suggested that the fifth toe was morphologically on the outside of the leg due to a forward position of the knee

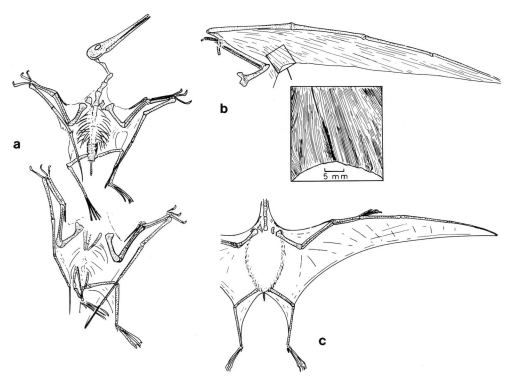

Fig. 11.17. a Two specimens of *Pterodactylus kochi. Top* the Vienna specimen, from a drawing by Wellnhofer (1987), indicating that the wing membrane may have been attached to the knee. *Bottom* the Munich specimen 1937.1.18, from a photograph by Wellnhofer (1983), showing sign of a tail membrane, b Sketches of wing of *Rhamphorhynchus muensteri*, from photographs by Wellnhofer (1983), with orientation of fine "fibres". The fourth digit is 38 cm long. c Possible wing reconstruction of *Pterodactylus kochi*

217

and bipedal locomotion (see the next section), but proposed no function for the fifth toe. Pennycuick (1988) maintained that an intermediate position of the leg with the femur splayed to a moderate extent would be possible and result in the fifth toe being directed outwards and up. He suggested that the toe could have been an anchor for a trailing-edge tendon of the wing membrane and the high mobility at the metatarso-phalangeal joint would have allowed the fifth toe to be swivelled around to point in various directions to control a downward curl in the trailing edge.

Wellnhofer (1975b, 1987) and Padian (1983) suggested that the pterosaur wing was reinforced internally by a system of fine, stiff, intercalated structural "fibres" of some hard material, whose function was analogous to the keratinous feather shafts in birds. These fibres ran nearly parallel to the wing finger and radiated through the wing in a closely spaced pattern with no reinforcement occurring along the trailing edge of the wing. Wellnhofer and Padian based their conclusions on a wing imprinting of the pterosaur *Rhamphorhynchus* (Fig. 11.17b). Pennycuick (1986, 1988) noted, however, that there are difficulties with this view, for the "fibres" did not transverse the full width of the membrane and are not visible at the proximal part of the membrane or connected to the skeleton. Furthermore, given the huge wing finger, it is unclear what loads these "fibres" would have been required to carry. Cross sections of the fourth digit suggest that it was subjected to large bending forces in the wing plane. The metacarpal was dorsoventrally compressed, and the first phalanx was triangularly shaped with the longest diameter in the plane of the wing (Bramwell and Whitfield 1974, Fig. 2). The "fibres" could have been wrinkles in the skin caused by contraction of elastic fibres when the wing was relaxed, as can be seen in relaxed bat wings. Pennycuick (1988) assumed that a trailing-edge tendon could have run from the tip of the wing finger to the fifth toe, supporting the posterior margin of the wing membrane and the posterior attachment for the main group of elastic fibres.

New studies, by David Unwin (of the University of Reading) and Nataly Bakhurina (of the Lower Tetrapod Laboratory in Moscow), are based on the best-preserved pterosaur fossils known in the world, unearthed in Kazakhstan and western Mongolia and collected by Soviet palaeontologists in the past 20 years. Their findings are reviewed in the New Scientist (1988, 1629:34–35), and they found that the pterosaurs were hairy and that their wing membranes were attached to their legs. Most pterosaur fossils from Kazakhstan are from the pigeon-sized *Sordes pilosus*. Its hairs, black filaments about 0.5 cm long, were probably made of keratin, while its wings were naked and leathery with fibre-like structures. They had also a membrane between their legs that was not attached to the tail. Fossils of the wrist bones of a large pterosaur, called *Phobetor*, show that it could fold its wings backward and downward, useful not only in rest but also in flight by permitting it to arch its wings and improve its flight performance. Unwin and Bakhurina conclude that pterosaurs were a diverse group with the same sorts of lifestyles as birds.

Figure 11.17c shows a possible reconstruction of a pterodactyl wing (such as *Pterodactylus kochi*) with hypothetical elastic fibres and muscles of the membrane indicated. If the wing had functioned as a bird wing with stiff elements carrying the loads, the handwing would not have needed to have this heavy leading edge skeleton. The bird handwing consists mainly of the light feathers, but each bird feather carries its own load while with a continuous membrane there will inevitably be tension forces in the membrane.

Figure 11.18 a and c shows the principal forces in a bat and pterosaur wing, respectively. The membrane of the bat is trisected by two elements, digits four

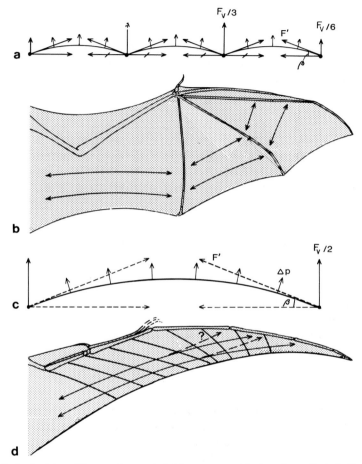

Fig. 11.18. Mechanical principles of the membrane-skeletal arrangement in the wings of bats (*a,b*) and pterosaurs (*c,d*). *a* Force diagram of a cross section of a schematic wing (bat type) extended from the body and by three digits forming equal interjacent areas. The lateral components of tension forces on each side of the middle digits cancel out, and the distal digit (*far right*) needs to carry the load (tension force F') from the outer membrane part only. F_v is the vertical force component acting on the digit and Δp is the local air pressure on the wing membrane. *b* The digits act to alter the direction of tensional forces, relieving the wingtip from large forces (Pennycuick 1971c). *c* Force diagram of a cross section of a schematic wing (pterosaur type) extended between the body and one digit. The digit (*far right*) has to take up bending forces F' transformed from the entire wing membrane (no canceling of lateral forces of interjacent digits as in bats). Symbols as in *a*. *d Arrows* indicate the assumed direction of tensional forces in a pterosaur wing. *Lines* indicate direction of elastic fibres, which may also occur along the trailing edge (After Norberg 1972a)

219

and five, and supported by these digits, the third digit and the body. The three membrane parts are assumed to be of equal area. The pterosaur membrane is undivided and supported by two elements, the body and the fourth digit. If the horizontal projections of the trisected (bat) and undivided (pterosaurs) membranes are equal, along with the angle β between the tangent to the membrane and horizontal, then the force per unit of length on one of the supporting members of the trisected membrane becomes a third of the corresponding force that the undivided membrane exerts on the single, distal, supporting member. This is because the area is reduced to a third, and because force F is proportional to the area of the supported membrane (Fig. 11.18c). The distal part of the wing's leading edge is exposed to unilateral tension forces transformed only from the adjacent patagium and tending to bend the digit.

The fourth and fifth digits act as compression members that alter the direction of the tensional forces (Pennycuick 1971c), so the distal part of the wing's leading edge is relieved of large tensional forces and the need to be thick with a powerful extensor muscle. If the wing membrane were outstretched only by one digit, as in pterosaurs, the leading edge (= fourth) digit and the hind limb would have to resist tensional forces transformed from aerodynamic pressure on the entire wing membrane posterior to the wing skeleton. The resulting tension would be much larger on the outer part of the fourth digit than on the inner part so the leading-edge digit would have to be much stronger and thicker, and controlled by much stronger muscles than the leading-edge digit in bats.

In addition to the digits, the elastin fibres determine the directions of the tension in the membrane. An appropriate structure of the membrane of pterosaurs, such as elastic strands along the paths indicated by lines in Fig. 11.18d, the distal part of the leading edge may have been relieved from some forces. On the other hand, if the wing had been composed of independent intercalated, stiff, structural fibres, there would have been no large tension transformed to the fourth digit, reducing the need for a thick leading-edge digit in the handwing.

11.3.4 Were Pterosaurs Quadrupedal or Bipedal?

If pterosaurs held their feet rotated outward-backward behind the body, they would have moved quadrupedally on the ground and, like bats, probably rested upside down on trees or cliffs. From analyses of the wing and hindlimb skeletons and from comparisons with the bird hindlimb skeleton, Padian (1983) proposed that pterosaurs could not have walked quadripedally by moving the forelimbs parasagitally over the ground, but that the hindlimb was designed for bipedal, parasagittal locomotion, as in birds. Padian noted that when the wing finger was flexed, it was oriented more down and back, not up and back according to Bramwell and Whitfield (1974); the down-and-back orientation would have made it difficult to walk on the forelimbs. The pelves of pterosaurs are usually preserved crushed or pressed flat in a lateral or dorsoventral plane and the prepubes (distal halves of the pubes) are normally separated from the rest of the girdle and into right and left halves. Different interpretations have been drawn from the preservations, so there are different views on the positions of the legs.

220

Wellnhofer (1975a) suggested that the acetabulum in fossils of *Rhamphorhynchus* points slightly upwards, while Padian (1983) argued that it points slightly downwards. With a downward position the femur could not be directed obliquely outward to clear the pelvis, and Padian also noted that the medial condyle of the distal end of the femur is larger than the lateral one, meaning that the tibia was the main bearer of the weight and that the movement at this joint was restricted to the anteroposterior plane as it is in birds. A reduced fibula also leads to this interpretation.

With a simple weight balance system, Pennycuick (1986) showed that pterosaurs may not have been able to support their weight on the hindlegs without help from the wings, so they must have walked or clambered quadrupedally. Animals with a long tail, including bipedal dinosaurs, *Archaeopteryx*, and kangaroos can balance their body weight about the hip joint (Fig. 11.19a). But in modern birds with no long tail the acetabulum is too far back from the centre of mass of the body and so birds have a tendency to topple forward about their hip joints. This has to be resisted by tonic muscles pulling the ischium towards the femur (T in Fig. 11.19b), and the long stiffened synsacrum allows the downward pull on the posterior end to hold up the anterior end. The early rhamphorhynchs had tails that could have balanced the body weight, and their pelves were about the same size as those of *Archaeopteryx*. But the later pterodactyls had no tail and no expanded synsacrum comparable with that of birds (although there was fusion of pelvic elements and vertebrae) so they lacked any obvious adaptation to allow moments to be balanced about the hip joint by tonic muscles in bipedal position. From these arguments pterosaurs appear to have been quadrupedal.

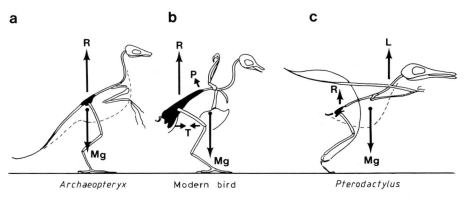

a   b   c

Archaeopteryx   Modern bird   *Pterodactylus*

Fig. 11.19a-c. Standing balance, where the *downward arrows* represent the body weight (*Mg*) and the *upward arrows* the reaction at the hip (*R*) and wing lift (*L*). *a Archaeopteryx*. The long tail would have allowed the body weight to act along approximately the same line as the upward reaction from the hip joint. *b* Modern birds have lost their tail, and the body weight therefore acts ahead of the hip joint, tending to topple the body forward. This is resisted by the enormously enlarged synsacrum, which works as a lever. The posterior end of the synsacrum (ischium) is pulled towards the femur by tonic muscles (*T*), which holds up the anterior part of the body (*P*). *c* The late, tailless, pterosaurs (here a *Pterodactylus*) had no obvious adaptations which could have allowed moments to be balanced about the hip joint by tonic muscles in bipedal standing. A running take-off might have been possible if part of the weight was supported by wing lift. (Pennycuick 1986)

11.4 Feather Structure and Function

11.4.1 Main Structure

Birds have a weakly keratinized skin and their bodies are covered by insulating soft down feathers and heavily keratinized contour feathers. The wings and tail are made up by enlarged and stiffened contour feathers, and the lower legs and toes by horny scales. Feathers can be considered as complicated horny scales and there are structural and developmental evidence that feathers evolved from the epidermal scales of their reptilian ancestors. In modern birds, the feathers have two physiological functions, thermoregulation and flight, and the evolution of feathers may have been influenced by one and/or the other purpose. Other functions of the feathers also have been proposed (see further Chap. 13).

Feather structure is described in most textbooks on vertebrate morphology so only a brief description is given here. A typical feather consists of a stiff axial shaft (rachis) with hollow and semitransparent proximal quill (calamus) and a distal vane. The vane is composed of a series of parallel barbs on each side, and each barb bears two rows of small barbules. The barbules on the distal (outer) side bear hooklets which hold together the adjacent barbs, and the net result is a strong, light and flexible web covering the body and shaping the wings and tail. The wing feathers (remiges) are divided into two groups; the handwing feathers (primaries) attaching to the manus and the armwing feathers (secondaries) attaching to the lower arm. The first (anterior) primaries run along the wingspan axis while the secondaries run perpendicular to it.

The shape and size of the cortex (outer wall) of the feather shaft account for the majority of bending behaviour. The calamus is hollow and elliptical in cross section (largest diameter in the vertical plane), while the rachis is rectangular in cross section, because ridges of cortical material run along two thirds of the length of the interior dorsal surface. In the pigeon, the first primary is as stiff laterally as dorso-ventrally, whereas the other primaries are stiffer dorso-ventrally (= as stiff as the first primary) than laterally (Purlsow and Vincent 1978).

Black primary feathers show less wear than white ones (Averill 1923; Lee and Grant 1986), and Voitkevich (1966) suggested that the presence of melanin is associated with an increased amount of keratin. The importance of the black pigment for durability has been confirmed experimentally (Burtt 1979).

The anatomy of the feather attachment permits both primaries and secondaries to rotate in the nose-up sense through about 90° from the normal position (see Fig. 9.7b). A nose-down rotation of the vane is impossible without deformation (twisting) of the shaft. When the wing spreads, as in the downstroke, the feathers are hooked together by the interlocking barbs, and they form a streamlined, more or less cambered, surface that prevents air from penetrating it. During the upstroke, and particularly in slow flight and hovering, the feathers rotate as a rigid unit when the wing flexes and air spills through the wing with reduction of drag. The mechanism of this is described below.

11.4.2 Vane Asymmetry and Feather Curvature

The body feathers have symmetrical or almost symmetrical vanes, while strongly asymmetric vanes appear in the wing and tail feathers. The asymmetry is most pronounced in the primaries but is present in the secondaries and in all tail feathers except for the central pair. In the asymmetric feathers the anterior (or outer) vane is narrower than the posterior (or inner) one. The wing feathers are also bent, which is most pronounced in the secondaries.

The interaction between feather structure and dynamics is very complicated. R.Å. Norberg (1985) developed a mechanical and aerodynamic theory on the function of vane asymmetry and feather curvature, and showed that the structural asymmetry of bird flight feathers is essential for their bending and twisting behaviour throughout the wingbeat cycle. His theory explains how the angles of attack of the feathers are adjusted automatically throughout the wingbeat cycle, despite continuously varying directions and velocities of the relative wind. His theory is summarized here.

Upstroke. In the upstroke the relative wind meets the wing primarily on the dorsal side from obliquely in front at some negative angle of incidence (see Sect. 9.3.2). The direction of the relative wind varies along the wing and changes throughout the upstroke. Figure 11.20 shows the dorsal view of a pied flycatcher (*Ficedula hypoleuca*) with right wing extended laterally as in the middle of the downstroke. The broken lines of primary feather no. 4 and secondary feather no. 2 are the longitudinal axes through the feather base (calamus) at the attachment. The mechanics in the upstroke refers to the secondary feather.

The aerodynamic force F'_2, generated by the distal part of the feather, acts downwards through the chordwise centre of pressure C.P. The centre of pressure lies near the mid-chord point of the feather whenever the relative wind is more or less perpendicular to the wing plane. When the relative wind meets the leading edge at some angle less than $90°$, the centre of pressure moves ahead of the airfoil's mid-chord point and the force F'_2 acts with moment arm l_1 about the axis of rotation of the feather. An initial moment $F'_2 \times l_1$ therefore tends to rotate the feather in the nose-up sense. So the section C.P. moves along a circular arc about the axis of rotation through the feather base (bottom right in Fig. 11.20), and the entire feather rotates in the nose-up sense. The feather's angle of attack then becomes smaller than $90°$, resulting in a lift force component making the resultant aerodynamic force F''_2 rotate laterally (counter-clockwise in Fig. 11.20, lower right). The section C.P. moves forwards to a position between the mid-chord and quarter-chord points, and the moment arm changes from l_1 to l'_1. As the feather chord approaches a "feathered" position along the direction of the relative wind (the angle of attack becoming smaller), the resultant force F'_2 diminishes and the torsion moment vanishes.

When the feather has rotated into a downwind position relative to its axis of rotation, there is no longer any torsion arm l_1 upon which the aerodynamic drag can act, but some friction drag will remain. With this nose-up rotation of the flight feathers, air spills through the wing producing little or no useful aerodynamic forces so the upstroke is thus primarily a recovery stroke in slow flight and hovering; the downstroke produces the power.

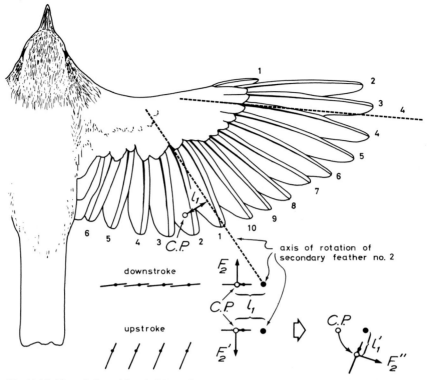

Fig. 11.20. Dorsal view of the pied flycatcher (*Ficedula hypoleuca*) with right wing extended as in the middle of the downstroke. The *broken lines* indicate the longitudinal axes for the fourth primary feather and second secondary feather through the feather bases (calamus) at their attachments to the wing. The wing feathers are not subjected to any aerodynamic load, but exhibit the curvature typical of the resting state. The *lower diagrams* show end-on views of secondary feathers as seen from behind in the plane of the upper, planform figure. *C.P.* is centre of pressure, F_2 is aerodynamic force generated during the downstroke, F_2' and F_2'' are forces generated during the upstroke, and l_1 and l_1' are moment arms. See text for further explanation (R.Å. Norberg 1985)

Downstroke. As in the upstroke, both the direction and magnitude of the resultant relative wind vary throughout the downstroke for every position along the wingspan (Fig. 9.6, Sect. 9.3.1). The most vertical direction of the relative wind is at the wingtip in the middle of the downstroke but since a wing operates with the most efficiency at a certain angle of incidence, there must be continuous adjustment of wing and/or feather twisting throughout the wingbeat cycle. The strong nose-down orientation of the tips of the first primaries in relation to the wing-base chord partly arises from their forward location in the anteroposteriorly cambered wing, the twisting of the entire wing, and the twisting of the separated feathers.

Referring to the same secondary as for the upstroke, the aerodynamic force F_2, generated during the downstroke, acts through the same chord-wise centre of pressure C.P. as force F_2' during the upstroke (Fig. 11.20). The moment arm l_1 about the feather's axis of rotation is the same, so there will be a moment $F_2 \times$

224

l_1 during the downstroke, which tends to rotate the feather in the nose-down sense.

Because of the feather shaft curvature and the overlap between adjacent feathers, the aerodynamic pressure on the vane of one feather is taken up largely by the nearest inner neighbour feather. Therefore, in their zone of overlap, adjacent feathers will be pressed together much more than if each feather were straight. A straight feather would carry its own aerodynamic load rather than much of its neighbour's load. The pressing together of adjacent feathers acts together with the hooklets to prevent the feathers from slipping apart forming a closed wing surface.

The mechanics of the first primary feathers differs strongly from those of the secondary feathers in the downstroke. The feather shafts of the first primaries are oriented about 90° to the direction of the relative wind and separated near their tips so each feather acts as an airfoil of its own. The wingtip separation is due partly to spreading of the feathers, but also to emargination of some feathers, both in their anterior and posterior vane. The overall aerodynamic function of separated primary feathers in bird wings is to increase lift by the leading-edge slat effect and possibly to reduce drag (see Sect. 11.5.1).

R.Å. Norberg (1985) treated three specific combinations of feather curvature and flexural stiffness in the downstroke mechanics:

1. A curved feather with equal flexural stiffness dorsoventrally and anteroposteriorly would bend in the direction of the aerodynamic force and the feather section would rotate in the nose-down sense. But high air velocities with ensuing strong aerodynamic forces may reduce the angle of attack below optimal values. This is refuted by the arrangement described below (2).

2. In a curved feather, which is more resistant to dorsoventral than to anteroposterior bending (as observed in the pigeon), the aerodynamic force on a distal feather section would result in feather bending so that the angle of attack may be maintained at near-optimal values despite large variations of the relative air speed. This mechanism tends towards self-adjustment of the angle of incidence at the outer, free parts of the anteriormost primary feathers, and the adjustment of the angle of incidence is dynamically stable. This feather anatomy is the one observed in real feathers.

3. In a straight, or near-straight feather which is more resistant to dorsoventral than to anteroposterior bending [the obvious alternative to (2)] the aerodynamic force would cause the feather to flip over in the nose-up sense, bringing the vane into a vertical attitude with the posterior vane downwards and aerodynamic disaster in the downstroke. So there is an urgent need for the first primary feathers to be curved if they are stiffer dorsoventrally than laterally, and they must be more resistant to dorsoventral bending than to lateral bending for the automatic maintenance of near-optimal angles of incidence, despite continuous variations of the direction and magnitude of the aerodynamic forces.

In summary, the inherent aeroelastic stability of the feather is achieved passively by an appropriate combination of the three main feather characteristics discussed above: (1) vane asymmetry, (2) the curved shaft, and (3) the greater flexural stiffness dorsoventrally than anteroposteriorly. Dorsoventral stiffness is achieved partly by a ventral furrow of the feather shaft.

11.4.3 Flight Feathers of *Archaeopteryx*

Advanced feather characteristics were present in *Archaeopteryx*, suggesting that it was adapted to flapping, powered flight. Feduccia and Tordoff (1979) were the first to draw attention to vane asymmetry in *Archaeopteryx*, and the first-discovered flight feather is asymmetrical and the feathers of the London and the Berlin specimens are about as asymmetrical as those in modern birds. Backward curvature also occurred in their flight feathers. According to Rietschel (1985) there is also a longitudinal furrow along the ventral side of flight feathers in *Archaeopteryx*, indicating they may have been stiffer dorsoventrally than anteroposteriorly, as in modern birds. These characters which are essential for flapping flight have beneficial effects also for gliding, so alone they cannot be used to conclude that *Archaeopteryx* was only a glider or capable of flapping flight as well. R.Å. Norberg (1985) noted, however, that it seems unlikely for flight feathers to have been so strongly curved in a purely gliding animal. The leading edge feathers in *Archaeopteryx* show the curvature that is a prerequisite for the automatic adjustment of angles of incidence at the primaries throughout the downstroke, provided that they separate. Four primary feathers of increasing length made up the leading edge in *Archaeopteryx*, an ideal arrangement for their separation towards the tip to function as leading-edge slats. Such slats are high-lift devices, important at take-off when the speed is so low that most of the relative air speed must be achieved by flapping. So, the striking similarity in feather structure between *Archaeopteryx* and modern birds strongly suggests that *Archaeopteryx* used flapping flight, and these features provide no grounds for rejecting the hypothesis that it flew with true powered flight (R.Å. Norberg 1985).

11.4.4 Silent Flight

The wings of flying birds generate some aerodynamic noise which can be disadvantageous to birds of prey, such as owls, which hunt by ear and whose prey have acute hearing. Owls have achieved a silent flight by three features in feather structure, that were first described by Graham (1934). First, a stiff comb-like fringe on the front margin of the leading edge feathers of the handwing reduces noise by affecting the pressure distribution in the boundary layer behind the leading edge. Second, soft, fringed feather margins at the trailing edge of the wing probably suppress trailing edge noises, for the barbs in the fringe are very flexible and are free to separate. Third, the soft downy covering of the upper surface of the wing feathers makes the feathers slide soundlessly on one another when the wing extends and flexes. Graham (1934) recognized a fourth flight noise reducing character in owls, namely their low wing loading that enables them to maintain small angles of attack and low speeds at the distal wing parts throughout the entire wingstroke.

Kroeger et al. (1972) measured the flight noise of owls and concluded that both the leading-edge comb and the soft downy compliant surfaces of the wing feathers were the main mechanisms for reducing the noise generated by turbulence in the boundary layer and by unsteady lift (cf. Sect. 11.5.3). Thorpe and

Griffin (1962) showed that the flight of owls is silenced not only in the frequency range audible to man but also in the ultrasonic range above 15,000 Hz, where small mammals are sensitive. Fishing owls lack the structural characteristics associated with silent flight and also generate flight noise (Graham 1934; Thorpe and Griffin 1962).

Watson (1973) and Blick et al. (1975) tested the effect of leading-edge barbs in the great horned owl (*Bubo virginianus*) on lift and drag characteristics of wings. They inserted the leading-edge barbs on model-airplane plywood and tested various combinations of spacings, widths and chord lengths of the barbs, and found that the effect of the barbs on stall is dramatic because they keep the flow attached near the leading edge even at large angles of attack and so prevent the wings from complete stalling. Instead of a sharp drop in lift at the stall angle, the lift coefficient remained more or less constant. The barbs tended to increase drag slightly at low angles of attack but they decreased drag above the original stall angle.

11.5 Wing Adaptations Enhancing Flight Performance

Animal wings must have special properties to produce enough lift at low flying speeds, but at one flight speed a particular wing profile is suitable and at another speed a different profile is better. The wings of birds and bats are moveable in different ways permitting the animal to change the planform, geometry, and aerodynamic characteristics of the wing to control the motion, or to improve the flight performance in some desired manner.

11.5.1 Wing Camber

Bats have the ability to vary the camber (anteroposterior curvature) of the wing, and in this respect they surpass the birds. An increase of the camber causes an increase of the lift coefficient (but along with an increase of the drag coefficient). In birds, slight camber is automatically produced by the feathers when the wings extend, while in bats, camber is produced mainly by downward pull of the first and fifth digits, and the downward inclination of the unit formed by the second and third digits (dactylopatagium minus). When the thumb is lowered it inclines the membrane parts anterior to the arm (propatagium) and second digit (dactylopatagium brevis) forward-downward. Contraction of the occipito-pollicalis muscle, which extends along the leading edge of these patagia, keep them taut. Bat wings, particularly of those species with wings of low aspect ratio, are aerofoils of high camber, which is efficient at low-speed flight. Broad-winged bats (long chord, low aspect ratio), such as members of the Megachiroptera, usually have relatively broader patagia anterior to the arm and third digit than long- and narrow-winged bats. In general, megachiropteran bats also have shorter metacarpals in relation to the total length of the digits than have microchiropteran bats, so the metacarpophalangeal joint of the fifth digit is situated more proximally on the digit. This wing structure of megachiropteran bats enhances their ability to camber the wing strongly. The relatively short metacarpals also contribute to giving a small handwing area meeting the airstream during the later part of the upstroke to reduce drag, when the third, fourth and fifth digits are flexed (Fig. 11.21a). The short metacarpals of megachiropteran bats have been suggested to be a primitive, non-adaptive character separating Megachiroptera from Microchiroptera (Pettigrew and Jamieson 1987), but it may as well be a highly adaptive character that evolved in larger bats with broad wings (as discussed above).

11.5.2 Wing Flaps

Leading edge flaps can delay stalling, and have been developed for airplane wings as high-lift devices. These flaps keep the flow laminar over the wing at higher angles of attack, permitting higher lift coefficients without flow separation (Figs. 11.21b and 11.22a-c), particularly for thin section wings with sharp leading edges (Abbott and von Doenhoff 1949). In bats the patagia (propatagium and dactylopatagii brevis and minus) together may function as a leading edge flap when lowered by the thumb and the second digit. These patagia are especially large (broad) in slow-flying bats with low aspect ratios (Fig. 11.22) and the efficiency of leading edge flaps increases with decreasing thickness of the leading edge. The leading edge is very thin in bats; in *Rousettus aegyptiacus* (body mass = 0.14 kg, wingspan = 0.57 m) it is about 0.5 mm thick at the armwing and about 0.1 mm at the handwing. The leading edge membrane and the pteroid bone in pterosaurs probably have functioned in the same way.

The flight feather layers and wing coverts of a bird wing form a *split-flap* configuration in many birds in slow flight, such as in landing (see Fig. 9.17a). The raised feathers prevent backflow of the turbulent air (forward flow along the wing's upper surface) and so suppress stalling, and this configuration generates high lift but also high drag (von Holst and Küchemann 1941). Multi-layered feathers can act also as a primitive *"blown-flap"* arrangement so that in extreme manoeuvres when the wing has a very large angle of incidence ($> 40°$) the feathers can be spread to provide a large surface area, and there can be a considerable flow through the wing. The high-pressure air flows up through the downy under-surface and when it is deflected by the flight feathers and forced out essentially parallel to the upper surface it tends to counteract the stalling tendency (Kuethe 1975).

11.5.3 Turbulence Generators

For a simple nonpermeable wing the maximum lift coefficient obtainable is between 1 and 2. When a critical stalling angle of attack is exceeded, the airstream separates from the wing's upper surface with a sudden fall of C_L and increase of C_D. But the stalling can be delayed and C_L increased up to a value of about 3,

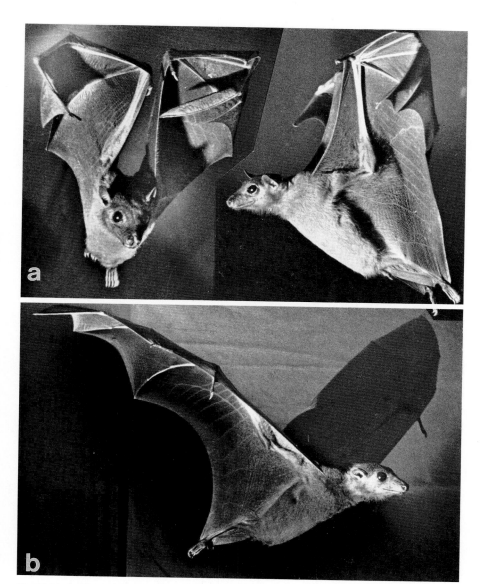

Fig. 11.21a,b. The megachiropteran bat *Rousettus aegyptiacus. a* The phalanges of digits three to five are strongly flexed to reduce drag during the later part of the upstroke. (Norberg 1972b; by courtesy of Springer-Verlag) *b* The membrane parts anterior to the arm and third digit form a leading-edge flap lowered by the thumb and second digit and by pronation of the arm. Photographs by the author. (Norberg 1972a)

229

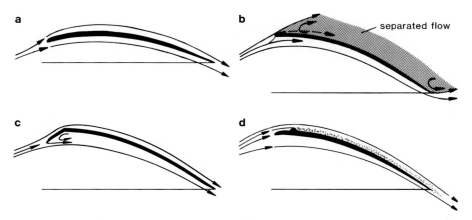

Fig. 11.22. a Laminar flow over a wing profile. *b* Airflow separation over the upper surface of the wing. *c* Leading-edge flap, keeping the airflow attached to the surface. *d* Transition from laminar to turbulent flow of the boundary layer behind a protruding structure, a turbulence generator. (Norberg 1985a; by courtesy of Harvard University Press)

which is beneficial particularly at slow speeds when the accompanying increase in drag has less effect than at faster speeds.

Very near the wing surface, the air movement is retarded due to friction, and this very thin layer (the boundary layer) can be either laminar, turbulent, or laminar anteriorly and turbulent posteriorly on the wing. One way to delay stalling is to make this boundary layer turbulent. Every pressure increase in the direction of flow is unfavourable for keeping the boundary layer laminar, especially at high Reynolds numbers. The transition from laminar to turbulent boundary layer flow occurs at the upper surface of the wing approximately at the location of the pressure peak (at about the highest point above the wing chord). But if the pressure peak is large enough, the laminar flow separates from the surface permanently so that the wing is stalling as the lift force rapidly declines. In a very interesting aerodynamic range (when the Reynolds number is less than about 10^5; Schmitz 1960) there is a critical value of the Reynolds number, Re_{crit}, below which (subcritical Re) the laminar boundary layer is very stable, and above which (supercritical Re) there is a transition from laminar to turbulent flow at the pressure peak. The lower the Reynolds number below critical, the more stable is the laminarity of the boundary layer. With a turbulent boundary layer there is a constant interchange of momentum between the rapid outer layers and the slow inner layers, so the layers close to the surface receive kinetic energy from the free external flow. For this reason, a turbulent boundary layer is much better than a laminar one to flow away backward against the pressure increase at the rear of the wing. Thus, a turbulent boundary layer can remain attached to the surface at higher angles of attack than a laminar boundary layer without wing stalling.

When the Reynolds number lies below the critical value, which is different for different profiles, the aerodynamic lift coefficient can be improved by

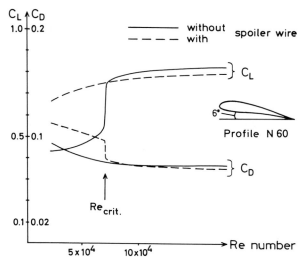

Fig. 11.23. Diagram showing the relations between the lift and drag coefficients with and without induced turbulence of the boundary layer of a particular wing profile. When the Reynolds number becomes less than the critical value, below which the boundary layer is laminar, the lift coefficient falls rapidly. When the boundary layer is made turbulent by a spoiler wire (and, in animals, by the arm, digits and, sometimes, concentration of hairs), the decrease of the lift coefficient below Re_{crit} is slight. Norberg 1972a, after Schmitz 1960)

induced turbulence of the boundary layer (Fig. 11.22d). Schmitz (1960) illustrated the variation of the lift and drag coefficients versus the Reynolds number with and without a spoiler wire (which acts as a turbulence generator near the leading edge of the wing) on a particular wing profile (Fig. 11.23). When the Reynolds number becomes less than the critical value (in the above example, $Re_{crit} = 6.3 \times 10^4$), the lift coefficient falls rapidly by about 50%, even when the wing has not stalled. When turbulence is established artificially (by a spoiler wire) in the range below the critical Reynolds number, the fall of C_L is very slight.

The Reynolds numbers of bird and bat wings lie in this interesting range. In the long-eared bat (*Plecotus auritus*) the Reynolds number is about 2×10^4 at the middle of the wing in flapping flight at a speed of 8.5 m s⁻¹, and in the fruit-bat *Rousettus aegyptiacus* the corresponding value is 7.8×10^4 at $V = 10$ m s⁻¹. To obtain a higher lift coefficient without stalling in birds and bats, a subcritical Reynolds number can be increased to the critical value, and/or boundary layer turbulence can be induced by morphological features. In manoeuvering and circling flight, when the flight speed has to be low, a larger chord would promote a higher Reynolds number, and a relatively rough surface would produce boundary layer turbulence and both would contribute to a high maximum lift coefficient.

In bats, projection of the digits and arm above the upper wing surface, and concentration of hairs on the skin at the arm, roughening the upper wing surface near the leading edge, may act as turbulence generators of the boundary layer

231

(Pennycuick 1971c; Norberg 1972b). In many microchiropteran bats the digits project more on the upper surface of the wing than on the lower surface, and the projections may be very sharp, which may have evolved to roughen the upper surface of the wing. Vaughan and Bateman (1980) compared the wing surfaces of slow-flying bats with those of fast-flying molossids and they found that the cross-sectional shape of the forearm in broad-winged slow-flying bats is almost rounded and the arm projects largely dorsally, while it is dorsoventrally flattened contributing to a smooth profile in the fast-flying molossid bats.

Molossids have unusually thick and leathery membranes that, when extended, retain a greater degree of corrugation by the elastic fibres. Vaughan and Bateman (1980) suggested that because the corrugation tends to be perpendicular to the long axis of the wing it may reduce the loss of lift due to wingtip vortices by limiting the span-wise movements of air towards the tip.

11.5.4 Wing Slots

A delay in stall as the angle of attack increases can be achieved by wing slots in birds (lifting of the alula and separation of wingtip feathers). The function of split alula (first described by Graham 1932 and later by Nachtigall and Kempf 1971) is to permit through-wing suction and so to prevent flow separation at the upper surface of the wing parts behind the slot (Fig. 11.24a). The slotted wingtips might not work as a high-lift device in this way, because the distance between the separated feathers is too large (especially at their distal parts) and the flow separation starts at the inner parts of the wings with rounded tips (Hummel 1980).

One function of separated wingtip feathers is to reduce the drag of the wings (Graham 1932) but splitting the wingtips leads to an increase of wing profile drag

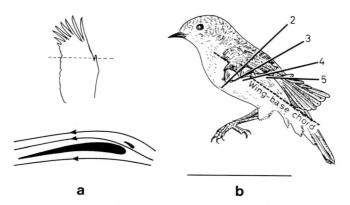

a   **b**

Fig. 11.24. a The alula forms a leading-edge slat at the wrist. *b* The multi-slotted wingtip feathers act as separate aerofoils. The full lines show the chord orientations of the individual feathers number *2* to *5* in their separated region near the tip, which have quite different inclinations from that of the wing-base chord (*broken line*). (*b* is modified from R.Å. Norberg 1985)

(the friction drag component). The friction drag coefficient depends on the Reynolds number ($C_{Df} \propto (Re)_{-1}{}^{\mu}{}_2$; cf. Sect. 2.3.1), which is smaller for the separate feathers because of the smaller chords, but wingtip splitting also leads to an increase of the induced drag if the tip feathers all lie in the same plane (Newman 1958). However, if the split tips are staggered in height, the induced drag will be reduced and this effect has been described by Prandtl (1923) and later by Hummel (1980). Cone (1962a,b) suggested that the bent primary feathers may act as endplates, inhibiting flow of air around the wingtip and reducing the effect of the shed vorticity. The first primary is bent upwards and forwards by aero-dynamic forces, whereas the bending becomes progressively reduced in the other primaries which are partly depressed by the partly overlapping feather in front. The tip feathers of birds with slotted tips are emarginated to reduce feather area and increase the slot widths.

The effects of staggered winglets at the wingtips have been tested in wind tunnels by the College of Aeronautics at Cranfield, and during a series of full-scale trials with Cranfield Institute of Technology research aircraft (reviewed in Hofton 1978). The greatest reductions in induced drag were achieved with three or four short winglets, where the anteriormost one pointed slightly upwards and the posteriormost one slightly downwards. Each winglet had a length of 41% and a root chord of 16% of the wing chord just proximal to the winglets. Reductions in induced drag of up to 29% were measured during take-off and on the landing approach, while the lift to drag ratio increased from 12.5 to 15.8.

However, this decrease of induced drag may not have any resulting improvement in slow-flight performance in smaller birds up to thrush size, since the Reynolds number of the flow past their separated primaries may be low enough to cause an important increase of friction drag (Oehme 1977). A moderate reduction of the total drag and an increase of the overall L/D may thus occur only in larger birds.

A more important function of separated primaries may be that the feathers act as individual aerofoils (Kokshaysky 1977; Oehme 1977), each producing lift and each twistable individually in the nose-down direction under aerodynamic load (Fig. 11.24b). As a result of the increasingly larger flapping velocity from base to tip of the wing, the air meets the wing increasingly more from below towards the tip in the downstroke. Since the chords of the separated feathers are much shorter than those of the unsplit wing, the Reynolds numbers of the primaries are smaller, perhaps below the critical value, for a longer time than in the unsplit wing. The separated primaries can be twisted more than the entire wingtip if the feathers were not separated, so higher local angles of attack occur at the inner parts of the wing before the outer parts reach stalling angles. Separated primaries, therefore, could increase the average lift coefficient of the entire wing.

A nose-down pitch moment of a primary feather results if its chordwise aerodynamic center of pressure is located behind the pitch (torsion) axis. As mentioned in Section 11.4.2, this is achieved by the asymmetrical location of the rachis near the leading edge of the primaries, and the primaries nearest to the leading edge are the most asymmetrical. The feather curvature results in the torsion axis of the feather base being located far behind local chordwise positions of centers of pressure.

The main benefit of the wingtip slotting may be a combination of increased lift and reduced induced drag. The same drag-reducing effect that can be obtained with the split, staggered primaries can be achieved by a slight enlargement of the wingspan, e.g. swifts, swallows, and seabirds all have long, high-aspect-ratio wings with pointed wingtips. Birds, that are limited in wingspan use the system of split, staggered wingtips, e.g. medium and large-sized birds, such as crows, hawks, bustards, and pheasants with rather short wings. Many soaring birds have both long wings and split tips (e.g. cranes, storks, pelicans), and they may achieve considerable savings in induced drag. These birds use reduced wingspan and closed tips during high-speed gliding. Bats and pterosaurs lack all kinds of wing slots.

Theoretical and experimental investigations have shown that wingtip slots also increase longitudinal stability as a planform effect, and this may be used by birds (Hummel 1980). This is an effect of the considerable rearward shift of the local aerodynamic centre in the slotted wingtip-region.

11.5.5 Energy-Saving Elastic Systems

Inertial power is considered to be small at medium and fast flight speeds, but of some importance during hovering and slow flight. One uncertainty is the extent to which inertial power is converted into useful aerodynamic power in flapping flight. Work must be done to accelerate the wing at the beginning of the downstroke, but at the lower reversal point in fast flight the kinetic energy of the wing can be transferred to the air and provide lift. In hovering and in very slow flight this transfer of energy is not as easily achieved because the relative airspeed at the turning points is very small, resulting from the small or absent forward speed component. Hence, the loss of inertial power should be important in hovering and in very slow flight, unless kinetic energy can be removed and stored by some other means.

Insects have a highly elastic, enery-saving mechanism at the wing hinge derived from the sclerotized cuticle of the thorax and the rubberlike ligaments whose main component is the protein resilin (Weis-Fogh 1960, 1965). Without this elastic system, many insects would not be able to produce the power necessary to fly. Vertebrates lack a comparable elastic wing-hinge system but possess structures that may be energy-saving.

In birds there are some possibilities for energy storage. Schaefer (1975) suggested that the clavicle may store and release energy at the top and bottom of the wingstroke. High-speed X-ray movies of flying starling in a wind tunnel demonstrate that the U-shaped furcula acts as a spring (Jenkins et al. 1988; Goslow et al. 1989; see Sect. 11.3.2) bending laterally during the downstroke, and recoiling during the upstroke. The distance between the dorsal ends reached 18.7 mm during the downstroke, an increase of 47% over the resting distance (12.3 mm), while by the end of the upstroke the distance was 13.5 mm (Fig. 11.25). Jenkins et al. determined that the force required to bend fresh excised starling

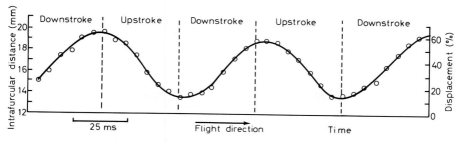

Fig. 11.25. Distance between the dorsal ends of the clavicles (furcula) during two wingbeats of *Sturnus vulgaris* in level flight. The *right scale* illustrates the percentage of displacement relative to resting distance. (Slightly modified from Goslow et al. 1989)

furculae through their normal range of excursion is 0.6 to 0.8 N, rising almost linearly with increasing excursion. They suggest, however, that the main function of the furcular bending may be to assist the bird's breathing (see Sect. 11.3.2), and whether or not the furcula can store and release this energy or, instead, if the lateral bending costs extra energy remains unknown.

The clavicles of some birds may not act in this manner. In soaring birds, such as raptors, the furcula probably is too robust and too much less compliant, while in barbets, toucans and some parrots the clavicles are unfused. In flightless birds the clavicles are either vestigial or absent. The condition observed in the starling may be common to most birds using powerful flapping flight. Jenkins et al. suggested that it is unlikely that the boomerang-shaped furcula of *Archaeopteryx* acted as a spring because its transversely flattened clavicles would resist bending in the desired direction.

Pennycuick and Lock (1976) described a mechanism by which the primary feathers might increase the efficiency of transfer of the wing's kinetic energy to the air toward the end of the downstroke when the feathers unbend (after being bent by aerodynamic loads in the beginning of the stroke). They proposed that although the mechanism exists it cannot transfer all the wing's kinetic energy to the air in hovering, so the mechanism would become fully effective only at some forward speed.

The wing membranes of bats (and possibly of pterosaurs) are highly elastic structures containing the protein elastin. Furthermore, in most bats the tips of the third, fourth, and fifth digits are cartilagineous and flexible, as is the tail tip of some species (such as species of family Nycteridae) with a large tail membrane. These features might absorb kinetic energy and then release elastic energy at the top and bottom of the wingstroke.

11.6 Tail and Feet

The Tail. Although the tail in birds may be used for longitudinal or directional control, particularly in rapid manoeuvres, the principal function appears to be analogous to that of trailing edge flaps on airplane wings (Pennycuick 1972a). At very low speeds, especially at take-off and landing, the tail is typically spread and depressed, increasing the supporting area and thus the lift. The action of the tail also helps to suck air downwards over the wing base, increasing the maximum lift coefficient of the wing itself. The effect of this was demonstrated by Pennycuick (1968a) in the pigeon gliding in a wind-tunnel. Birds with long forked tails, such as some swifts, swallows, terns, frigate birds, and the fork-tailed falcon (*Elanoides furcatus*), use the spread tail as a long flap on each side to help deflect air over the main wing surface and keep the flow attached at higher angles of attack than otherwise would be possible (von Holst and Küchemann 1941; Pennycuick 1975). The two tail halves act as aspirators, accelerating the flow near the trailing edge so that stall is delayed with no significant increase in drag (Kuethe 1975). Such birds are specialized for very slow flight and manoeuvrability, although their inexpensive flight also allows them to fly rapidly.

In bats the tail membrane is connected to the wing membrane via the legs, providing a supplementary wing area. In both bats and birds the tail is moved up and down synchronously with the wings in slow flight and in hovering. The tail membrane is large in most microchiropteran bats, but consists of only small flaps along the legs in Megachiroptera and most fruit bats within Microchiroptera. The inclusion of the hind legs in the wing and tail membranes adds greatly to the control of stability, wing camber and wing twisting. Species with particularly large tail membranes, such as megadermatids, nycterids and some vespertilionids, are specialized for hovering and slow manoeuvrable flight.

The long-tailed rhamphorhynch pterosaurs may have used their enlarged tail-tips for steering, while the smaller tail flaps at the legs in both rhamphorhynchs and pterodactyls functioned as auxilary membrane surfaces.

The Feet. Birds usually carry their feet near the body in flight so that they generate little or no drag and those with reasonably large feet also use them as airbrakes by lowering and spreading them. Lowering the feet decreases the lift/drag ratio (by increasing the parasite drag) and steepens the angle of glide and it is used by birds before landing, when trying to soar at a constant height in strong lift, and sometimes (auks) large webbed feet are used as auxilary gliding areas when stretched backwards at either side of the tail (Pennycuick 1968a, 1971b, 1975).

Chapter 12

Flight and Ecology

12.1 Introduction

The combination of morphological, ecological and behavioural attributes most benefiting a flying animal is related to the type of habitat it lives in and to its way of exploiting it (Fig. 12.1). The animals are adapted to occupy different niches, so they often have to fly in different ways associated with different wing morphology. Some birds and bats hawk insects in the air, in open spaces or within vegetation, while others fly continuously during foraging and still others perch between foraging bouts. Some birds forage in trees, bushes or on the ground, by climbing, hanging or walking, while some achieve low searching costs by soaring over large areas in search of food; still others use high cost hovering in front of flowers to drink nectar. Widely differing wing and leg structures are required for minimum cost of transport and for different locomotion types (e.g. U.M. Norberg 1979, 1981a,b, 1986a; Norberg and Rayner 1987; Rayner 1988).

The observed combination of morphological, behavioural and ecological traits is usually assumed to represent a near-optimal solution maximizing the fitness of the individual. Optimal foraging models are often built to maximize net energy intake or to minimize foraging time, optimization criteria substituted for

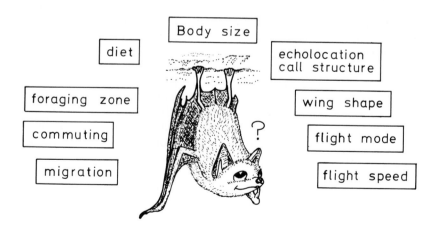

What combination?

Fig. 12.1. What combination of ecological, behavioural and morphological factors an animal should have depends on the type of habitat it lives in and its way of exploiting the habitat

fitness (e.g. R.Å. Norberg 1977, 1981b, 1983; Pyke et al. 1977; Krebs 1978), so part of optimal foraging theory is relevant to understanding the morphology of ecologically important structures (ecological morphology). The structure and function of the locomotor organs are closely related to the way an animal exploits its environment.

12.2 Predictions on Wing Shape and Flight Behaviour

Certain combinations of wing loadings and aspect ratios permit a flying animal to exhibit only certain kinds of behaviour and ecology and we can predict the predominant flight mode and flight performance of an animal from body mass, wingspan and wing area.

Since flight is very expensive, there should be strong selection to minimize the mechanical power required to fly. Low total power is attainable with a high aspect ratio particularly when this is combined with a low body mass and low wing loading, for this suite of characters permits slow flight with low profile and parasite powers. So slow-flying species should have long, high-aspect-ratio wings (of large area to enable slow flight), while fast-flying ones benefit from short wings (small wing area to reduce profile power), of high aspect ratio (Norberg 1987).

Pursuit of flying insects requires high manoeuvrability (the ability to make swift rolls and tight turns) as well as enduring and fairly slow flight to minimize distance travelled during the reaction time, from prey detection to initiation of a roll (R.Å. Norberg, personal communication). Slow flight and tight turning radius are achieved by a low wing loading (Sects. 9.3 and 9.7.1).

Migratory species should have wings of high aspect ratio for enduring flight (Pennycuick 1975), and their wings should be rather short (small wing area, high speed) if time is an important factor, but long if it is not. Long-distance migrants should benefit from long, narrow wings more than those migrating shorter distances.

Flying within vegetation puts demands on slow flight and short wings that have to be broad (long chord) to compensate for their shortness and give enough area to allow slow flight, a set of features conflicting with low power. So hovering and slow flight within vegetation present selection forces: towards long wings for low induced power but towards short wings for manoeuvrability among clutter. The wings are short or of average length in most birds and bats foraging within vegetation.

Species taking heavy prey should have a large wing area (low wing loading) so that they can carry the extra weight. This also is important for lactating bats that sometimes carry their young in flight.

The combination of aspect ratio and a wing-loading index that is independent of body mass can reveal a pattern for the understanding of the relationship between wing shape and locomotion modes in birds and bats from various families and with different foraging strategies. For scaling reasons, large birds and bats have higher wing loadings than have smaller ones, so for comparing species of different sizes a measure of relative wing loading must be used. For many birds the data points for wing loading versus body mass lie approximately along a straight line of slope $1/3$ (Lighthill 1977), as expected for geometrically similar species, although this is not always the case for species within the various families or orders of flying animals (Greenewalt 1962, 1975; U.M. Norberg 1981b; Norberg and Rayner 1987). Still, it would be convenient to use the slope for geometric similarity to calculate the relative wing loading for birds as $(Mg/S)/M^{1/3}$, which remains constant irrespective of the animal size, provided that they are geometrically similar. Aspect ratio is non-dimensional and independent of body mass for geometrically similar birds and can be used for direct comparison between differently sized species. Any deviation from the rule of geometric similarity in both characters may represent adaptations to different foraging behaviours.

Norberg and Rayner (1987) used a Principal Components Analysis for a similar investigation of bat-wing shapes and used body mass, wingspan and wing area to obtain indices for wing loading (PC 2) and aspect ratio (PC 3). This procedure gives a result comparable to the one described above, but with the disadvantage that the principal components are not quite independent of body mass (see Sect. 12.6) although they correct for deviations from geometric similarity within bat groups. In the method described above the birds are assumed to be geometrically similar.

12.3 Wing Design in Birds

Bird wing morphology as related to flight pattern and foraging behaviour has been considered by many workers (e.g. Greenewalt 1962, 1975; Pennycuick 1975; Norberg 1979; Andersson and Norberg 1981; Rayner 1988; U.M. Norberg and R.Å. Norberg 1989). Figure 12.2 shows the variation in aspect ratio and relative wing loading for 141 extant bird species selected from various families with different foraging strategies and choice of food. The flightless *Gallirallus australis* is included for comparison. Aspect ratio varies across the species with a factor of about 4 (ranging from 4.4 to 17.2) and relative wing loading with a factor of about 7 (from 26.8 to 194.1 in flying species, but it is 270 in the flightless one). Different foraging groups, as defined in Table 12.1, are encircled in Fig. 12.2, and since each such group may consist of different taxa, it does not necessarily represent a monophyletic, systematic entity. The various foraging categories are characterized by a more or less limited range of wing loading and aspect ratio, reflecting specific flight modes. The relationships among foraging strategy, flight performance and wing shape of the various birds are given in Table 12.1, and the grouping below refers to Table 12.1 and to Fig. 12.2. The groupings are arbitrarily made on the basis of wingshape and foraging mode, and are used here only to facilitate the understanding of relationships between morphology and ecology.

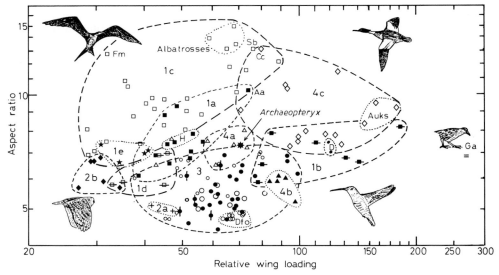

Fig. 12.2. Aspect ratio (b^2/S) versus relative wing loading $(Mg/S)/M^{1/3}$ in various birds. The *encircled* foraging groups are defined solely by the similarity of the flight and foraging modes of their members, not by systematic affinity. Each group may contain several systematic entities, and a systematic group may have representatives distributed into several of the *encircled* foraging groups. The numbers refer to the foraging groups in Table 12.1: ■, 1a; ▬, 1b; □, 1c; ⊟, 1d; ★, 1e; +, 2a; ♦, 2b; ◆, 3a; ●, 3b; ○ ○ ○, 3c; △, 4a; ▲, 4b; ◊, 4c; *Aa* = *Apus apus*; *Cc* = *Cygnus cygnus*; *D* = diving petrels; *Df* = Darwin's finches; *Fm* = *Fregata magnificens*; *Ga* = *Gallirallus australis*; *H* = hirundines; *Sb* = *Sula bassana*. Data from sources compiled in U.M. Norberg and R.Å. Norberg (1989) and from Pennycuick (1987a). (After U.M. Norberg and R.Å. Norberg 1989)

12.3.1 Continuous Foraging Flights

Birds with long wings and high aspect ratios have low wing loadings, and their flight is slow and inexpensive (long wings reduce the induced power which is large in slow flight). These birds use continuous foraging flights in open spaces and included are many seabirds and the swifts and the swallows (groups 1a and 1c in Table 12.1 and Fig. 12.2). Many seabirds and birds of prey use soaring flight, which further reduces the flight costs. Their low wing loadings give low sinking speeds during gliding and permit long periods of soaring (Pennycuick 1975). The long soaring wings of albatrosses, frigate birds and gulls (group 1c) are adapted to inexpensive flight, while the shorter and broader soaring wings (of low aspect ratios) of vultures, buzzards, storks and eagles (groups 1d and 1e) provide still lower wing loadings, enabling the birds to use narrow thermals (since they can make tight turns; Sect. 6.3.1) and to take off and land more easily (Pennycuick 1971a, 1975). For a given wing area, a long and narrow wing gives less induced power than a shorter and broader one, so the take-off requirements probably dictate the low aspect ratio in soaring land birds (Pennycuick 1975). Soaring in thermals is used during foraging and during migration. Terns, ospreys and harriers (group 1c) have low wing loadings and rather high aspect ratios and they

Table 12.1. Relationships between flight mode, wing characters and flight performance in various birds appearing in Fig. 12.2

Flight and foraging modes	Observed among	Observed wing characters	Flight performance predicted for the observed wing characters, based on aerodynamic theory
1. Continuous flight			
a) Cruising and hawking in open spaces	Swifts, hirundines, *Falco, Accipiter*	Long or average wingspan, high or rel. high aspect ratio, rel. high wing loading	Relatively slow and enduring flight (low flight costs), rel. high manoeuvrability
b) Hovering	Hummingbirds	Average to high aspect ratio, short wings, high wing loading	Fast enduring flight, high manoeuvrability
c) Crusing flight in search of prey or carcass on ground or in water in open spaces	*Diomedea, Milvus, Circus, Pandion, Larus, Sterna, Fregata, Pelecanus*	High or average aspect ratio, long wings, low or average wing loading	Slow enduring flight, soaring ability
d) Soaring or cruising flight in search of prey, often perching	Vultures, *Buteo*, eagles	low aspect ratio (broad wings), low wing loading	Slow flight, soaring ability
e) Foraging on foot on ground or shallow water, soaring during migration	Cranes, storks, herons	Low aspect ratio, low wing loading	Slow flight, soaring ability
2. Perching			
a) Perching and hawking within vegetation	Flycatchers	Low aspect ratio, short wings, low wing loading	Slow, highly manoeuvrable, but rather expensive flight
b) Perching and striking prey on the ground	Owls, many raptors	Low aspect ratio, low wing loading,	Slow, expensive silent flight, high load-carrying capacity
3. Locomotion among vegetation			
a) Short flights and climbing	Certhiidae, Sittidae, Picidae	Low wing loading, wings of average length, low aspect ratio	Slow, highly manoeuvrable, but expensive flight
b) Short flights and foraging in trees and bushes	Paridae, Sylvidae, Aegithalidae, Troglodytidae	Average wing loading low aspect ratio, short or rather short wings	Rather slow, manoeuvrable, but expensive flight

Table 12.1. (continued)

Flight and foraging modes	Observed among	Observed wing characters	Flight performance predicted for the observed wing characters, based on aerodynamic theory
c) Short flights and foraging on the ground and among vegetation	*Emberiza, Passer, Sturnus,* wagtails, Darwin's finches, corvids, thrushes,	Low (corvids) or average wing loading, low aspect ratio, average (large corvids, wagtails) or short wings	Slow, rather non-manoeuvrable flight
4. Foraging on ground or in water			
a) Ground foraging in open spaces	*Charadris, Calidris, Tringa, Limosa,*	Above average aspect ratio, rather high wing loading, rather short wings	Fast and relatively cheap flight (adaptations to commuting and migration)
b) Ground foraging within vegetation	Galliformes	Rather high aspect ratio, short wings, very high wing loading, large flight muscle mass	Fast flight, ability of vertical take-off
c) Foraging in open water	Alcidae, Anatidae, *Gavia, Mergus, Anser, Cygnus*	High aspect ratio, short wings, high wing loading small flight muscle mass	Rapid enduring flight (adaptations to migration and commuting), low manoeuvrability; most of these species need taxiing runs for taking-off

can use slow continuous and inexpensive flight during foraging while birds with short soaring wings and lower aspect ratios often perch between foraging bouts (see below).

Hummingbirds are exceptional, for they use more or less continuous flight (including long hovering sequences) within vegetation during foraging. But they also perch frequently. They also are exceptional in their wing stroke kinematics: during the upstroke (= backstroke) the fully extended and twisted, inverted, wings also produce lift, and together with their higher-than-average aspect ratios, making their flights between flowers and their hovering relatively inexpensive. Their pointed wingtips and extremely short armwings reduce the inertial power (U.M. Norberg 1981b), which may otherwise become large during hovering, and their small size is an advantage for flight among vegetation. Hummingbirds' high wing loadings might be an adaptation for rapid flights between foraging sites.

12.3.2 Perching

Birds flying near or within vegetation, such as many flycatchers and owls, are perchers with lower aspect ratios so their flight is more expensive than that of the open-space foragers. Flycatchers (group 2a in Table 12.1 and Fig. 12.2) hawk insects in the air and perch between foraging bouts and their small sizes and low wing loadings make their flight highly manoeuvrable and well adapted for insect-hawking (Norberg 1986a). Most owls (group 2b) forage in forests and have low aspect ratios and for these, continuous flight is more expensive than for birds of similar sizes but with higher aspect ratios although their low wing loadings somewhat reduce flight power (by making possible slow flight with low profile and parasite powers). Owls therefore mostly hunt from perches rather than from continuous flight and those hunting in open areas (such as *Tyto* and *Asio* species) usually have longer wings and higher aspect ratios than those hunting within vegetation, and use continuous flight more often. The low wing loadings in owls permit them to fly slowly and silently and to increase their load-carrying capacity.

Take-off and landing are costly and since birds with low aspect ratios should tend to search for food from perches rather than from continuous flight, they should make as few foraging flights as possible. Furthermore, to maximize the rate of net energy gain they should select the largest available prey. Insectivorous birds with very low aspect ratios do not usually fly much during foraging (see below).

12.3.3 Locomotion Among Vegetation

Foraging within vegetation is favoured by shorter wings, large wing area (broad wings of low aspect ratio) and low wing loading for slow manoeuvrable flight. The very low aspect ratios of many smaller birds that forage near or within vegetation mean high flight costs, so these species spend much of their foraging time walking, climbing, clinging and hanging, which puts different demands on leg structure. This category includes many of the smaller passerines, such as the Darwin's finches (of the genera *Geospiza*, *Platyspiza*, *Camarhynchus* and *Certhidea*) on the Galapagos Islands, which are very similar in wing shape (short, broad wings of extremely low aspect ratio) but differ in leg proportions (U.M. Norberg and R.Å. Norberg, unpublished, but see also Norberg 1979).

The values of aspect ratio (7.0) and relative wing loading (70) for *Archaeopteryx* (represented by a star in Fig. 12.2) are based on data from Yalden (1971). Although this bird had somewhat higher aspect ratio and relative wing loading than the average values for arboreal birds, it was not exceptional, and its wing form and wing characters suggest that it was capable of flapping flight. The various wing types of extant birds could have evolved from this form.

12.3.4 Migratory Species

Long wings are advantageous for migratory birds, where the goal is to cover as great a distance as possible on a given amount of energy (Pennycuick 1969). Birds that need both sustained and fast flight (to be able to reach a distant destination in moderate time) benefit from both a high aspect ratio and a high wing loading. Several migratory species, however, do not have to fly fast, and many of them have rather low wing loadings.

Many small riparian and forest species are migratory and although several have rather low aspect ratios and high flight costs they must store large amounts of fat before migratory flights (Pennycuick 1975). Most of the birds with very low aspect ratios (two thirds of the included species with AR< 6) are sedentary. Most of the birds shown in Fig. 12.2 (except for hummingbirds) are marked in Fig. 12.3 according to their migratory habits, and most of the seabirds and other large birds that make long migratory flights (covering several tens of degrees of latitude) have very high aspect ratios. Among the smaller passerines the difference between residents and migrants is not so clear, but there is a tendency for the birds that make short migratory flights (covering 10–20° latitude) to have higher aspect ratios than the sedentary ones, while the long-distance fliers are scattered in the diagram. This situation can be clarified by considering the mean aspect ratio and relative wing loading of various categories. The mean aspect ratio is 8.22 (SD = 3.03) and the mean relative wing loading 53.7 (SD = 16.1) for 28 long-distance migratory birds, and the corresponding measurements are 7.38 (SD = 1.71) and 71.5 (SD = 35.5) for 50 short-distance and partly migratory birds and 5.54 (SD = 0.95) and 59.5 (SD = 15.2) for 51 non-migratory birds. The long-distance fliers have low relative wing loadings, meaning that their optimal flight speeds during migration (maximum range speeds) should be higher than for the short-distance fliers.

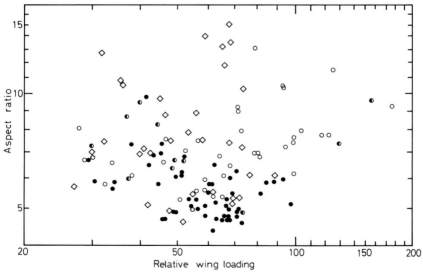

Fig. 12.3. Aspect ratio versus relative wing loading in long-distance migrants (*rhombs*), short-distance migrants (*open circles*), partly migratory birds (*half-filled circles*), and sedentary birds (*filled circles*). The birds are separated among those occurring in Fig. 12.2 (hummingbirds excluded). The majority of the long-distance migrants have a high aspect ratio (mean 8.2), whereas the majority of the sedentary species have a low aspect ratio (mean 5.4)

244

12.3.5 Foraging on Ground or in Water

Birds with high wing loading, short wings and high aspect ratio include loons, mergansers, geese, swans, ducks and auks (group 4c in Table 12.1 and Fig. 12.2). They are capable of fast and, for their speeds, rather economical flight, used for commuting and migration (short wings reduce profile power which is a large power component in fast flight). Among these birds, mergansers and auks have the highest wing loadings, and their flight muscle masses, and those of the loons, are small or relatively small (7–19% of body mass; cf. Magnan 1922; cited in Greenewalt 1962) and these birds must run (skitter) along the water to achieve the speed necessary for take-off (R.Å. Norberg and U.M. Norberg 1971; Fig. 9.16). Gallinaceous birds (group 4b) also have very high wing loadings but low aspect ratios and hence very expensive flight. Their flight muscle masses are usually large (22–29%; cf. Magnan 1922), however, and they can take off almost vertically and with high acceleration despite their short wings. The very short wingspan in these species, most of which are big, is an adaptation for flight within dense vegetation and for practical take-offs from the ground.

Most waders (group 4a) have above average aspect ratios and wing loadings and so achieve rather fast and inexpensive flight. They usually forage on the ground and their wings are adapted for migration and commuting.

Some birds use their wings not only for flight in air but also for propulsion, "flight", in water in pursuit of prey. These include the diving petrels (Pelecanoididae, Procellariiformes) in the southern oceans, the alcids (Alcidae, Charadriiformes) in the northern oceans, and the dippers (e.g. *Cinclus cinclus*, Cinclidae, Passeriformes) in running water. The optimal span and area of wings used for flight in air and in water is quite different from those used for flight only, or water only (e.g. penguins), because of the enormous difference in density between the two media. When flight is used in both media there must be compromise adaptations and when adaptation for "flight" in water is driven very far, the ability for flight in air can be lost; e.g. penguins in the southern hemisphere and the great auk (*Alca impennis*) in the northern hemisphere. To some extent, some of the birds flying in both media (such as *Cinclus cinclus*; Goodge 1959) can adjust their wings behaviourally to the higher density of water by keeping the wings only partly open during the wingbeat.

12.4 Wing Shape and Foraging Energetics of Hummingbirds at Different Altitudes

Hummingbirds are the most accomplished hoverers among birds and the most specialized nectar-feeders, and they exhibit a wide array of foraging behaviours. Some species are highly territorial and show interference competition (by agressive encounters), others are non-territorial and show exploitation competition (competitive utilization of resources), while others are intermediate. Feinsinger and Chaplin (1975) found that, in one community, territorial species had shorter wingspans than non-territorial species. A shorter span is better for manoeuvrability than a long one, and this is important for access in territory conflicts. Short wings, however, increase the power required for hovering flight, but ready access to abundant nectar supplies permits the territorial humming-birds to refuel quickly, saving energy by reducing hovering time; instead they spend more time for territory defence. Non-territorial hummingbirds forage among dispersed or nectar-poor flowers (Colwell 1973), and they spend much time in foraging flight (hovering), and hardly any time in defence of food. They apparently compete through exploitation (Wolf et al. 1976) and may therefore be subjected to stronger selection for long wings than are territorial hum-mingbirds, because long wings reduce the wing-disk loading ($Mg/S_d = 4Mg/\pi b^2$) and the power used for hovering [cf. Eq. 8.6].

Because air density decreases with increased elevation ($\rho = 1.22$ kg m^{-1} at sea level, 1.00 kg m^{-1} at 2000 m, and 0.66 kg m^{-1} at 6000 m), the power required for hovering by a given bird increases with increasing elevation. But some optimum wingspan must minimize the total energetic cost for hovering (induced power decreases with increasing wing length but inertial power should increase; hence some optimal length), and the wingspan should increase with the elevation so the power for hovering remains the same (Feinsinger et al. 1979). With increasing elevation, Feinsinger et al. suggested that the overall pattern would be one of sequential replacement of non-territorial species by other non-territorial ones with even lower wing-disk loading (longer wingspan), while territorial (short-winged) species would be continuously replaced by other territorial species with somewhat lower wing-disk loading, as the former could no longer obtain sufficient energy to produce the power required for hovering (Fig. 12.4). They tested the four interrelated predictions: (1) A given hummingbird species will become increasingly territorial at higher elevations because the power it uses for hovering increases with increasing elevation due to decreasing air density. At higher elevations the bird should reduce its hovering time. (2) The limits on power for hovering should remain substantially constant with elevation. (3) Mean wing-disk loading for hummingbirds as a group will decrease with increased elevation. (4) Mean power for hovering for hummingbirds as a group will not vary with elevation.

Feinsinger et al. found that *Colibri thalassinus* is non-territorial at an elevation of 1400 m, but territorial at 3000 m. But, as stated by the authors, the first prediction was suggested by these findings and cannot be used to support it. There are no other data to test the first prediction, but field data verify predictions (2)–(4) so there are interrelationships among behaviour, morphology and energetics among hummingbirds.

12.5 Wing Design of Species in a Pariform Guild

Several other bird species, and some bats, hover, but not as proficiently as hummingbirds and not with the same kinematics (see Sects. 8.1 and 8.2). The goldcrest (*Regulus regulus*) uses asymmetrical hovering and often gathers in

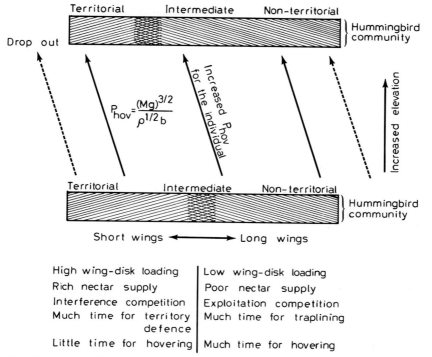

Mean P_{hov} constant for the community
Extremes of P_{hov} constant for the community
Mean wing-disk loading for the community decreases with elevation

Territorial	Intermediate	Non-territorial

) Hummingbird
) community

Drop out

$$P_{hov}=\frac{(Mg)^{3/2}}{\rho^{1/2}b}$$

Increased P_{hov} for the individual

Increased elevation

Territorial	Intermediate	Non-territorial

) Hummingbird
) community

Short wings ⟷ Long wings

High wing-disk loading	Low wing-disk loading
Rich nectar supply	Poor nectar supply
Interference competition	Exploitation competition
Much time for territory defence	Much time for traplining
Little time for hovering	Much time for hovering

Fig. 12.4. Hummingbird communities at different altitudes. For a given bird the power for hovering P_{hov} increases with increasing elevation because of decreasing air density ρ. A given hummingbird is therefore expected to reduce its hovering time and become more territorial the higher its elevational distribution is. The mean wingspan b for the entire community would be expected to increase and the wing-disk loading to decrease with increasing elevation, so that the mean power for hovering remains about the same at different altitudes. The model is based on Feinsinger et al. (1979). (U.M. Norberg 1981b; by courtesy of The Zoological Society of London)

mixed-species foraging flocks during the non-breeding season. These flocks often contain the coal tit (*Parus ater)*, the willow tit (*P. montanus)*, the crested tit (*P. cristatus*), and the treecreeper (*Certhia familiaris*). These species are sympatric over large regions of their distribution range, occur in the same habitat, and partly overlap in their choice of feeding station and food. Because they belong to the same foraging guild, they probably influence each other via interspecific competition (exploitation and interference competition) in an ecological, short time scale, and, therefore, also in the evolutionary perspective. Their co-adaptation may be expected to have involved some divergence in feeding behaviour and adaptive morphology and the use of different feeding stations (though overlapping) should produce different structural adaptations of wings and legs. The question is, how large are the differences that have evolved?

Figure 12.5 illustrates the feeding station selection in Norwegian spruce (*Picea abies*) by these birds. The goldcrest forages on the needled outer parts of the branches and is very good at manoeuvring and often hovers in front of, underneath, or among, the branches. It moves about within the mesh-work of subbranches, has a very agile food-searching behaviour, and uses its wings much more than the other species. Hovering ability and high manoeuvrability permit the goldcrest to find and collect food in places the other species cannot exploit as efficiently and economically. About 6% of its foraging time in autumn is spent hovering (R.Å. Norberg, personal communication).

The coal tit also uses the outer parts of branches, and both the goldcrest and the coal tit are more acrobatic than the other species. The coal tit also often hangs under branches and cones, while the crested tit mostly moves about with hops on main and secondary branches and on the ground. The willow tit also moves with hops on the branches, hangs under branches and cones, but also often clings to the trunk. The treecreeper climbs on the trunk in a vertical, head-up position, but sometimes also moves about on the main branches.

The goldcrest has several adaptations for slow, manoeuvrable and hovering flight, such as low mass, low wing loading and a short armwing, and it is superior in this respect to the other four species. The armwing should be short in hovering species because wingbeat frequency usually increases with decreasing flight speed as a compensation for the reduction of flight speed. The higher the wingbeat frequency, the larger the inertial loads on the wing skeleton and the larger the inertial power (power needed to oscillate the wings). Since angular accelerations are larger in hovering than in forward flight, the inertial loads on the wing skeleton are particularly high in hovering, so the supporting structures of the arm wing must be strong (= relatively heavy) if a certain load safety factor against breakage is to be maintained in a hovering animal. Therefore, it is important for a hovering animal to keep the wing's moment of inertia low by having a proximal location of the wing's mass. This may be achieved by reduced lengths of humerus and ulna and/or reduced angle between these bones. So the shorter the armwing is in relation to the total length of the wing, the less the inertial loads on the wing skeleton becomes and the lower the inertial power

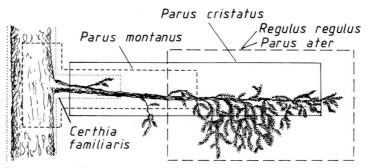

Fig. 12.5. Feeding station selection in Norway spruce (*Picea abies*) by three tits, the goldcrest and the treecreeper, which often gather in mixed-species foraging flocks during the non-breeding season. (Norberg 1979; by courtesy of the Royal Society, after Haftorn 1956)

(Norberg 1979). The handwing consists mainly of the primary feathers and has a low mass even when it is long.

We will now see how the wing morphology of the five birds is correlated to their respective flight behaviour. The differences in wing morphology are small but statistically significant in several respects (Norberg 1979). The goldcrest and the treecreeper both have shorter armwings in relation to the total length of the wings than any of the three tits (Table 12.2). Longer handwings compensate for the shortened armwings, but do not add much to the total wing mass.

The goldcrest is the lightest species, has low wing loading and the lowest wing-disk loading. Although the wingspan is not particularly long, the induced power during hovering is the lowest among the five species. Now, why is not the span longer to further reduce the power for hovering? We have seen that the induced power decreases with increasing wingspan, but the longer the wings the larger the inertial power and the less the manoeuvrability. Furthermore, long wings are not practical for flapping flight in dense vegetation, so the actual wing length is probably a near-optimal compromise between the following opposing tendencies: minimization of induced power (tending towards long span) and improvement of manoeuvrability and practicability among branches (tending towards short span).

The tree creeper should be rather good at hovering because it has very low wing and wing-disk loadings but it is a treetrunk forager adapted to vertical climbing (long tail, long toes, long curved claws, and proportionately short tibia; Norberg 1979; Norberg 1986). The bird climbs upwards on the treetrunk and then flies downwards to a lower level in the next tree and climbs upwards again and so saves energy during locomotion (R.Å. Norberg 1981a, 1983). A longer wingspan should be of little hindrance to a treecreeper foraging in this way compared to the tits, which fly among branches, and it has no need for particularly manoeuvrable flight. Furthermore, the treecreeper is partly migratory (as are the goldcrest and the coal tit) and so should benefit from long wings. These arguments may explain why opposing selection pressures balance at a relatively long wingspan in the treecreeper.

The coal tit has relatively low weight and long wings producing relatively low wing and wing-disk loadings and rather low induced power output for hovering.

Table 12.2. Body mass, wing characters of importance for hovering flight, and power and specific power for hovering in five coniferous forest passerines. (Norberg 1979)

	Regulus regulus	Certhia familiaris	Parus ater	Parus montanus	Parus cristatus
Body mass (g)	5.8 ± 0.11	9.1 ± 0.22	9.1 ± 0.08	10.9 ± 0.08	11.5 ± 0.37
Wing loading (N m^{-2})	12.9 ± 0.30	12.4 ± 0.60	14.2 ± 0.41	14.4 ± 0.22	16.2 ± 0.26
Wing disk loading (N m^{-2})	3.4 ± 0.07	3.5 ± 0.19	3.7 ± 0.10	4.0 ± 0.07	4.3 ± 0.15
Length of hand wing/length of arm wing	2.9 ± 0.09	3.0 ± 0.10	2.6 ± 0.10	2.3 ± 0.05	2.5 ± 0.11
Induced power P_{ind} (W)	0.07	0.10	0.11	0.14	0.15
Specific induced power P_{ind}/M (W kg^{-1})	1.19	1.13	1.21	1.28	1.31

It does not have especially short arm wings, and is adapted to slow, rather manoeuvrable, flight but not particularly to hovering. The willow tit has relatively large body mass, rather short span and high wing-disk loading, and it has the longest arm wing relative to the total wing length of the five species so it is less adapted to hovering and manoeuvrable flight. The crested tit has the highest wing and wing-disk loadings among the five species, so it is the least adapted to hovering and slow, manoeuvrable flight. It often hops about on the branches or on the ground when searching for food instead of flying about among the branches, as the other tits and particularly the goldcrest more often do. Its large mass (= larger volume:surface area) makes it better able to survive cold winter nights than the other species.

So the niche differentiation among the five species is associated with clear structural differentiation of the locomotor organs but to rather different degrees for various structures. The three tits, which are congeneric, are similar in the wing and leg skeletons, while the wing forms are not dependent on the wing skeleton, but mainly on the lengths of the feathers. The flight adaptations to different niches and different flight patterns during foraging have led to divergent evolution of wing feathers, and thus of wing forms, rather than of wing skeletons.

12.6 Wing Design in Bats

Bats exhibit almost similar relationships as birds between wing shape and foraging behaviour, and several authors have considered bat flight morphology in relation to flight pattern and foraging behaviour (for example, Vaughan 1970c; Findley et al. 1972; Kopka 1973; Lawlor 1973; U.M. Norberg 1981a, 1986b, 1987; Baagøe 1987; Norberg and Fenton 1988). Norberg and Rayner (1987) examined the wing morphology of bats in relation to flight performance and flight behaviour, to clarify the functional basis of the eco-morphological correlations in bats.

Figure 12.6 is based on a principal components analysis of wing morphology of 215 bat species from 16 families, and shows different aspect ratio and wing loading categories of bats (Norberg and Rayner 1987). These components emphasize the close scaling of wing proportions with size in bats. The wing loading WL' and aspect ratio AR' components are obtained from the following exponential functions,

$\longrightarrow$

Fig. 12.6. *a* Scatter plot of second and third principal components of wing morphology of 215 species of bats. The second component, along the *horizontal* axis, is identified as a size-independent wing size or loading, and the third component, along the *vertical* axis, as wing shape or aspect ratio. The *dotted lines* separate species with short, average and long wingspans, and indicate 8% deviation from the mean span. The *broken line* encircles the estimated components for seven fossil bats (After Norberg and Rayner 1987). *b* Wing shapes, flight modes and food choice in bats represented in *a*. Species with high wing loadings are fast fliers, while those with low wing loadings are slow fliers. Species with high aspect ratios can afford to use continuous flight, while those with low aspect ratios usually are perchers. The diagram axes are defined as in *a*. (Norberg 1989)

a

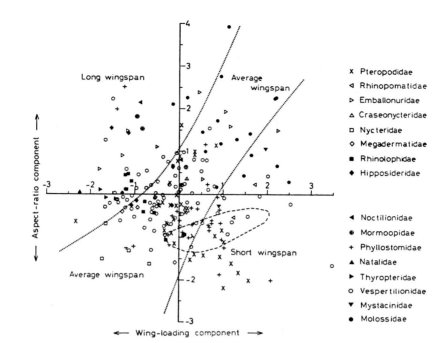

Long wingspan

Average wingspan

Average wingspan

Short wingspan

← Aspect-ratio component →

← Wing-loading component →

×   Pteropodidae
◁   Rhinopomatidae
▷   Emballonuridae
△   Craseonycteridae
□   Nycteridae
◇   Megadermatidae
■   Rhinolophidae
◆   Hipposideridae

◀   Noctilionidae
◉   Mormoopidae
+   Phyllostomidae
▲   Natalidae
▶   Thyropteridae
○   Vespertilionidae
▼   Mystacinidae
●   Molossidae

b

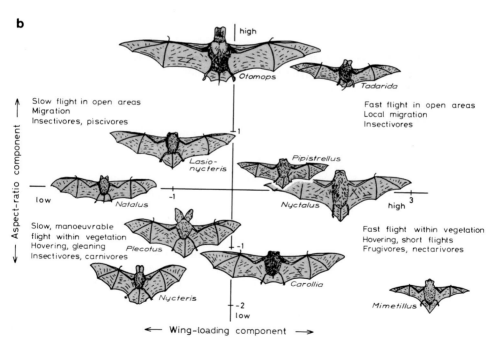

high

Otomops

Tadarida

Slow flight in open areas
Migration
Insectivores, piscivores

Fast flight in open areas
Local migration
Insectivores

↑ Aspect-ratio component ↓

Lasio-nycteris

Pipistrellus

low

Natalus

Nyctalus

high

Slow, manoeuvrable
flight within vegetation
Hovering, gleaning
Insectivores, carnivores

Plecotus

Fast flight within vegetation
Hovering, short flights
Frugivores, nectarivores

Nycteris

Carollia

Mimetillus

low

← Wing-loading component →

251

$$e^{WL'} = 3.77 \times 10^{-3} \, M^{3.02} \, b^{-2.08} \, S^{-3.71} \tag{12.1}$$

and

$$e^{AR'} = 1.81 \times 10^{-5} \, M^{-1.47} \, b^{14.6} \, S^{5.11}. \tag{12.2}$$

Equation (12.1) is rather close in form to wing loading Mg/S and Eq. (12.2) to aspect ratio b^2/S.

High wing loading and large aspect ratio are characteristic for bats with fast sustained flight, such as molossids, species in the emballonurid genus *Taphozous* and the vespertilionids *Tylonycteris* and some *Pipistrellus*. Most species in this category are insectivorous and hunt flying insects in the air in open spaces. Molossids have short or average wingspan with slightly rounded wingtips permitting them to make quick agile rolling manoeuvres, but their high wing loadings result in a large radius of turn.

Low wing loading and large aspect ratio occur in other emballonurids, in hipposiderids, noctilionids, and some vespertilionids (*Lasionycteris, Nyctophilus, Eptesicus*). These species have slow and enduring flight, often in open spaces, for their long wings would be a hindrance in dense vegetation. Some migratory species belong to this group. Noctilionids are partly piscivores, while the others are insectivores.

Bats with low aspect ratio and low wing loading can fly slowly within vegetation, and carry heavy prey. A large tail membrane and broad wings enable the bat to make tight turns in slow flight, and even to use the membrane as an insect-catching device. The wingspans are average and the wingtips are usually very rounded in these species, which is closely related to the bat's good manoeuvring ability. Nycterids, megadermatids, rhinolophids, and many vespertilionids belong to this group, and several of these species are insectivorous gleaners and hoverers or carnivores. Some of these bats have large ears for better auditory localization and the large ears facilitate the acoustically difficult task of detecting and localizing insects on surfaces, such as leaves. Only species with a low wing loading (= slow flight) can afford to have large, drag-producing ears (U.M. Norberg 1981a), and only species with low wing loading can manoeuvre to pick insects from vegetation.

Most frugivorous and nectarivorous species (families Pteropodidae and Phyllostomidae) have short wings with high wing loading and low aspect ratio, characteristics of fast but not enduring flight. The shapes of their wings are rather similar to those of the former group, but the tail membranes are very small in most of these species, and the wingspan is usually shorter relative to the body size. Since the tail membrane is included in the total wing area, the wing loading is higher than for the former group and the combination of short wingspan, low aspect ratio and high wing loading means that the cost of transport is high. Nectarivorous phyllostomids (glossophagines) often hover when feeding from flowers, whereas the larger nectarivorous pteropodids (macroglossines) more often perch. Hovering species should have long wings to minimize induced power, but these species often fly within vegetation putting much greater limits on the wing length than for non-hovering insectivorous species that hunt within vegetation, for the hovering ones fly closer to it. Muscle physiology of frugivorous bats shows that they are well suited to making short bursts of flight (Valdivieso et al. 1968). Less need to manoeuvre and to catch insects may be a reason why pteropodids and most phyllostomids have small tail membranes, while another may be climbing within vegetation for fruits which would be facilitated if the tail membrane is small. Most of those phyllostomids with larger tail membranes also eat insects. If nectar is sparse or if there are long distances between flowers or fruits, the bats would benefit from high wing loadings (=fast flight). Most phyllostomid bats have very long handwings with rounded wingtips, which may be adaptations for rapid acceleration and hovering.

Several species of bats (such as *Lasiurus cinereus, L. borealis, Lasionycteris noctivagans, Nyctalus noctula*, and *Tadarida brasiliensis*) migrate long distances to avoid dry or cold seasons, and others (*Tadarida* spp.) make long nightly commuting flights. As predicted for long-distance fliers, most of them have high or rather high aspect ratio wings and pointed wingtips. Commuting species that make long flights every night usually have high wing loadings and fast flights so they can cover long distances in short times.

The wing shapes of fossil bats are similar to those of several recent pteropodids, phyllostomids, rhinopomatids and vespertilionids (Norberg 1989). Compared with recent bats, the ancient ones had low aspect ratios and average to high relative wing loadings. Short wings and low aspect ratios mean that these ancient bats probably foraged or lived among vegetation and were perch-hunters. There is nothing in their wing shapes to indicate that these bats were poor fliers, although they may have had more expensive flight than most extant bats.

12.7 Wing Design and Echolocation Call Structure in Bats

Wing shape and echolocation call structure are adapted to the bat's foraging style, and there are usually associations between the three features (Fig. 12.7). There is a close association between wing design and call structure among bats adapted for fast-flying aerial-hawking, for which long-range detection is necessary. These bats use long, narrow-band echolocation calls of low frequency during searching and cruising, calls well suited for prey detection, but which give little information about the structure of the target and are not clutter-resistant (Simmons and Stein 1980). Calls of this design occur in bats with high aspect ratio and high wing loading. After detecting a target the bat must manoeuvre to catch it, and the high wing loading makes the turning radius large, so even though the reaction time is short a fast-flying bat will travel far before it can initiate a manoeuvre, making long-range detection essential (Norberg 1989).

Slow foraging flights, on the other hand, enable the bat to manoeuvre and catch prey detected at short ranges and echolocation calls designed for short-range detection among vegetation are usually broad-band and ultrasonic. The echolocation call and wing shape in bats foraging in clutter seem not to be as closely associated as in fast-flying aerial-hawking species, but different combinations of wing design and echolocation call structure may represent various

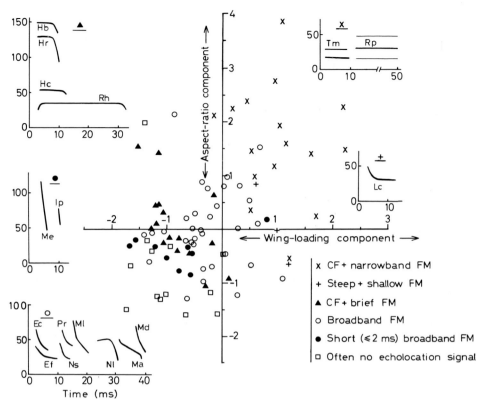

Fig. 12.7. Scatter plot of second and third principal components, identified as measures of wing loading and aspect ratio of bats represented in Fig. 12.6a and for which sonagrams of the echolocation sounds are known. The diagram axes are defined as in Fig. 12.6a. The sonagrams are based on Simmons et al. (1979), Fenton and Bell (1981), Thompson and Fenton (1982), Fenton et al. (1983) and Neuweiler (1984). *Ec = Eptesicus capensis; Ef = E. fuscus; Hb = Hipposideros bicolor; Hc = H. caffer; Hr = H. ruber; Ip = Idionycteris phyllotis; Lc = Lasiurus cinereus; Ma = Myotis adversus; Md = M. daubentoni; Me = M. evotis; Ml = M. lucifugus; Nl = Noctilio leporinus; Ns = Nycticeius schlieffeni; Pr = Pipistrellus rueppelli; Rh = Rhinolophus hildebrandtii; Rph = Rhinopoma hardwickei; Ta = Tadarida aegyptiaca; Tm = Taphozous mauritianus.* (Norberg and Rayner 1987; by courtesy of The Royal Society of London)

solutions to problems of foraging among vegetation of different density and structure.

The fossil bats *Icaronycteris, Archaeonycteris* and *Palaeochiropteryx* share special basicranial characters with microchiropteran species, suggesting comparable refinement for echolocation (Novacek 1985; Habersetzer and Storch 1988, 1989), so there are grounds for believing that their echolocation was well developed. The wing design (low aspect ratios and average to high relative wing loadings) in these bats indicates that they foraged close to or among vegetation, and they should therefore have benefited most from echolocation calls for short-range detection. However, using a special micro-X-ray method, Habersetzer and Storch (1988) found that the cochlea of *P. tupaiodon* shows no adaptation to constant or to high frequency echolocation sounds, which can rule out flutter detection but not echolocation.

12.8 Evolution of Wing Morphology

The distribution of species and foraging groups in Figs. 12.2 and 12.4 may have come about by one or both of the following evolutionary processes. If foraging behaviour and the associated flight mode dictate the flight morphology, then there should be large variation in aspect ratio and relative wing loading within a phylogenetic group, provided there have been opportunities for ecological diversification. Several phylogenetic groups should have their representatives spread over large regions in Figs. 12.2 and 12.4, assuming that there are no strong structural constraints.

If phylogeny severely constrains the potential for structural change of birds or bats within a phylogenetic group, there should be less differentiation into ecological roles. Morphological constraints, due to phylogenetic affinity, could limit structural and ecological differentiation within a phylogenetic group, making it relatively homogeneous and producing phylogenetic groups that are differently adapted and that occupy different, narrower, and well-defined regions in the diagrams.

The categories 1–4 of birds in Table 12.1 and Fig. 12.2 (encircled) were formed by grouping species with similar foraging behaviour and ecology, irrespective of their systematic position, and each group includes species from several families and orders. Some taxonomic groups are represented in different categories. For example, passeriform birds occur in five of the ecological groups (1a, 2a, 3a–c); aspect ratio varies between 4.4 (*Parus major*, group 3b) and 7.9 (*Delichon urbica*, group 1a) and relative wing loading between 38 (*Corvus frugilegus*, group 3c) and 99 (*Troglodytes troglodytes*, group 3b). Within the family Anatidae (group 4c) aspect ratio varies from 7.3 (*Anas platyrhynchos*) to 13.1 (*Cygnus cygnus*) and relative wing loading from 70.5 (*Anser anser*) to 128 (*Mergus merganser*). Pteropodid, vespertilionid and phyllostomid bats occur in all four quadrants in Fig. 12.6a, but most members of each family are gathered together.

On the other hand, members of several different orders and families of birds or bats show convergence in flight morphology and have similar foraging behaviour and ecology. For example, swifts, swallows, and bee-eaters *(Merops apiaster)* belonging to the orders Apodiformes, Passeriformes and Coraciiformes have very similar foraging modes and show similarities in wing morphology. The bee-eater weighs about 18 g (Greenewalt 1962; further data is lacking) and its wing form is similar to that of swallows. Furthermore, the "seabirds" in foraging group 1c originate from seven different families within the orders Accipitriformes, Procellariiformes, Pelecaniformes and Charadriiformes. Similar patterns emerge among the animal-eating bats (see above).

Summarizing, in some phylogenetic groups ecological diversification has obviously dictated morphological changes and enabled adaptation to diverse life styles. Such evolutionary changes have arisen by gradual, mutual co-adaptation between behaviour, ecology and structure. Structural constraints due to phylogenetic affinity obviously have not prevented anatomical radiation, but some phylogenetic groups are relatively homogeneous in flight morphology. This can occur in very specialized groups, such as swifts and molossid bats, where small structural changes do not permit shifting into other ecological roles. Another reason could be that the ecological roles into which a species might evolve by relatively small structural changes are already occupied by others.

However, what are recognized as taxonomic groups on the same hierarchical level may mean different things for different groups: within some groups all members may be very similar, while in others they may be more diverse in morphology, all depending upon the characters on which the taxonomic classification is based.

The selection pressure for a diversification appears to be different in different taxonomic and ecological groups. But even small morphological variations within a family or genus can reflect different behaviour and ecology (for example, contrast owls foraging in open habitat and in forest). Flight morphology is strongly selected and formable, but it is strongly controlled by behaviour, such as foraging mode, habitat selection and flight mode.

Chapter 13

Evolution of Flight

13.1 The Major Theories

There are two major conflicting hypotheses about the origin of powered flight in vertebrates, the arboreal ("trees-down") and cursorial ("ground-up") theories. According to the arboreal theory, powered flight in birds, bats and pterosaurs evolved via gliding in tree-living animals, while the cursorial theory holds that flapping flight in birds evolved in ground-running and jumping animals without a gliding intermediate.

13.1.1 Trees-Down Theory

Darwin (1859) was probably the first to propose the arboreal theory for the origin of flight in bats, and Marsh (1880) did so for birds, while a major summary of both theories was made by Feduccia (1980). The pre-flying pterosaurs, birds and bats may have been tree-dwellers that used gliding as a main type of locomotion during foraging; the transition from a ground-dwelling habit to a tree-climbing one opened up a new niche dimension. Ground-living insectivorous proto-birds and omnivorous proto-pterosaurs and proto-bats might have begun to forage among bushes and then to climb trees.

An argument against the trees-down theory has been the lack of trees in the Solnhofen Limestones (Upper Jurrasic) where all five fossils of *Archaeopteryx* have been found (e.g. Viohl 1985). The Solnhofen Limestones are marine backreef sediments with clay minerals and the thickest remains of branches do not exceed 3 to 4 cm in diameter. The conifers *Brachyphyllum* and *Palaeocyparis* were the most frequent land plants; they were stem succulents, with the same structures as extant cacti, and must have been shrubs not exceeding 3 m (Jung 1974). However, the absence of larger trees does not preclude the evolution of flight from arboreal proto-birds, for flight evolution may have been initiated much earlier when the flora was different, or in other areas. Furthermore, there is strong evidence from its anatomy that *Archaeopteryx* was already a good flier (see, for example, Sect. 11.4.3 and R.Å. Norberg 1985). The gliding phase could also have been initiated from jumps from cliffs or mountain slopes (Peters 1985; Norberg 1985b) with the same evolutionary stages as in the trees-down model.

Bock (1965, 1983, 1986) suggested that the proto-birds might have begun to climb trees to escape from predators, to roost at night, or to nest, and he identified the following evolutionary stages. Endothermy and insulating feathers could have been adaptations for life in trees where it is colder than at ground level.

Climbing and clinging among branches and leaves would require better control of movements with concomitant improvements in sense organs, neuro-muscular control and external morphology. Development of elongated feathers was assumed to have reduced the rate of fall and promoted safe landings. Elongated feathers on the forelimbs could then have acted as gliding surfaces, which eventually developed into a flapping wing for increase of the flight lengths. Finally, powered flight originated. Another variation of this theory is that feathers evolved directly for flight (Heilmann 1926; Feduccia 1980).

Maximization of net energy gain during foraging in trees or among cliffs might have been a reason for strong selection for increased gliding performance. R. Å. Norberg (1981a, 1983) showed that gliding from one tree to another and climbing upwards during foraging is a way to maximize net energy gain by modern birds (Fig. 13.1a,b). Once a glide surface had evolved, the animal's energy and time demands for locomotion during foraging might have been drastically reduced and their foraging efficiency increased. The first glides must

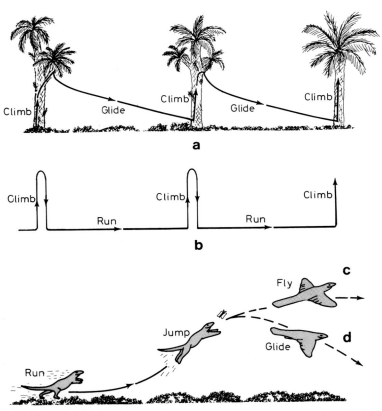

Fig. 13.1. a Energy-saving mode of locomotion. It costs less for an animal to climb a tree and then glide to the next tree than *b* to climb up and down in a tree and then run to the next, as modelled by R.Å. Norberg (1981a, 1983). *c* The run-jump-fly scenario in birds. *d* The same as *c* but with a gliding stage before the flying one

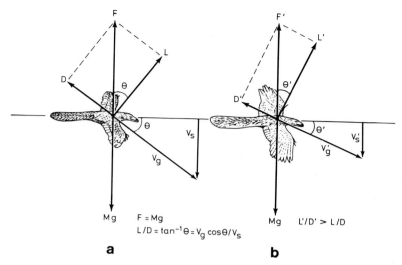

Fig. 13.2. Aerodynamics of gliding. *a* The glide angle θ is determined by the lift/drag ratio L/D. F is the resultant force which must balance body weight Mg. V_g is gliding speed and V_s is sinking speed. *b* An increase of aspect ratio (longer wings) results in larger L/D ratio and hence in a smaller glide angle and lower sinking speed. Starting from a given height the animal will cover more ground in a glide the better its L/D ratio (glide ratio) is. A decrease of wing loading, as a result of increased wingspan and area, reduces the minimum gliding speed. (Norberg 1985c)

have been steep, since the wings' aspect ratio and the entire glide surface were necessarily small in the original stages, resulting in low lift to drag ratios. But even steep parachuting jumps from trees reduced the time and energy required for foraging, while increased wing area decreased the wing loading and the gliding speed, allowing safer landings (Fig. 13.2). Parachuting for escape might also have been important for even squirrels without a glide membrane leap among branches and trees and spread whatever they have (legs, tail) to glide on. This behaviour, before a glide surface occurs, strongly promotes every incipient skin area for gliding purposes. Gliding must have been used only for commuting and not for insect-catching, which would require high manoeuvrability, which did not evolve until true flight was well established.

13.1.2 Ground-Up Theory

The cursorial theory includes a variety of ideas on the intermediary forms. Nopsca (1907, 1923) stated that flapping forelimbs first increased speed in the running proto-bird and later provided lift for short glides. Ostrom (1974) proposed the insect-net theory, which suggests that the animal used enlarged forelimbs for insect catching, although he has now reassessed this hypothesis (Ostrom 1986). Inspired by Ostrom's ideas, Caple et al. (1983) made calculations on stability during running in a proto-bird model and showed that extended lifting devices, such as enlarged forelimbs, could act as stabilizers in running, jumping, and in flight. Their study indicates that small increments of lift will dramatically increase body orientation control, and they assumed that the proto-birds evolved from small, bipedal, cursorial, insectivorous dinosaurs, which jumped for flying insects but had no lift capabilities. The forelimbs were extended and moved for stability and control of the body, and better stability and lift increment increased the animal's ability to capture insects. Limb motions were similar to wing strokes in powered flight of modern birds. Rapid movements and twisting of the forelimbs for control and stability eventually led to the production of more lift and thrust (Fig. 13.1c).

Jepsen (1970) and Caple et al. (1983) suggested that proto-bats foraged for insects by jumping from trees or cliffs (Fig. 13.3). But insects evolved flight long before the bats did and were probably highly aerobatic at the time proto-bats might have foraged for them. Without good manoeuvrability, jumping proto-bats might then have had great difficulties catching flying insects, and, if they did manage, might have got only one prey per jump. They would then have had to climb a tree or cliff again before the next attack. Such a foraging would be hopelessly time- and energy-consuming and ecologically impossible. Ballistic jumps to a lower level on a nearly vertical surface, as suggested by Caple et al. (1983), would require high control of the movements, and could be performed only by a well-developed flier. These power and control arguments might identify formidable barriers to the evolution of flight from a jumping behaviour without an intermediary gliding stage.

13.2 Transition from Gliding to Flapping Flight, an Aerodynamic Model

A main argument against the arboreal theory has been that lift could not be produced when a gliding animal begins to flap its proto-wings (e.g. Caple et al. 1983; Balda et al. 1985), but, as demonstrated in my aerodynamic model (Norberg 1985b,c, 1986a), a transition from gliding to active flight is mechanically and aerodynamically quite feasible. I showed that for every step along the hypothetical route from gliding, through stages of incipient flapping, to fully powered flight, there would have been an advantage over previous stages in terms of length and control of the flight path.

The selection pressure for good control of movements must have been high already on the tree-climbing proto-fliers, and higher still on the early gliders. Better control would have allowed the glider to change course to reach a particular destination. Good stability and control of movements were probably achieved before powered flight evolved, and may have evolved progressively along with the ability to glide. There might have been no difficulties for a gliding proto-flier to achieve stability and control of movements by simple wing movement coordination (twisting, retraction, control of dihedral angle etc.). From the beginning slight flapping was probably used for stability and for manoeuvres in turning and landing. When flapping became more powerful it could also

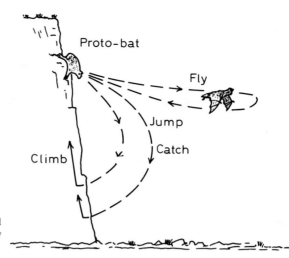

Fig. 13.3. The ballistic jump and fly scenario in bats, as modelled by Caple et al. (1983)

have been used to increase the glide path length and this would have made locomotion among vegetation more efficient both in terms of speed and energy cost.

My model, based on quasi-steady-state aerodynamics, shows that a net thrust force can be produced even during very slight flapping in a gliding animal while the necessary vertical lift is still produced resulting in a shallower glide path. The model is constructed so that, regardless of the amount of flapping, the different flight variables always combine so that the vertical lift produced during one complete wing stroke in partially powered flight always equals the weight of the animal (as in gliding with no flapping). Then,

$$\int_0^{t/2} L_{v,d}\, dt + \int_{t/2}^{t} L_{v,u}\, dt = \int_0^{t} Mg\, dt, \tag{13.1}$$

where $L_{v,d}$ and $L_{v,u}$ are the vertical lift forces produced during the downstroke and upstroke, respectively, t the time for a whole wing stroke, and Mg the body weight. With this requirement always prevailing, the model explores whether there is any net horizontal thrust force, over and above that needed to balance body drag, that is, whether

$$\int_0^{t/2} T\, dt - \int_{t/2}^{t} D_h\, dt > 0, \tag{13.2}$$

where T is the horizontal thrust produced during the downstroke, and D_h the horizontal drag produced during the upstroke. These forces are named thrust and drag although they may change sign at the extreme positions of the wings for large amplitudes. The outcome is that a net thrust always will be generated during gliding with incipient flapping (partially powered flight), given certain flight variables, and any thrust can be used to make the flight path shallower than it would be in equilibrium gliding for the same animal.

The initial stage is a hypothetical proto-flier in stable gliding at angle θ to the horizontal. The vertical lift is then identical with the resultant force F ($= Mg$) of

261

the lift and drag forces, L and D. Drag D ($= Mg \sin\theta$) is the sum of the wing drag, D_w, and body drag, D_b, and the horizontal component of the body drag is $D_{b,h}$ (Fig. 13.4a). The resultant force is the sum of the vertical components of the wings' resultant force, F_w, and the body drag, and are denoted $L_{v,w}$ and $D_{b,v}$, respectively. Similarly the horizontal drag, D_h, is the sum of the horizontal components of wing drag, $D_{w,h}$, and body drag, $D_{b,h}$. Treating the forces acting on wings and body as separate units, the Eqs. (13.1) and (13.2) can be written as

$$\int_0^{t/2} L_{v,d} \, dt + \int_{t/2}^t L_{v,u} \, dt + \int_0^t D_{b,v} \, dt = \int_0^t F \, dt \tag{13.3}$$

and

$$\int_0^{t/2} T \, dt > \int_{t/2}^t D_{w,h} \, dt + \int_0^t D_{b,h} \, dt. \tag{13.4}$$

(Fig. 13.4b,c).

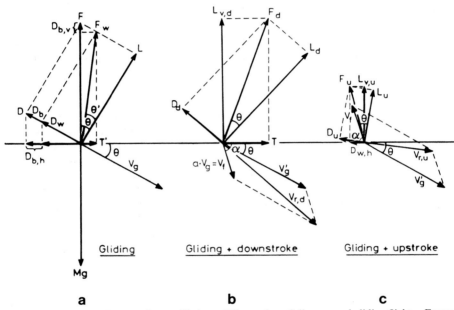

Fig. 13.4a-c. Force diagrams for equilibrium gliding and partially powered gliding flight. *a* Forces acting on a hypothetical proto-flier during gliding at angle θ to the horizontal. The resultant force F of the lift L and drag D forces equals body weight Mg. The total drag D is the sum of wing drag D_w and body drag D_b. F_w is the resultant of the forces acting on the wings only. $D_{b,h}$ and $D_{b,v}$ are the horizontal and vertical components, respectively, of F_w. T' is the horizontal component of F_w and equals the horizontal component of body drag, $T' = D_{b,h}$. V_g is gliding speed. *b* Forces acting on the wings during a downstroke in a hypothetical proto-flier flapping at angle α to the horizontal in partially powered gliding flight. $L_{v,d}$ is the lift component, perpendicular to the long wing axis, of the instantaneous resultant force F_d, and T is the horizontal thrust component of F_d. V'_g is the gliding speed vector, V_f the instantaneous flapping speed, and $V_{r,d}$ their resultant. *c* Forces acting on the wings during an upstroke in a hypothetical proto-flier flapping at angle α to the horizontal in partially powered gliding flight. $L_{v,u}$ is the lift component, perpendicular to the long wing axis, of the instantaneous resultant force F_u, and $D_{w,h}$ is the horizontal drag component of F_u. $V_{r,u}$ is the resultant of V'_g and V_f. (Norberg 1985c)

262

The lift/drag ratio of the wings, L/D, is assumed to equal that in gliding and to remain constant throughout the entire wingstroke. Furthermore, the downstroke and upstroke are assumed to be of equal duration, and the wings to supinate during the upstroke so that the air meets the wings' lower sides during the entire wingbeat cycle which makes the wingtip trailing vortex always rotate in the same direction producing some positive lift (directed upwards) on the upstroke.

The model accounts for the dihedral and anhedral attitudes of the wings in various phases of the wingbeat cycle changing the lift and drag force components of the wings and the local gliding and flapping speed vectors with the instantaneous wingbeat angle (see Norberg 1985b). Moreover, the vertical and horizontal forces obtained during flapping vary with time and wing's positional angle during the wingstroke. The lift force index, $L_v/(1/2)\rho C_L S V_g^2 t$, and the net horizontal force index, $(T-D_{w,h}-D_{b,h})/(1/2)\rho C_L S V_g^2 t$, where V_g is gliding speed, were calculated for different positions of the wings and for different glide angles. The vertical lift produced during flapping must equal that produced during gliding so that the shaded area above the force index line for gliding must equal that below it in Fig. 13.5. Furthermore, the net horizontal thrust produced during the downstroke must exceed the horizontal drag produced during the upstroke meaning that the hatched area above the zero line must exceed that below it in Fig. 13.5 for the animal to flatten out the glide.

The outcome of the model is that a horizontal net thrust force will be generated during partially powered flight, given certain flight variables. Incipient wing flapping in a gliding proto-flier would have carried larger benefits (in terms of flattened flight paths) the better the L/D ratio of the wings. Therefore, no matter how small the wing flapping speed, it will provide a net thrust force and the lift necessary to balance the body weight, provided that the stroke amplitude is small enough and the flapping frequency exceeds a threshold value. A particular flapping frequency must be attained for the animal to obtain thrust and the lift necessary to balance body weight, but even very small flapping speeds and amplitudes are sufficient. For a proto-bird the size of *Archaeopteryx* with a wingspan of 60 cm and a glide speed of about 7 m s⁻¹ or a proto-bat the size of *Palaeochiropteryx* with a wingspan of about 26 cm and a glide speed of about 2.9 m s⁻¹, the wingbeat frequency must be about 6.5 strokes s⁻¹ with vertical wingstrokes and about 6 strokes s⁻¹ when the stroke plane is tilted forward relative to the vertical as it is in modern birds and bats, given that the duration of the downstroke equals that of the upstroke. If the animal had beaten its wings faster during the downstroke than during the upstroke, the stroke frequency could be reduced to 2 strokes per second (Norberg 1985b).

Increased aspect ratio of the wings (longer span) and more efficient wing profiles, that gave larger L/D ratio and thrust, as well as increased abilities for wing flexion and movement coordination eventually led to horizontal flight. More sophisticated wing characters evolved increasing manoeuvrability and permitting the animals to forage in the air. Radiation to different flight habits led to different wing forms specialized for different foraging and flight modes.

In terms of vortex patterns, there would have been no problem with the transition from gliding to flapping flight. The intermediary stage with gliding and

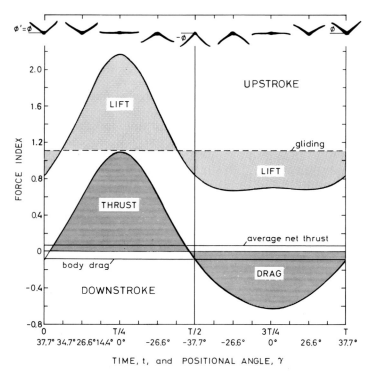

Fig. 13.5. Diagram showing how the lift force index $L_v /(1/2)\rho C_L SV^2_g t$ and net horizontal force index $(T-D_{w.h}-D_{b.h})/(1/2)\rho C_L SV^2_g t$ vary with time t and the wing's positional angle γ during one complete wingstroke in a gliding-flapping hypothetical proto-flier, when the glide angle θ is 30°, wingstroke plane angle is 90°, the mean flapping speed V_f is 0.5 times the gliding speed V_g, and the ratio (wing drag)/(body drag) is 5. For these values the vertical lift produced during flapping (*solid curves*) equals that produced during gliding when half the wingstroke amplitude Ø is 37.7° [$\gamma_{max} = (1/2)\emptyset$; *shaded areas above and below the broken line* for gliding being equal]. The thrust produced during the downstroke is larger than the drag produced during the upstroke (*hatched area* above the zero line larger than that below). This net thrust can be used to flatten out the glide. (Norberg 1985c, after Norberg 1985b)

slight flapping could have been associated with a wake consisting of a pair of continuous undulating vortex tubes, like those observed in kestrels in cruising flight at low wingbeat amplitude and frequency (Spedding 1987b). Undulating vortex tubes might then have formed closed loops in more aerodynamically active wingstrokes (Fig. 13.6).

In gliding and in partially powered flight along a more or less descending path, the animal is assisted by using part of the potential energy gained during a previous climb, energy that may have been produced over a long period of time, i.e. with a low power output. Therefore, partially powered flight is relatively inexpensive in energetic terms, since much of the lift production is obtained by converting some of the animal's potential energy (resulting in height loss).

264

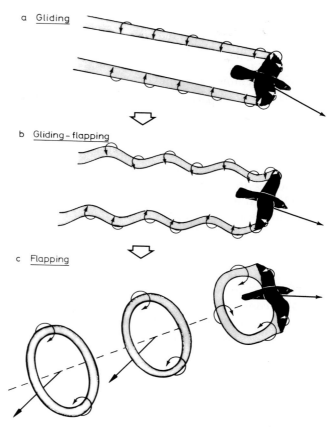

Fig. 13.6a-c. Probable changes in vorticity during the transition from gliding to active flight in the evolution of flight. *a* Trailing wingtip vortices produced during gliding. *b* Vortex wake, a pair of undulating vortex tubes that might have been produced by the ancient gliders with slight flapping. *c* Vortex rings produced during more active flight. The undulating vortex tubes in *b* might have been the intermediary form between the straight vortices produced during gliding and the closed loops produced by more active wingstrokes. (Norberg 1985b, by the courtesy of The University of Chicago Press)

13.3 The Ground-Running and Jumping Scenario, a Discussion

The step from a ground-running and jumping mode of life to powered flight as envisaged by the ground-up theory (Fig. 13.1c) seems very difficult because the animal had to work *against* gravity. In a running hypothetical proto-flier the drag on body and wings would increase with speed squared, as in flight, and would retard the animal, meaning that still more work would be required to reach the speed needed to generate the lift and thrust required for take-off. Longer forelimbs and feathers would further have increased drag and the work necessary for the same speed. The protoflier would have had to produce not only the power necessary for flight but also the power needed to run near take-off speed during the transient evolutionary stages. So it would have been very difficult for the animal to produce the power required for flapping flight. This power argument identifies a formidable obstacle to the evolution of flight from a cursorial mode of life. A gliding animal does not suffer from this negative feed-back system, since it works *with* gravity, and hence reaches a high forward speed almost free of cost before it begins to flap.

But fast running with *retracted* wings punctuated by regular leaps and shallow glides on extended wings is a commuting mode that is energetically cheaper than continuous running with the same average speed, but the savings are much less than those realized in the climbing-gliding scenario (Rayner 1985c; Fig. 13.1d). It might be unlikely that a running proto-bird had to move so fast except for escape from predators, and modern cursorial animals reach only half the speed necessary for take-off. Jump-and-glide also involves some deceleration and would be slower than continuous running. Taken together, these points make the gliding theory a much more feasible hypothesis than the running-jumping-leaping one, both in terms of energy- and time-saving arguments, as well as for aerodynamic reasons.

I am convinced that gliding must have preceded flapping, either from some elevation, such as trees or cliffs, or following leaps from the ground. Such leaps were not used for foraging in air, but for commuting. My aerodynamic model (Norberg 1985b) applies to the transition from gliding to powered flight regardless of whether gliding occurred from some elevation or followed upon leaps in fast runs.

13.4 The Climbing Ability in Proto-Fliers

An important question for the understanding of the evolution of flight in birds from a climbing ancestor is that it is unclear whether or not the proto-bird could climb or not. The free claws on the hands in early birds suggest that they were bush-, tree-, or cliff-climbers, and it is probable that the near ancestors were. The only modern bird with claws on free digits, the young of the Hoatzin (Opisthocomidae), are bush- and tree-climbers, and *Archaeopteryx* should have been able to climb trees (Yalden 1985). Those familiar with the basilisc (*Basiliscus vittatus*) know that it is a very fast-running quadrupedal reptile that can run over water surfaces to escape predators *and* climb trees! The first pterosaurs were probably small, quadrupedal animals that used their free, clawed digits for climbing, just like bats.

Chapter 14

Concluding Remarks

I will begin by citing Charles Darwin (1859): "When we see any structure highly perfected for any particular habit, as the wings of a bird for flight, we should bear in mind that animals displaying early transitional grades of the structure will seldom have survived to the present day, for they will have been supplanted by their successors, which were gradually rendered more perfect through natural selection."

Natural selection in flying animals is assumed to result in some near-optimal combination of various parameters, such as body size, wing design, flight mode and flight energetics, which are closely related to the type of habitat in which a flying animal lives and its way of exploiting the habitat (its ecological niche). There also should be mutual interrelationships among optimal foraging behaviour, flight characteristics and wing design. Optimal foraging behaviour is often likely to shift in a seasonal environment, which may also lead natural selection to act in conflicting directions. Moreover, the flight cost is minimized in different ways in slow and in fast flights; long wings are beneficial in slow flight and small wings in fast flight, but long wings would not be practical in dense vegetation or for take-offs from the ground or for manoeuvrability. There are different demands on wing design in activities such as gliding, hovering, horizontally flying, soaring, and migrating species, and in animals of different size. The optimal flight speed varies with the type and abundance of food, so selection should produce a wing design that minimizes the power required to fly at the speed and in the manner optimal for the animal.

A problem with a book like this is that aerodynamic theory seems to consider flying animals as perfectly adapted flying machines. But there often have to be compromises, and various constraints and trade-offs operate so that the observations often differ from the predictions. Understanding how natural selection may work on flying animals means knowing something about flight mechanics and energetics. The energetics of flight have been estimated from theory and by physiological measurements in a number of birds and bats. Aerodynamic models contain some approximations and empirical coefficients and they have become rather complicated for flapping wings, but can still provide quite exact estimates of energetic requirements. The conversion of results from physiological measurements into estimates of power output also involve many assumptions about the most favourable circumstances, and the results often refer to measurements of the total amount of energy consumed during a time period. But one must know if the flight had been steady or unsteady, continuous or intermittent, or if the animal has been forced to fly. The doubly labelled water method has frequently been used in the past few years, and it gives estimates of the animal's total metabolism over a limited time. If the different activities can be timed, the energy requirements for each can be obtained by multiple regression. But flight often involves manoeuvres, as well as take-offs and landings, whose cost may be difficult to separate from other flight activities. So the doubly labelled water method may be good for determining energy budgets, but may be difficult to use for estimating the cost for each different flight type. The picture is further complicated because we often do not know how birds are stressed by the measurement techniques and how this may affect their behaviour and metabolism, and hence energy consumption. This means that comparisons between physiological estimates and those obtained by aerodynamic theory should be cautious, also because physiological methods are not obviously more reliable than methods based on kinematics, morphology and aerodynamics.

Archaeopteryx posessed several advanced morphological characters for flapping flight, and so is considered to have been a true flyer. Its forerunners were probably gliders, and optimization of net energy gain might have been a reason for strong selection for improved gliding performance. Increased wing area reduced wing loading and gliding speed, allowing safer landings, and elongation of the wings increased aspect ratio and L/D ratio, resulting in shallower glides. Slight flapping resulted in the production of thrust, used to flatten out the glide and this, and the ability to coordinate the movements, eventually led to fully powered horizontal flight. Radiation to different habitats led to different wing forms, and the evolution of more sophisticated wing characters improved the aerodynamic performance.

The same steps were probably taken by the proto-pterosaurs and proto-bats. The controversial opinions about the morphology, locomotion and evolution of flight in pterosaurs can only be resolved by functional approaches, e.g. by biomechanical treatments and aerodynamic models. The behaviour of fossil animals and the evolution of flight will probably always be a subject of contention, and although we may never know for certain if we have found the right answers, we can always distinguish the possible from the impossible, the probable from the improbable.

References

Abbott JH, Doenhoff AE von (1949) Theory of wing sections. McGraw-Hill in Aeron Sci, New York

Aldridge HDJN (1986) Kinematics and aerodynamics of the greater horseshoe bats, *Rhinolophus ferrumequinum*, in horizontal flight at various flight speeds. J Exp Biol 126:479–497

Aldridge HDJN (1987a) Body accelerations during the wingbeat in six bat species: the function of the upstroke in thrust generation. J Exp Biol 130:275–293

Aldridge HDJN (1987b) Turning flight of bats. J Exp Biol 128:419–425

Alerstam T (1982) Fageltflyttning (Bird migration). Signum, Lund (in Swedish)

Alerstam T (1985) Strategies of migratory flight, illustrated by arctic and common terns, *Sterna paradisaea* and *Sterna hirundo*. In: Rankine MA (ed) Migration: mechanics and adaptive significance. Contrib Mar Sci Suppl 27:580–603

Alerstam T (1987) Bird migration across a strong magnetic anomaly. J Exp Biol 130:63–86

Alerstam T, Högstedt G (1983) The role of the geomagnetic field in the development of bird's compass sense. Nature (Lond) 306:463–465

Alexander RMcN (1977) Flight. In: Alexander RMcN, Goldspink G (eds) Mechanics and energetics of animal locomotion. Chapman and Hall, London, pp 249–278

Alexander RMcN (1982) Optima for animals. Arnold, London

Alexander RMcN (1983) Animal mechanics, 2nd edn. Blackwell, Oxford

Altenbach JS (1979) Locomotor morphology of the vampire, *Desmodus rotundus*. Spec Publ Am Soc Mammal 6

Altenbach JS, Hermanson JW (1987) Bat flight muscle function and the scapulohumeral lock. In: Fenton MB. Racey PA, Rayner JMV (eds) Recent advances in the study of bats. Univ Press, Cambridge, pp 100–118

Andersson M, Norberg RÅ (1981) Evolution of reversed sexual size dimorphism and role partitioning among predatory birds, with a size scaling of flight performance. Biol J Linn Soc 15:105–130

Archer RD, Sapuppo J, Betteridge DS (1979) Propulsion characteristics of flapping wings. Aeronaut J 83:355–371

Armstrong RB, Ianuzzo CD, Kunz TH (1977) Histochemical and biochemical properties of flight muscle fibers in the little brown bat *Myotis lucifugus*. J Comp Physiol 119:141–154

Aschoff J, Pohl H (1970) Der Ruheumsatz von Vögeln als Funktion der Tageszeit und der Körpergrösse. J Ornithol 111:38–47

Averill CK (1923) Black wing tips. Condor 25:57–59

Baagøe HJ (1987) The Scandinavian bat fauna: adaptive wing morphology and free flight behaviour in the field. In: Fenton MB, Racey PA, Rayner JMV (eds) Recent advances in the study of bats. Univ Press, Cambridge, pp 57–74

Badgerow JP, Hainsworth FR (1981) Energy savings from formation flight? A re-examination of the Vee formation. J Theor Biol 93:41–52

Balda RP, Caple G, Willis WR (1987) Comparison of the gliding to flapping sequence with the flapping to gliding sequence. In: Hecht MK, Ostrom JH, Viohl G, Wellnhofer P (eds) The beginnings of birds. Proc Int *Archaeopteryx* Conf Eichstätt 1984. Freunde des Jura-Museums Eichstätt, Willibaldsburg, pp 267–277

Baudinette RV, Schmidt-Nielsen K (1974) Energy cost of gliding flight in Herring Gulls. Nature (Lond) 248:83–84

Beck W (1984) The influence of the earth's magnetic field on the migratory behaviour of pied flycatchers (*Ficedula hypoleuca*). In: Varju D, Schnitzler H-U (eds) Localization and orientation in biology and engineering. Port Aransas: Contrib Mar Sci Suppl 27:544–552

Bell GP, Bartholomew GA, Nagy KA (1986) The roles of energetics, water economy, foraging behavior, and geothermal refugia in the distribution of the bat, *Macrotus californicus*. J Comp Physiol B156:441–450

Bennett PM, Harvey PH (1987) Active and resting metabolism in birds: allometry, phylogeny and ecology. J Zool Lond 213:327–363

Berger M (1972) Formationsflug ohne Phasenbeziehungen der Flügelschäge. J Ornithol Lpz 113:161–169

Berger M (1974a) Energiewechsel von Kolibris beim Schwirrflug unter Höhenbedingungen. J Ornithol 115:273–288

Berger M (1974b) Oxygen consumption and power of hovering hummingbirds at varying barometric and oxygen pressures. Naturwissenschaften 61:407

Berger M (1978) Ventilation in the hummingbirds *Colibri coruscans* during altitude hovering. In: Piiper J (ed) Respiratory function in birds, adult and embryonic. Springer, Berlin Heidelberg New York, pp 85–88

Berger M (1981) Aspects of bird flight respiration. Proc 17th Int Ornithol Congr 1978, 1:365–369

Berger M, Hart JS (1972) Die Atmung beim Kolibri *Amazilia fimbriata* während des Schwirrfluges bei verschiedenen Umgebungstemperaturen. J Comp Physiol 81:363–380

Berger M, Hart JS (1974) Physiology and energetics of flight. In: Farner DS, King JR (eds) Avian biology, vol 4. Academic Press, London New York, pp 415–477

Berger M, Hart JS, Roy OZ (1970a) Respiration, oxygen consumption and heart rate in some birds during rest and flight. Z Vergl Physiol 66:201–214

Berger M, Roy OZ, Hart JS (1970b) The co-ordination between respiration and wing beats in birds. Z Vergl Physiol 66:201–214

Berger M, Hart JS, Roy OZ (1971) Respiratory water and heat loss of the black duck during flight at different temperatures. Can J Zool 49:767–774

Bernstein MH (1976) Ventilation and respiratory evaporation in the flying crow, *Corvus ossifragus*. Respir Physiol 26:371–382

Bernstein MH, Thomas SP, Schmidt-Nielsen K (1973) Power input during flight of the fish crow, *Corvus ossifragus*. J Exp Biol 58:401–410

Betteridge DS, Archer RD (1974) A study of the mechanics of flapping wings. Aeronaut Q 25:129–142

Betz A (1963) Applied airfoil theory. In: Durand WF (ed) Aerodynamic theory, vol 4. Dover, New York, pp 1–129

Betz E (1958) Untersuchungen über die Korrelation des Flugmechanismus bei den Chiropteren. Zool Jahrb Abt Anat 77:491–526

Bibby CJ, Green RE (1981) Autumn migration strategies of Reed and Sedge Warblers. Ornis Scand 12:1–12

Biesel W, Nachtigall W (1987) Pigeon flight in a wind tunnel. IV. Thermoregulation and water homeostasis. J Comp Physiol 157:117–128

Bilo D (1971) Flugbiophysik von Kleinvögeln. I. Kinematik und Aerodynamik des Flügelabschlages beim Haussperling (*Passer domesticus* L.). Z Vergl Physiol 71:382–454

Bilo D (1972) Flugbiophysik von Kleinvögeln. II. Kinematik und Aerodynamik des Flügelauf-schlages beim Haussperling (*Passer domesticus* L.). Z Vergl Physiol 76:426–437

Blake RW (1983) Mechanics of gliding in birds with special reference to the influence of the ground effect. J Biomech 16:649–654

Blake RW (1985) A model of foraging efficiency and daily energy budget in the black skimmer (*Rynchops nigra*). Can J Zool 63:42–48

Blick EF, Watson D, Belie G, Chu H (1975) Bird aerodynamic experiments. In: Wu TY, Brokaw CJ, Brennen C (eds) Swimming and flying in nature, vol 2. Plenum Press, New York, pp 939–952

Bock WJ (1965) Tte role of adaptive mechanisms in the origin of higher levels of organization. Syst Zool 14:272–287

Bock WJ (1983) On extended wings: another view of flight. Sciences 23:16–20

Bock WJ (1986) The arboreal origin of avian flight. In: Padian K (ed) The origin of birds and the evolution of flight. Calif Acad Sci San Francisco 8:57–72

Bond J (1948) Origin of the bird fauna of the West Indies. Wilson Bull 60:207–229

Bowman RE (1961) Morphological differentiation and adaptation in the Galapagos finches. Univ Calif Publ Zool 58:1–302

Bramwell CD (1971) Aerodynamics of *Pteranodon*. Biol J Linn Soc 3:313–328

Bramwell CD, Whitfield GR (1974) Biomechanics of *Pteranodon*. Phil Trans R Soc Lond B 267:503-592

Brower JC, Veinus J (1981) Allometry in pterosaurs. Univ Kansas Palaeont Contrib Pap 105:1-32

Brown RHJ (1948) The flight of birds. The flapping cycle of the pigeon. J Exp Biol 25:322-333

Brown RHJ (1953) The flight of birds. II. Wing function in relation to flight speed. J Exp Biol 30:90-103

Brown RHJ (1963) The flight of birds. Biol Rev 38:460-489

Burke RE (1978) Motorunits: physiological/histochemical profiles, neural connectivity and functional specialization. Am Zool 18:127-134

Burtt EH Jr (1979) Tips on wings and other things. In: Burtt EH Jr (ed) The behavioral significance of color. Garland STPM, New York, pp 75-110

Butler PJ (1981) Respiration during flight. In: Hutas I, Debreczeni LA (eds) Advances in physiological sciences. Pergamon Press, Oxford, pp 155-164

Butler PJ (1985) New techniques for studying respiration in free flying birds. Proc 18th Int Ornithol Congr 1982, 2:995-1004

Butler P, Woakes AJ (1980) Heart rate, respiratory frequency and wing beat frequency of free flying barnacle geese, *Branta leucopsis*. J Exp Biol 85:213-226

Butler PJ, Woakes AJ (1985) Exercise in normally ventilation and apnoeic birds. In: Gilles R (ed) Circulation, respiration, and metabolism. Springer, Berlin Heidelberg New York Tokyo, pp 39-55

Butler PJ, West NH, Jones DR (1977) Respiratory and cardiovascular responses of the pigeon to sustained, level flight in a wind tunnel. J Exp Biol 71:7-26

Calder WA III (1968) Respiratory and heart rates of birds at rest. Condor 70:358-365

Calder WA III (1974) The consequences of body size for avian energetics. In: Paynter RA Jr (ed) Avian energetics. Nuttall Ornithol Club, Cambridge Mass 15:86-157

Calder WA III (1984) Size, function, and life history. Harvard Univ Press, Cambridge Mass

Calder WA III, King JR (1974) Thermal and caloric relations of birds. In: Farner DS, King JR (eds) Avian biology. Academic Press, London New York, pp 259-413

Caple G, Balda RP, Willis WR (1983) The physics of leaping animals and the evolution of preflight. Am Nat 121:455-467

Carpenter RE (1969) Structure and function of the kidney and the water balance of desert bats. Physiol Zool 42:288-302

Carpenter RE (1975) Flight metabolism of flying foxes. In: Wu TY-T, Brokaw CJ, Brennen C (eds) Swimming and flying in nature. Plenum Press, New York, pp 883-890

Carpenter RE (1985) Flight physiology of flying foxes. J Exp Biol 114:619-647

Carpenter RE (1986) Flight physiology of intermediate-sized fruit bats (Pteropodidae) J Exp Biol 120:79-103

Carpenter RE, Graham JB (1967) Physiological responses to temperature in the long-nosed bat, *Leptonycteris sanborni*. Comp Biochem Physiol 22:709-722

Chandra-Bose DA, George JC (1965a) Studies on the structure and physiology of the flight muscles of birds. 13. Characterization of the avian pectoralis. Pavo 3:14-22

Chandra-Bose DA, George JC (1965b) Studies on the structure and physiology of the flight muscles of birds. 14. Characterization of the avian supracoracoideus. Pavo 3:23-28

Childress S (1981) Mechanics of swimming and flying. Univ Press, Cambridge

Clutton-Brock TH, Harvey PH (1979) Comparison and adaptation. Proc R Soc Lond B 205:547-565

Colwell RK (1973) Competition and coexistence in a simple tropical community. Am Nat 107:737-760

Cone CD (1962a) Thermal soaring of birds. Am Sci 50:180-209

Cone CD (1962b) The soaring flight of birds. Sci Am 206:130-139

Cone CD (1968) The aerodynamics of flapping bird flight. Spec Sci Rep Va Inst Mar Sci 52

Costa DP, Prince PA (1987) Foraging energetics of grey-headed albatrosses *Diomedea chrysostoma* at Bird Island, South Georgia. Ibis 129:149-158

Coues E (1872) On the mechanism of flexion and extension in birds' wings. Proc Am Assoc Advmt Sci (1871) pp 278-284

Csicsáky M (1977) Aerodynamische und ballistische Untersuchungen an Kleinvögeln. Ph.D. Thesis, University of Hamburg

Cutts A (1986) Sacromere length changes in the wing muscles during the wing beat cycle of two bird species. J Zool Lond A 209:183–185

Dalton S (1982) Caught in motion. Weidenfeld and Nicolson, London

Darwin C (1859) The origin of species. J Murray, London

Dathe HH, Oehme H (1980) Kinematik und Energetik des Rüttelfluges mittelgrosser Vögel. Proc 17th Int Ornithol Congr 1978, 1:384–390

Dawson DM, Goodfriend TL, Kaplan NO (1964) Lactic dehydrogenases: Functions of the two types. Science 143:929–933

Dial KP, Kaplan SR, Goslow GE Jr, Jenkins FA Jr (1988) A functional analysis of the primary upstroke and downstroke muscles in the domestic pigeon (*Columba livia*) during flight. J Exp Biol 134:1–16

Dolnik VR, Blyumental TI (1967) Autumnal premigratory and migratory periods in the chaffinch (*Fringilla coelebs coelebs*) and some other temperate-zone passerine birds. Condor 69:435–468

Dolnik VR, Gavrilov VM, Ezerskas LJ (1963) Energy losses of small birds to natural migrational flight. In: Tezisy Dokladov Pyatoi Pribaltiiskoi Ornitologicheskoi Konferentsii. Akad Nauk Inst Zool Bot, pp 65–67 (in Russian)

Duellman WE (1970) The hylid frogs of Middle America. Monogr Mus Nat Hist Univ Kansas 1

Duncker H-R (1971) The lung air sac system in birds. Ergeb Anat Entwichlungsgesch 45 (6)

Durand WF (1963) Aerodynamic theory, vol 4. Dover, New York

Ellington CP (1978) The aerodynamics of normal hovering: three approaches. In: Schmidt-Nielsen K, Bolis L, Maddrell SHP (eds) Comparative physiology: water, ions, and fluid mechanics. Univ Press, Cambridge, pp 327–345

Ellington CP (1980) Vortices and hovering flight. In: Nachtigall W (ed) Instationäre Effekte an schwingenden Tierflügeln. Steiner, Wiesbaden pp 64–101

Ellington CP (1984a) The aerodynamics of hovering insect flight. I. The quasi-steady analysis. Phil Trans R Soc Lond B 305:1–15

Ellington CP (1984b) The aerodynamics of hovering insect flight. II. Morphological parameters. Phil Trans R Soc Lond B 305:17–40

Ellington CP (1984c) The aerodynamics of hovering insect flight. III. Kinematics. Phil Trans R Soc Lond B 305:41–78

Ellington CP (1984d) The aerodynamics of hovering insect flight. IV. Aerodynamic mechanisms. Phil Trans R Soc Lond B 305:79–113

Ellington CP (1984e) The aerodynamics of hovering insect flight. V. A vortex theory. Phil Trans R Soc Lond B 305:115–144

Ellington CP (1984f) The aerodynamics of hovering insect flight. VI. Lift and power requirements. Phil Trans R Soc Lond B 305:145–181

Ellington CP (1984g) The aerodynamics of flapping animal flight. Am Zool 24:95–105

Emlen ST (1975) The stellar-orientation system of a migratory bird. Sci Am 233:102–111

Feduccia A (1980) The age of birds. Harvard Univ Press, Cambridge Mass

Feduccia A, Tordoff HB (1979) Feathers of *Archaeopteryx*: asymmetric vanes indicate aerodynamic function. Science 203:1021–1022

Feinsinger P, Chaplin SB (1975) On the relationship between wing disc loading and foraging strategy in hummingbirds. Am Nat 109:217–224

Feinsinger P, Colwell RK, Terborgh J, Chaplin SB (1979) Elevation and the morphology, flight energetics, and foraging ecology of tropical hummingbirds. Am Nat 113:481–497

Fenton MB, Bell GP (1981) Recognition of species of insectivorous bats by their echolocation calls. J Mammal 62:233–243

Fenton MB, Merriam HG, Holroyd GL (1983) Bats of Kootenay, Glacier, and Mount Revelstoke National Parks in Canada: identification by echolocation calls, distribution and biology. Can J Zool 61:2503–2508

Findley JS, Studier EH, Wilson DE (1972) Morphologic properties of bat wings. J Mammal 53:429–444

Fisher HI (1957) Bony mechanism of automatic flexion and extension in the pigeon's wing. Science 126:446

Foehring RC, Hermanson JW (1984) Morphology and histochemistry of free-tailed bats, *Tadarida brasiliensis*. J Mammal 65:388–194

272

Frey E, Riess J (1981) A new reconstruction of the pterosaur wing. Neues Jahrb Geol Paläontol Abh 161:1-27

Froude RE (1889) On the part played in propulsion by differences of fluid pressure. Trans Inst Nav Arch 30:390

Fung YC (1969) An introduction to the theory of aeroelasticity. Dover, New York

Gans C, Darevski I, Tatarinov LP (1987) *Sharovipteryx*, a reptilian glider? Paleobiology 13:415-426

George JC (1965) The evolution of the bird and bat pectoral muscles. Pavo 3:131-142

George JC, Berger AJ (1966) Avian myology. Academic Press, London New York

George JC, Jyoti D (1955) Histological features of the breast and leg muscles of bird and bat and their physiological and evolutionary significance. J Anim Morphol Physiol 2:31-36

George JC, Naik RM (1957) Studies on the structure and physiology of the flight muscles of bats. 1. The occurrence of two types of fibres in the pectoralis major muscle of the bat (*Hipposideros speoris*), their relative distribution, nature of fuel store and mitochondrial content. J Anim Morphol Physiol 4:96-101

George JC, Naik RM (1958a) The relative distribution and chemical nature of the fuel store of the two types of fibres in the pectoral major muscle of the pigeon. Nature (Lond) 181:709-710

George JC, Naik RM (1958b) The relative distribution of the mitochondria in the two types of fibres in the pectoralis major muscle of the pigeon. Nature (Lond) 181:782-783

Gilbert EG, Parsons MG (1976) Periodic control and the optimality of aircraft cruise. J Aircraft 13:828-830

Goldspink G (1977a) Design in muscles in relation to locomotion. In: Alexander RMcN, Goldspink G (eds) Mechanics and energetics of animal locomotion. Chapman and Hall, London, pp 1-22

Goldspink G (1977b) Muscle energetics and animal locomotion. In: Alexander RMcN, Goldspink G (eds) Mechanics and energetics of animal locomotion. Chapman and Hall, London, pp 57-81

Goldspink G (1977c) Mechanics and energetics of muscle in animals of different sizes, with particular reference to the muscle fiber composition of vertebrate muscle. In: Pedley (ed) Scale effects in animal locomotion. Academic Press, London New York, pp 37-55

Goldspink G, Larson RE, Davies RE (1970) Fluctuations in sacromere length in the chick posterior latissimus dorsi muscle during isometric contraction. Experientia 26:16-18

Goodge WR (1959) Locomotion and other behaviour of the dipper. Condor 61:4-17

Goodhart HCN (1965) Glider performance: A new approach. Org Sci Tech Vol Voile 144 (2)

Goslow GE Jr (1985) Neural control of locomotion. In: Hildebrand M, Bramble DM, Liem KF, Wake DB (eds) Functional vertebrate morphology. Harvard Univ Press, Cambridge Mass, pp 338-365, refs pp 410-413

Goslow GE Jr, Dial KP, Jenkins FA Jr (1989) The avian shoulder: an experimental approach. Am Zool 29:287-301

Gould JL (1982) The map sense of pigeons. Nature (Lond) 296:205-211

Graham RR (1932) Safety devices in wings of birds. J R Aeronaut Soc 36:24-58

Graham RR (1934) The silent flight of owls. J R Aeronaut Soc 38:837-843

Greenewalt CH (1960) Hummingbirds. Doubleday, New York

Greenewalt CH (1962) Dimensional relationships for flying animals. Smithson Misc Collect 144:1-46

Greenewalt CH (1975) The flight of birds. Trans Am Phil Soc 65:1-67

Grubb BR (1983) Allometric relations of cardiovascular function in birds. Am J Physiol (Heart Circ Physiol) 245:H567-H572

Guidi G (1938) Osservazioni sul volo ad ala battente dei piccione. Aerotecnica, Roma 18:954-966. (A study of the wing-beats of pigeons in flight. J R Aeronaut Soc 42:1104-1115, 1938)

Habersetzer J, Storch G (1987) Klassifikation und funktionelle Flügelmorphologie paläogener Fledermäuse (Mammalia, Chiroptera). Cour Forsch Inst Senckenberg Frankfurt a M 91:117-150

Habersetzer J, Storch G (1988) Grube Messel: akustische Orientierung der ältesten Fledermäuse. Spektrum Wiss 7:12-14

Habersetzer J, Storch G (1989) Ecology and echolocation of the Eocene Messel bats. In: Hanák V, Horáček I, Gaisler J (eds) European bat research. (in press)

Haftorn S (1956) Contribution to the food biology of tits especially about storing surplus food. IV. A comparative analysis of *Parus atricapillus* L., *Parus cristatus* L. and *Parus ater* L. K Nor Vidensk Selsk Skr 4:1-54

Hails CJ (1979) A comparison of flight energetics in hirundines and other birds. Comp Biochem Physiol 63A:581–585

Hails CJ, Bryant DM (1979) Reproductive energetics of a free-living bird. J Anim Ecol 48:471–482

Hainsworth FR (1986) Why hummingbirds hover: A commentary. Auk 103:832–833

Hainsworth FR (1987) Precision and dynamics of positioning by Canada geese flying in formation. J Exp Biol 128:445–462

Hainsworth FR, Wolf LL (1969) Resting, torpid, and flight metabolism of the hummingbird *Eulampis jugularis*. Am Zool 9:1100–1101

Hamilton WJ III (1967) Orientation in waterfowl migration. In: Storm RM (ed) Animal orientation and navigation. Oregon State Univ Press, Corvallis

Hankin EH (1913) Animal flight. Iliffe, London

Hart JS, Berger M (1972) Energetics, water economy and temperature regulation during flight. Proc 15th Int Ornithol Congr, pp 189–199

Hart JS, Roy OZ (1966) Respiratory and cardiac responses to flight in pigeons. Physiol Zool 39:291–305

Hartman FA (1961) Locomotor mechanisms of birds. Smithson Misc Collect 143

Haubenhofer M (1964) Die Mechanik des Kurvenfluges. Schweiz Aero-Rev 39:561–565

Heilmann G (1926) The origin of birds. Witherby, London

Heppner FH (1974) Avian flight formations. Bird-Banding 45:160–169

Hermanson JW (1979) The forelimb morphology of the pallid bat, *Antrozous pallidus*. Ms thesis, Northern Arizona Univ Flagstaff, 84 pp

Hermanson JW (1981) Functional morphology of the clavicle in the pallid bat, *Antrozous pallidus*. J Mammal 62:801–805

Hermanson JW, Altenbach JS (1981) Functional anatomy of the primary downstroke muscles in the pallid bat, *Antrozous pallidus*. J Mammal 62:795–800

Hermanson JW, Altenbach JS (1983) The functional anatomy of the shoulder of the pallid bat, *Antrozous pallidus*. J Mammal 64:62–75

Hermanson JW, Altenbach JS (1985) Functional anatomy of the shoulder and arm of the fruit-eating bat *Artibeus jamaicensis*. J Zool Lond A 205:157–177

Hermanson JW, Foehring RC (1988) Histochemistry of flight muscles in the Jamaican Fruit Bat, *Artibeus jamaicensis*: implications for motor control. J Morphol 196:353–362

Hertel H (1966) Structure–form–movement. Reinhold, New York

Herzog K (1968) Anatomie und Flugbiologie der Vögel. Fischer, Stuttgart

Hesse R (1921) Das Herzgewicht der Wirbeltiere. Zool Jahrb Abt Allg Zool Physiol Tiere 38:242–364

Heusner AA (1982) Energy metabolism and body size. I. Is the 0.75 mass exponent of Kleiber's equation a statistical artifact? Respir Physiol 48:1–12

Higdon JJL, Corrsin S (1978) Induced drag of a bird flock. Am Nat 112:727–744

Hill AV (1938) The heat of shortening and the dynamic constants of muscle. Proc R Soc Ser B 126:136–195

Hill AV (1950) The dimensions of animals and their muscular dynamics. Sci Prog London 38:209–230

Hofton A (1978) How sails can save fuel in the air. New Sci 20 April: 146–147

Holbrook KA, Odland GF (1978) A collagen and elastic network in the wing of the bat. J Anat 126:21–36

Holst E von, Küchemann D (1941) Biologische und aerodynamische Probleme des Tierfluges. Naturwissenschaften 29:348–362 (Biological and aerodynamical problems of animal flight. J R Aeronaut Soc 46:39–56, 1942 (abridged) and NASA Tech Memor 75337, 1980)

Houghton EL, Brock AE (1960) Aerodynamics for Engineering Students, 2nd edn. Arnold, London

Hudson DM, Bernstein MH (1981) Temperature regulation and heat balance in flying white-necked ravens, *Corvus cryptoleucus*. J Exp Biol 90:267–281

Hudson DM, Bernstein MH (1983) Gas exchange and energy cost of flight in the white-necked raven, *Corvus cryptoleucus*. J Exp Biol 103:121–130

Hummel D (1980) The aerodynamic characteristics of slotted wing-tips in soaring birds. Proc 17th Int Ornithol Congr 1:391–396

Hussell DJT (1969) Weight loss of birds during nocturnal migration. Auk 86:75–83

Jenkins FA Jr, Dial KP, Goslow GE Jr (1988) A cineradiographic analysis of bird flight: the wishbone in starlings is a spring. Science 241:1495–1498

Jepsen GL (1970) Bat origins and evolution. In: Wimsatt WA (ed) Biology of bats, vol 1. Academic Press, London New York, pp 1–64

Johnston DW, McFarlane RW (1967) Migration and bio-energetics of flight in the Pacific golden plover. Condor 69:156–168

Johnston IA (1985) Sustained force development: specializations and variation among the vertebrates. J Exp Biol 115:239–251

Jung W (1974) Die Konifere *Brachyphyllum nepos* Saporta aus den Solnhofener Plattenkalken (unteres Untertithon), ein Halophyt. Mitt Bayer Staatssamml Paläonthol Hist Geol München 14:49–58

Kaiser CE, George JC (1973) Interrelationship amongst the avian orders Galliformes, Columbiformes, and Anseriformes as evinced by the fiber types in the pectoralis muscle. Can J Zool 51:887–892

Kanwisher JW, Williams TC, Teal JM, Lawson KO (1978) Radiotelemetry of heart rates from free-ranging gulls. Auk 95:288–293

Kármán T von, Burgers JM (1935) General aerodynamic theory – perfect fluids. In: Durand W (ed) Aerodynamic theory, vol 2. Div E. Springer, Berlin

Kiepenheuer J (1982) The effect of magnetic anomalies on the homing behaviour of pigeons: an attempt to analyse the possible factors involved. In: Papi F, Wallraff HG (eds) Avian navigation. Springer, Berlin Heidelberg New York Tokyo, pp 120–128

Kiessling K-H (1977) Muscle structure and function in the goose, quail, pheasant, guinea hen, and chicken. Comp Biochem Physiol 57 B:287–292

King JR, Farner DS (1961) Energy metabolism, thermoregulation and body temperature. In: Marshall AJ (ed) Biology and comparative physiology of birds, vol 2. Academic Press, London New York, pp 215–288

Kleiber M (1932) Body size and metabolism. Hilgardia 6:315–353

Kleiber M (1961) The fire of life. Wiley, New York

Kleiber M (1967) Der Energiehaushalt von Mensch und Haustier. (The fire of life). Parey, Hamburg

Kokshaysky NV (1977) Some scale dependent problems in aerial animal locomotion. In: Pedley TJ (ed) Scale effects in animal locomotion. Academic Press, London, New York, pp 421–435

Kokshaysky NV (1979) Tracing of the wake of a flying bird. Nature (Lond) 279:146–148

Kopka T (1973) Beziehungen zwischen Flügelfläche und Körpergrösse bei Chiropteren. Z Wiss Zool 185:235–284

Kovtun MF (1970) Morfofunkcional'nyj analiz myšc pleča letučich myšej v svjazi s ich poletom (Morpho-functional analysis of the shoulder muscles in bats in relation to their flight). Vest Zool 1:18–22

Kovtun MF (1981) Sravnitel 'naja morfologija i évoljucija organov lokomocii rukokrylych (Comparative morphology and evolution of locomotor organs in bats). D Sc Thesis, Acad Sci Ukrainian SSR, Kiev

Kovtun MF (1984) Strojenje i evolucja organow lokomocji rukokrylych. (Structure and evolution of the locomotor organs in bats). Nauk Dumka, Kiev

Kovtun MF, Moroz VF (1974) Isslyedovaniye bioelektrichyeskoi aktivnosti myishts plyechovo poyasa u *Eptesicus serotinus* Schreb. (Chiroptera). (Investigation of bio-electrical activity of the shoulder girdle muscles in *Eptesicus serotinus*.) Dokl Akad Nauk SSSR 210:1481–1484

Krebs JR (1978) Optimal foraging: decision rules for predators. In: Krebs JR, Davis NB (eds) Behavioural ecology: an evolutionary approach. Blackwell, Oxford, pp 23–63

Kreithen ML, Keeton WT (1974) Detection of polarized light by homing pigeon *Columba livia*. J Comp Physiol 89:83–92

Kroeger RA, Grushka HD, Helvy TC (1972) Low speed aerodynamics for ultra-quiet flight. AF FDL-TR-71–75, Air Force Flight Dynamics Laboratory, Dayton, Ohio

Kuethe AM (1975) Prototypes in nature. The carry-over into technology. TechniUM, Spring 1975: 3–20

Kunz TH, Nagy KA (1987) Methods of energy budget analysis. In Kunz TH (ed) Ecological and behavioral methods for the study of bats. Smithson Inst, Baltimore, pp 277–302

Kuroda N (1961) A note on the pectoral muscle of birds. Auk 78:261–263

Langston W Jr (1981) Pterosaurs. Sci Am 244:92–102

Larkin RP, Sutherland PJ (1977) Migrating birds respond to Project Seafarer's electromagnetic field. Science 195:777–779

Lasiewski RC (1962) The energetics of migrating hummingbirds. Condor 64:324

Lasiewski RC (1963a) The energetic cost of small size in hummingbirds. Proc 13th Int Ornithol Congr 1962, pp 1095–1103

Lasiewski RC (1963b) Oxygen consumption of torpid, resting, active, and flying hummingbirds. Physiol Zool 36:122–140

Lasiewski RC (1972) Respiratory function in birds. In: Farner DS, King JR (eds) Avian biology, vol 2. Academic Press, London New York, pp 288–342

Lasiewski RC, Calder WA (1971) A preliminary allometric analysis of respiratory variables in resting birds. Respir Physiol 11:152–166

Lasiewski RC, Dawson WR (1967) A re-examination of the relation between standard metabolic rate and body weight in birds. Condor 69:13–23

Lawlor TE (1973) Aerodynamic characteristics of some neotropical bats. J Mammal 54:71–78

Lawson DA (1975) A pterosaur from the latest Cretaceous of West Texas: discovery of the largest flying creature. Science 187:947–948

Lee DS, Grant GS (1986) An albino Greater Shearwater: feather abrasion and flight energetics. Wilson Bull 98:488–490

LeFebvre EA (1964) The use of D2O18 for measuring energy metabolism in *Columba livia* at rest and in flight. Auk 81:403–416

Lifson M, McClintock RM, (1966) Theory and use of the turnover rates of body water for measuring energy and material balance. J Theor Biol 12:46–74

Lifson M, Gordon GB, Visscher NB, Nier AO (1949) The fate of utilized molecular oxygen of respiratory CO_2, studied with the aid of heavy oxygen. J Biol Chem 180:803–811

Lighthill MJ (1975) Aerodynamic aspects of animal flight. In: Wu T Y-T, Brokaw CJ, Brennen C (eds) Swimming and flying in nature, vol 2. Plenum Press, New York, pp 423–491

Lighthill MJ (1977) Introduction to the scaling of aerial locomotion. In: Pedley TJ (ed) Scale effects in animal locomotion. Academic Press, London New York, pp 365–404

Lissaman PBS, Shollenberger CA (1970) Formation flight of birds. Science 168:1003–1005

Lundgren BO (1988) Catabolic enzyme activities in the pectoral muscle of migratory and non-migratory goldcrests, great tits, and yellowhammers. Ornis Scand 19:190–194

Lundgren BO, Kiessling K-H (1985) Seasonal variation in catabolic enzyme activities in breast muscle of some migratory birds. Oecologia 66:468–471

Lundgren BO, Kiessling K-H (1986) Catabolic enzyme activities in the pectoral muscle of premigratory and migratory juvenile Reed Warblers *Acrocephalus scirpaceus* (Herm.). Oecologia 68:529–532

Lundgren BO, Kiessling K-H (1988) Comparative aspects of fibre types, areas, and capillary supply in the pectoralis muscle of some passerine birds with differing migratory behaviour. J Comp Physiol B 158:165–173

Lyuleeva DS (1970) Energy of flight in swallows and swifts. Dokl Akad Nauk SSR 190:1467–1469

Magnan A (1922) Les caractéristiques des oiseaux suivant la mode de vol. Ann Sci Nat Sér Zool 5:125–334

Maresca C, Favier D, Rebont J (1979) Experiments on an aerofoil at high angle of incidence in longitudinal oscillations. J Fluid Mech 92:671–690

Marey EJ (1887) Physiologie du mouvement. Le vol des oiseaux. Masson, Paris

Marsh OC (1880) Odontornithes: a monograph on the extinct toothed birds of North America. Rep US Geological Exploration of the 40th Parallel 7. Government Printing Office, Washington DC

Matsumoto Y, Hoekman T, Abbott BC (1973) Heat measurements associated with isometric contraction in fast and slow muscles of the chicken. Comp Biochem Physiol 46A:785–797

Maxworthy T (1979) Experiments on the Weis-Fogh mechanism of lift generation by insects in hovering flight. Part 1. Dynamics of the 'fling'. J Fluid Mech 93:47–63

McGowan C (1982) The wing musculature of the brown kiwi (*Apteryx australis mantelli*) and its bearing on ratite affinities. J Zool 197:173–219

McMahon TA (1984) Muscles, reflexes, and locomotion. Univ Press, Princeton

McMasters JH (1976) Aerodynamics of the long pterosaur wing. Science 191:899

Miller RS (1985) Why hummingbirds hover. Auk 102:722–726

Milne-Thomson LM (1958) Theoretical aerodynamics, 4th edn. Dover, New York

Mises R von (1959) Theory of flight. Dover, New York. First published 1945

Moore FR (1977) Geomagnetic disturbance and the orientation of nocturnally migrating birds. Science 196:682–684

Moreau RE (1961) Problems of Mediterranean-Saharan migration. Ibis 103:373–427, 580–623

Morgado E, Günther B, Gonzales U (1987) On the allometry of wings. Rev Chilena Hist Nat 60:71–79

Muller BD, Baldwin J (1978) Biochemical correlates of flying behaviour in bats. Aust J Zool 26:29–37

Nachtigall W (1970) Phasenbeziehungen der Flügelschläge von Gänsen während des Verbandsflugs in Keilformation. Z Vergl Physiol 67:414–422

Nachtigall W (1979) Gleitflug des Flugbeutlers *Petaurista breviceps papuanus*. II. Filmanalysen zur Einstellung von Gleitbahn und Rumpf sowie zur Steuerung des Gleitflugs. J Comp Physiol 133:89–95

Nachtigall W (1980) Bird flight: kinematics of wing movements and aspects on aerodynamics. Proc 17th Int Ornithol Congr 1:377–383

Nachtigall W, Kempf B (1971) Vergleichende Untersuchungen zur flugbiologischen Funktion des Daumenfittichs (*Alula spuria*) bei Vögeln. I. Der Daumenfittich als Hochauftriebserzeuger. Z Vergl Physiol 71:326–341

Nachtigall W, Rothe HJ (1978) Eine Methode, Tauben für den freien Flug im Windkanal zu trainieren. Naturwissenschaften 65:266–267

Nagy KA (1975) Water and energy budgets of free-living animals: measurement using isotopically labelled water. In: Hadley NF (ed) Environmental physiology of desert organisms: Dowden, Hutchinson and Ross, Stroudsburg, pp 227–245

Nagy KA (1980) CO_2 production in animals: analysis of potential errors in the doubly labeled water method. Am J Physiol 238:R466–R473

Nagy KA (1987) Field metabolic rate and food requirement scaling in mammals and birds. Ecol Monogr 57:111–128

Nair KK (1954) A comparison of the muscles in the forearm of a flapping and a soaring bird. J Anim Morph Physiol 1:26–34

Neuweiler G (1984) Foraging, echolocation and audition in bats. Naturwissenschaften 71:446–455

Newman BG (1958) Soaring and gliding flight of the black vulture. J Exp Biol 35:280–285

Newman BG, Savage SB, Schouella D (1977) Model tests on a wing section of an *Aeschna* dragonfly. In: Pedley TJ (ed) Scale effects in animal locomotion. Academic Press, London New York, pp 445–477

Nisbet ICT, Drury WH, Baird J (1963) Weight loss during migration. I. Deposition and consumption of fat by the blackpoll warbler *Dendroica striata*. Bird-Banding 34:107–138

Noll UG (1979) Body temperature, oxygen consumption, noradrenaline response and cardiovascular adaptations in the flying fox, *Rousettus aegyptiacus*. Comp Biochem Physiol 63 A:79–88

Nopsca F (1907) Ideas on the origin of flight. Proc Zool Soc Lond 1907:223–236

Nopsca F (1923) On the origin of flight in birds. Proc Zool Soc Lond 1923:463–477

Norberg RÅ (1975) Hovering flight of the dragonfly *Aeschna juncea* L., kinematics and aerodynamics. In: Wu TY-T, Brokaw CJ, Brennen C (eds) Swimming and flying in nature, vol 2. Plenum Press, New York, pp 763–781

Norberg RÅ (1977) An ecological theory on foraging time and energetics and choice of optimal food-searching method. J Anim Ecol 46:511–529

Norberg RÅ (1981a) Why foraging birds in trees should climb and hop upwards rather than downwards. Ibis 123:281–288

Norberg RÅ (1981b) Optimal flight speed in birds when feeding young. J Anim Ecol 50:473–477

Norberg RÅ (1983) Optimum locomotion modes for birds foraging in trees. Ibis 125:172–180

Norberg RÅ (1985) Function of vane asymmetry and shaft curvature in bird flight feathers; inference on flight ability of *Archaeopteryx*. In: Hecht MK, Ostrom JH, Viohl G, Wellnhofer P (eds) The beginnings of birds. Proc Int *Archaeopteryx* Conf Eichstätt 1984. Freunde des Jura-Museums Eichstätt, Willibaldsburg, pp 303–318

Norberg RÅ (1986) Treecreeper climbing; mechanics, energetics, and structural adaptations. Ornis Scand 17:191–209

Norberg RÅ, Norberg UM (1971) Take-off, landing, and flight speed during fishing flights of *Gavia stellata* (Pont.). Ornis Scand 2:55–67

Norberg UM (1969) An arrangement giving a stiff leading edge to the hand wing in bats. J Mammal 50:766–770

Norberg UM (1970a) Hovering flight of *Plecotus auritus* Linnaeus. Bijdr Dierk 40:62–66

Norberg UM (1970b) Functional osteology and myology of the wing of *Plecotus auritus* Linnaeus (Chiroptera). Ark Zool 22:483–543

Norberg UM (1972a) Functional osteology and myology of the wing of the dog-faced bat *Rousettus aegyptiacus* (E. Geoffroy) (Pteropidae). Z Morphol Tiere 73:1–44

Norberg UM (1972b) Bat wing structures important for aerodynamics and rigidity (Mammalia, Chiroptera). Z Morphol Tiere 73:45–61

Norberg UM (1975) Hovering flight of the pied flycatcher (*Ficedula hypoleuca*). In: Wu TY-T, Brokaw CJ, Brennen C (eds) Swimming and flying in nature, vol 2. Plenum Press, New York, pp 869–881

Norberg UM (1976a) Aerodynamics, kinematics, and energetics of horizontal flapping flight in the long-eared bat *Plecotus auritus*. J Exp Biol 65:179–212

Norberg UM (1976b) Aerodynamics of hovering flight in the long-eared bat *Plecotus auritus*. J Exp Biol 65:459–470

Norberg UM (1976c) Some advanced flight manoeuvres of bats. J Exp Biol 64:489–495

Norberg UM (1979) Morphology of the wings, legs and tail of three coniferous forest tits, the goldcrest, and the treecreeper in relation to locomotor pattern and feeding station selection. Phil Trans R Soc Lond B 287:131–165

Norberg UM (1981a) Allometry of bat wings and legs and comparison with bird wings. Phil Trans R Soc Lond B 292:359–298

Norberg UM (1981b) Flight, morphology and the ecological niche in some birds and bats. In: Day MH (ed) Vertebrate locomotion. Symp Zool Soc Lond 48:173–197. Academic Press, London New York

Norberg UM (1985a) Flying, gliding, and soaring. In: Hildebrand M, Bramble DM, Liem KF, Wake DB (eds) Functional vertebrate morphology. Harvard Univ Press, Cambridge Mass, pp 129–158, refs pp 391–392

Norberg UM (1985b) Evolution of vertebrate flight: an aerodynamic model for the transition from gliding to flapping flight. Am Nat 126:303–327

Norberg UM (1985c) Evolution of flight in birds: aerodynamic, mechanical and ecological aspects. In: Hecht MK, Ostrom JH, Viohl G, Wellnhofer P (eds) The beginnings of birds. Proc Int *Archaeopteryx* Conf Eichstätt 1984. Freunde des Jura-Museums Eichstätt, Willibaldsburg, pp 293–302

Norberg UM (1986a) On the evolution of flight and wing forms in bats. In: Nachtigall W (ed) Bat flight — Fledermausflug. Biona Report 5. Fischer, Stuttgart, pp 13–26

Norberg UM (1986b) Evolutionary convergence in foraging niche and flight morphology in insectivorous aerial-hawking birds and bats. Ornis Scand 17:253–260

Norberg UM (1987) Wing form and flight mode in bats. In: Fenton MB, Racey PA, Rayner JMV (eds) Recent advances in the study of bats. Univ Press, Cambridge, pp 43–56

Norberg UM (1989) Ecological determinants of bat wing shape and echolocation call structure with implications for some fossil bats. In: Hanák V, Horácèk I, Gaisler J (eds) European Bat Research (in press)

Norberg UM, Fenton MB (1988) Carnivorous bats? Biol J Linn Soc 33:383–394

Norberg UM, Norberg RÅ (1989) Ecomorphology of flight and tree-trunk climbing in birds. Proc Int 19th Ornithol Congr 1986, 2:2271–2282

Norberg UM, Rayner JMV (1987) Ecological morphology and flight in bats (Mammalia; Chiroptera): wing adaptations, flight performance, foraging strategy and echolocation. Phil Trans R Soc Lond B 316:335–427

Novacek MJ (1985) Evidence for echolocation in the oldest known bats. Nature (Lond) 315:140–141

Obrecht III HH, Pennycuick CJ, Fuller MR (1988) Wind tunnel experiments to assess the effect of back-mounted radio transmitters on bird body drag. J Exp Biol 135:265–273

Odum EP, Connell CE, Stoddard HL (1961) Flight energy and estimated flight ranges of some migratory birds. Auk 78:515–527

Oehme H (1968) Der Flug der Mauersseglers (*Apus apus*). Biol Zbl 87:287–311

Oehme H (1977) On the aerodynamics of separated primaries in the avian wing. In: Pedley TJ (ed) Scale effects in animal locomotion. Academic Press, London New York, pp 479–494

Oehme H (1985) Über die Flügelbewegung der Vögel im schnellen Streckenflug. Milu, Berlin 6:137–156

Oehme H (1986) Vom Flug des Habichts *Accipiter gentilis* (L.). Ann Naturhist Mus Wien 88/89 B:67–81

Oehme H, Kitzler U (1974) Über die Kinematik des Flügelschlages beim unbeschleunigten Horizontalflug. Untersuchungen zur Flugphysiologie der Vögel I. Zool Jahrb Physiol 78:461–512

Oehme H, Dathe HH, Kitzler U (1977) Research on biophysics and physiology of bird flight. IV. Flight energetics in birds. Fortschr Zool 24:257–273

Ogilvie MA (1978) Wild geese. Poyser, Berkhamsted

Ohtsu R, Uchida TA (1979a) Correlation among fiber composition and LDH isozyme patterns of the pectoral muscles and flight habits in bats. J Fac Agr Kyushu Univ 24:145–155

Ohtsu R, Uchida TA (1979b) Further studies on histochemical and ultrastructural properties of the pectoral muscles of bats. J Fac Agr Kyushu Univ 24:157–163

Osborne MFM (1951) Aerodynamics of flapping flight with application to insects. J Exp Biol 28:221–245

Ostrom JH (1974) *Archaeopteryx* and the origin of flight. Q Rev Biol 49:27–47

Ostrom JH (1986) The cursorial origin of avian flight. In: Padian K (ed) The origin of birds and the evolution of flight. Calif Acad Sci San Francisco 8:73–81

Padian (1983) A functional analysis of flying and walking in pterosaurs. Paleobiology 9:218–239

Paynter RA Jr (1974) Avian energetics. Nuttall Ornithol Club, Cambridge Mass

Pearson OP (1950) The metabolism of hummingbirds. Condor 52:145–152

Pearson OP (1964) Metabolism and heat loss during flight in pigeons. Condor 66:182–185

Pennycuick CJ (1960) Gliding flight in the fulmar petrel. J Exp Biol 37:330–338

Pennycuick CJ (1968a) Power requirements for horizontal flight in the pigeon. J Exp Biol 49:527–555

Pennycuick CJ (1968b) A wind-tunnel study of gliding flight in the pigeon *Columba livia*. J Exp Biol 49:509–526

Pennycuick CJ (1969) The mechanics of bird migration. Ibis 111:525–556

Pennycuick CJ (1971a) Gliding flight of the white-backed vulture *Gyps africanus*. J Exp Biol 55:13–38

Pennycuick CJ (1971b) Control of gliding angle in Rüppell's griffon vulture *Gyps rüppellii*. J Exp Biol 55:39–46

Pennycuick CJ (1971c) Gliding flight of the dog-faced bat *Rousettus aegyptiacus* observed in a wind tunnel. J Exp Biol 55:833–845

Pennycuick CJ (1972a) Animal flight. Arnold, London

Pennycuick CJ (1972b) Soaring behaviour and performance of some East African birds, observed from a motor-glider. Ibis 114:178–218

Pennycuick CJ (1975) Mechanics of flight. In: Farner DS, King JR (eds) Avian biology, vol 5. Academic Press, London New York, pp 1–75

Pennycuick CJ (1982) The flight of petrels and albatrosses (Procellariiformes), observed in south Georgia and its vicinity. Phil Trans R Soc Lond B 300:75–106

Pennycuick CJ (1983) Thermal soaring compared in three dissimilar tropical bird species, *Fregata magnificens, Pelecanus occidentalis* and *Coragyps atratus*. J Exp Biol 102:307–325

Pennycuick CJ (1986) Mechanical constraints on the evolution of flight. In: Padian K (ed) The origin of birds and the evolution of flight. Calif Acad Sci San Francisco 8:83–98

Pennycuick CJ (1987a) Flight of auks (Alcidae) and other northern seabirds compared with southern Procellariiformes: ornithodoloite observations. J Exp Biol 128:335–347

Pennycuick CJ (1987b) Flight of seabirds. In: Croxall JP (ed) Seabirds: feeding ecology and role in marine ecosystems. Univ Press, Cambridge, pp 43–62

Pennycuick CJ (1988) On the reconstruction of pterosaurs and their manner of flight, with notes on vortex wakes. Biol Rev 63:299–331

Pennycuick CJ (1989) Bird flight performance: a practical calculation manual. Univ Press, Oxford

Pennycuick CJ, Bartholomew GA (1973) Energy budget of the lesser flamingo (*Phoeniconaias minor* Geoffroy). E Afr Wildl J 11:199–207

Pennycuick CJ, Lock A (1976) Elastic energy storage in primary feather shafts. J Exp Biol 64:677–689

Pennycuick CJ, Parker GA (1966) Structural limitations on the power output of the pigeon's flight muscles. J Exp Biol 45:489–498

Pennycuick CJ, Rezende MA (1984) The specific power output of aerobic muscle, related to the power density of mitochondria. J Exp Biol 108:377–392

Pennycuick CJ, Alerstam T, Larsson B (1979) Soaring migration of the Common Crane *Grus grus* observed by radar and from aircraft. Ornis Scand 10:241–251

Pennycuick CJ, Obrecht III HH, Fuller MR (1988) Empirical estimates of body drag of large waterfowl and raptors. J Exp Biol 135:253–264

Peters DS (1985) Functional and constructive limitations in the early evolution of birds. In: Hecht MK, Ostrom JH, Viohl G, Wellnhofer P (eds) The beginnings of birds. Proc Int *Archaeopteryx* Conf Eichstätt 1984. Freunde des Jura-Museums Eichstätt, Willbaldsburg, pp 243–249

Peters RH (1983) The ecological implications of body size. Univ Press, Cambridge

Pettigrew JD (1986) Flying primates? Megabats have the advanced pathway from eye to midbrain. Science 231:1304–1306

Pettigrew JD, Jamieson BGM (1987) Are flying-foxes (Chiroptera: Pteropodidae) really primates? Aust Mammal 10:119–124

Phlips PJ, East RA, Pratt NH (1981) An unsteady lifting line theory of flapping wings with application to the forward flight of birds. J Fluid Mech 112:97–125

Prandtl L (1923) Ergebnisse der Aerodynamischen Versuchsanstalt zu Göttigen, II. Lieferung, München

Prandtl, Tietjens (1957a) Fundamentals of hydro- and aeromechanics. Dover, New York

Prandtl, Tietjens (1957b) Applied hydro- and aeromechanics. Dover, New York

Prinzinger R, Hänssler I (1980) Metabolism-weight relationship in some small non-passerine birds. Experimentia 37:1299–1300

Purslow PP, Vincent JFV (1978) Mechanical properties of primary feathers from the pigeon. J Exp Biol 72:251–260

Pyke GH (1981) Why hummingbirds hover and honeyeaters perch. Anim Behav 29:861–867

Pyke GH, Pulliam HR, Charnov EL (1977) Optimal foraging: a selective review of theory and tests. Q Rev Biol 52:137–154

Quine DB, Kreithen ML (1981) Frequency shift discrimination: Can homing pigeons locate infrasounds by Doppler shifts? J Comp Physiol 141:153–155

Racey PA, Speakman JR (1987) The energy costs of pregnancy and lactation in heterothermic bats. Symp Zool Soc Lond 57:107–125

Raikow RJ (1985) Locomotor system. In: King AS, McLelland J (eds) Form and function in birds, vol 3. Academic Press, London New York, pp 57–148

Rankine WJM (1865) Mechanical principles of the action of propellers. Trans Inst Nav Arch 6:13–38

Raveling DG, LeFebvre EA (1967) Energy metabolism and theoretical flight range of birds. Bird-Banding 38:97–113

Rayner JMV (1977) The intermittent flight of birds. In: Pedley TJ (ed) Scale effects in animal locomotion. Academic Press, London New York, pp 437–443

Rayner JMV (1979a) A vortex theory of animal flight. I. The vortex wake of a hovering animal. J Fluid Mech 91:697–730

Rayner JMV (1979b) A vortex theory of animal flight. II. The forward flight of birds. J Fluid Mech 91:731–763

Rayner JMV (1979c) A new approach to animal flight mechanics. J Exp Biol 80:17–54

Rayner JMV (1980) Vorticity and animal flight. In: Elder HY, Trueman ER (eds) Aspects of animal movement. Sem Ser Soc Exp Biol 5. Univ Press, Cambridge, pp 177–199

Rayner JMV (1981) Flight adaptations in vertebrates. In: Day MH (ed) Vertebrate locomotion. Symp Zool Soc Lond, Academic Press, London New York 48:137–172

Rayner JMV (1985a) Vertebrate flight: a bibliography to 1985. Univ Press, Bristol

Rayner JMV (1985b) Bounding and undulating flight in birds. J Theor Biol 117:47–77

Rayner JMV (1985c) Cursorial gliding in protobirds. An expanded version of a discussion contribution. In: Hecht MK, Ostrom JH, Viohl G, Wellnhofer P (eds) The beginning of birds. Proc Int *Archaeopteryx* Conf Eichstätt 1984. Freunde des Jura-Museums Eichstätt, Willibaldsburg, pp 289–292

Rayner JMV (1986) Vertebrate flapping flight mechanics and aerodynamics, and the evolution of flight in bats. In: Nachtigall W (ed) Bat flight — Fledermausflug. Biona report 5. Fischer, Stuttgart, pp 27–75

Rayner JMV (1988) Form and function in avian flight. In: Johnston RF (ed) Current Ornithology, vol 5. Plenum Press, New York, pp 1–66

Rayner JMV, Aldridge HDJN (1985) Three-dimensional reconstruction of animal flight paths and the turning flight of microchiropteran bats. J Exp Biol 118:247–265

Rayner JMV, Jones G, Thomas A (1986) Vortex flow visualizations reveal change in upstroke function with flight speed in bats. Nature (Lond) 321:162–164

Reid EG (1932) Applied wing theory. McGraw-Hill, New York

Richardson WJ (1976) Autumn migration over Puerto Rico and the western Atlantic: a radar study. Ibis 118:309–332

Richmond CR, Langham WH, Trujillo TT (1962) Comparative metabolism of trititated water by mammals. J Cell Comp Physiol 59:45–53

Rietschel S (1976) *Archaeopteryx*-Tod und Einbettung. Natur Museum Frankfurt a M 99:1–8

Rietschel S (1985) Feathers and wings of *Archaeopteryx*, and the question of her flight ability. In: Hecht MK, Ostrom JH, Viohl G, Wellnhofer P (eds) The beginning of birds. Proc Int *Archaeopteryx* Conf Eichstätt 1984. Freunde des Jura-Museums Eichstätt, Willibaldsburg, pp 251–260

Rosser BWC, George JC (1984) Some histochemical properties of the fiber types in the pectoralis muscle of an emu (*Dromaius novaehollandiae*). Anat Rec 209:301–305

Rosser BWC, George JC (1986a) The avian pectoralis: histochemical characterization and distribution of muscle fiber types. Can J Zool 64:1174–1185

Rosser BWC, George JC (1986b) Slow muscle fibers in the pectoralis of the turkey vulture (*Cathartes aura*): an adaptation for soaring flight. Zool Anz 217:252–258

Rothe HJ, Nachtigall W (1980) Physiological and energetic adaptations of flying birds, measured by the wind tunnel technique. A survey. Proc 17th Int Ornithol Congr 1980, 1:400–405

Rothe HJ, Biesel W, Nachtigall W (1987) Pigeon flight in a wind tunnel. II. Gas exchange and power requirements. J Comp Physiol B 157:99–109

Russell AP (1979) The origin of parachuting locomotion in gekkonid lizards (Reptilia: Gekkonidae). Zool J Linn Soc 65:233–249

Salt WR (1963) The composition of the pectoralis muscles of some passerine birds. Can J Zool 41:1185–1190

Schaefer GW (1975) Dimensional analysis of avian forward flight. Symp on biodynamics of animal locomotion, Cambridge UK Sept 1975 (unpublished)

Scheid P (1979) Mechanisms of gas exchange in bird lungs. Rev Physiol Biochem Pharmacol 86:137–184

Schmeidler W (1934) Mathematische Theorie des Schwingenfluges. Z Angew Math Mech 14:163–172

Schmidt-Nielsen K (1972) Locomotion: energy cost of swimming, flying, and running. Science 177:222–228

Schmidt-Nielsen K (1984) Scaling: why is animal size so important? Univ Press, Cambridge

Schmitz FW (1960) Aerodynamik des Flugmodells. Tragflügelmessungen I und II bei kleinen Geschwindigkeiten. Lange, Duisberg

Scholey KD (1983) Developments in vertebrate flight: climbing and gliding of mammals and reptiles, and the flapping flight of birds. Ph D Thesis, University of Bristol

Scholey KD (1986) An energetic explanation for the evolution of flight in bats. In: Nachtigall W (ed) Bat flight – Fledermausflug. Biona report 5. Fischer, Stuttgart, pp 1–12

Scott NJ, Starrett A (1974) An unusual breeding aggregation of frogs, with notes on the ecology of *Agalychnis spurrelli* (Anura: Hylidae) Bull S Cal Acad Sci 73:86–94

Segrem NP, Hart JS (1967) Oxygen supply and performance in *Peromyscus*. Metabolic and circulatory responses to exercise. Can J Physiol Pharmacol 45:531–541

Shapiro J (1955) Principles of helicopter engineering. McGraw-Hill, New York

Sharov AG (1971) Novyiye Lyetayushchiye reptili iz myezozoya Kazakhstana i Kirgizii. (New flying reptiles of the Mesozoic from Kasachstan and Kirgisia). Akad Nauk SSSR Trudy Paläont Inst 130:104–113, Moscow

Short GH (1914) Wing adjustments of pterodactyls. Aeronaut J 18:336–343

Shyestakova GS (1971) Stroyeniye kryil'yev i myekhanika polyeta ptits. (The structure of wings and the mechanics of bird flight.) Nauka, Moscow

Simmons JA, Stein RA (1980) Acoustic imaging in bat sonar: echolocation signals and the evolution of echolocation. J Comp Physiol A 135:61–84

Simmons JA, Fenton MB, O'Farrell MJ (1979) Echolocation and the pursuit of prey by bats. Science 203:16–21

Smith JD (1977) Comments on flight and the evolution of bats. In: Hecht MK, Goody PC, Hecht BM (eds) Major patterns in vertebrate evolution. Plenum Press, New York, pp 427–437

Smith NG, Goldstein DL, Bartholomew GA (1986) Is long-distance migration possible for soaring hawks using only stored fat? Auk 103:607–611

Sokal RR, Rohlf FJ (1981) Biometry. 2nd edn. Freeman, New York

Speakman JR, Racey PA (1988) The doubly-labelled water technique for measurement of energy expenditure in free-living animals. Sci Prog Oxford 72:227–237

Spedding GR (1982) The vortex wake of birds: an experimental investigation. Ph D Thesis, University of Bristol

Spedding GR (1986) The wake of a jackdaw (*Corvus monedula*) in slow flight. J Exp Biol 125:287–307

Spedding GR (1987a) The wake of a kestrel (*Falco tinnunculus*) in gliding flight. J Exp Biol 127:45–57

Spedding GR (1987b) The wake of a kestrel (*Falco tinnunculus*) in flapping flight. J Exp Biol 127:59–78

Spedding GR, Rayner JMV, Pennycuick CJ (1984) Momentum and energy in the wake of a pigeon (*Columba livia*). J Exp Biol 111:81–102

Stahl WR (1962) Similarity and dimensional methods in biology. Science 137:205–212

Stahl WR (1967) Scaling of respiratory variables in mammals. J Appl Physiol 22:453–460

Steadman D (1982) The origin of Darwin's Finches (Fringillidae, Passeriformes). Trans San Diego Soc Nat Hist 19:279–296

Stegmann B von (1964) Die funktionelle Bedeutung des Schlüsselbeines bei den Vögeln. J Ornithol 105:450–463

Strickler TL (1978) Functional osteology and myology of the shoulder in Chiroptera. Contrib Vert Evol 4. Karger, Basel

Studier EH, Howell DJ (1969) Heart rate of female big brown bats in flight. J Mammal 50:842–845

Suthers RA, Thomas SP, Suthers BJ (1972) Respiration, wing-beat and ultrasonic pulse emission in an echo-locating bat. J Exp Biol 56:37–48

Suzuki A (1978) Histochemistry of the chicken skeletal muscles. II. distribution and diameter of three fibre types. Tohoku J Agric Res 29:38–43

Swan LW (1961) Ecology of high Himalayas. Sci Am 205:68

Swinebroad J (1954) A comparative study of the wing myology of certain passerines. Am Midl Nat 51:488–514

Sy M-H (1936) Funktionell-anatomische Untersuchungen am Vogelflügel. J Ornithol Lpz 84:199–296

Talesara CL, Goldspink G (1978) A combined histochemical and biochemical study of myofibrillar ATPase in pectoral, leg and cardiac muscle of several species of bird. Histochem J 10:695–710

Tatner P, Bryant DM (1986) Flight cost of a small passerine measured using doubly labeled water: implications for energetics studies. Auk 103:169–180

Taylor CR, Weibel E, Bolis L (1985) Design and performance of muscular systems. Comp Biol Limited, Cambridge

Teal JM (1969) Direct measurement of CO_2 production during flight in small birds. Zoologica 54:17–23

Thomas SP (1975) Metabolism during flight in two species of bats, *Phyllostomus hastatus* and *Pteropus gouldii*. J Exp Biol 63:273–293

Thomas SP (1981) Ventilation and oxygen extraction in the bat *Pteropus gouldii* during rest and steady flight. J Exp Biol 94:231–250

Thomas SP, Suthers RA (1970) Oxygen consumption and physiological responses during flight in an echolocating bat. Fedn Proc 29:265

Thomas SP, Suthers RA (1972) The physiology and energetics of bat flight. J Exp Biol 57:317–335

Thompson D, Fenton MB (1982) Echolocation and feeding behaviour of *Myotis adversus* (Chiroptera: Vespertilionidae). Aust J Zool 30:543–546

Thorpe WH, Griffin DR (1962) The lack of ultrasonic components in the flight noise of owls compared with other birds. Ibis 104:256–257

Torre-Bueno JR (1976) Temperature regulation and heat dissipation during flight in birds. J Exp Biol 65:471–482

Torre-Bueno JR (1978) Evaporative cooling and water balance during flight in birds. J Exp Biol 75:231–236

Torre-Bueno JR, Larochelle J (1978) The metabolic cost of flight in unrestrained birds. J Exp Biol 75:223–229

Tucker VA (1966) Oxygen consumption of a flying bird. Science 154:150–151

Tucker VA (1968a) Respiratory physiology of house sparrows in relation to high-altitude flight. J Exp Biol 48:55–66

Tucker VA (1968b) Respiratory exchange and evaporative water loss in the flying budgerigar. J Exp Biol 48:67–87

Tucker VA (1969) The energetics of bird flight. Sci Am 220:70–78

Tucker VA (1970) Energetic cost of locomotion in animals. Comp Biochem Physiol 34:841–846

Tucker VA (1972) Metabolism during flight in the laughing gull, *Larus atricilla*. Am J Physiol 222:237–245

Tucker VA (1973) Bird metabolism during flight: evaluation of a theory. J Exp Biol 58:689–709

Tucker VA (1974) Energetics of natural avian flight. In: Paynter RA (ed) Avian energetics. Nuttall Ornithol Club, Cambridge Mass, pp 298–334

Tucker VA (1975) Flight energetics. Symp Zool Soc Lond 35:49–63

Tucker VA (1976) Special review of 'The flight of birds' by CH Greenewalt. Auk 93:848–854

Tucker VA (1987) Gliding birds: The effect of variable wing span. J Exp Biol 133:33–58

Tucker VA (1988) Gliding birds: descending flight of the white-backed vulture, *Gyps africanus*. J Exp Biol 140:325–344

Tucker VA, Parrott GC (1970) Aerodynamics of gliding flight in a falcon and other birds. J Exp Biol 52:345–367

Utter JM (1971) Daily energy expenditures of free-living Purple Martins (*Progne subis*) and Mockingbirds (*Mimus polyglottos*), with a comparison of two northern populations of Mockingbirds. Dissertation. Rutgers University, New Brunswick, New Jersey

Utter JM, LeFebvre EA (1970) Energy expenditure for free flight by the Purple Martin (*Progne subis*). Comp Biochem Physiol 35:713–719

Utter JM, LeFebvre EA (1973) Daily energy expenditure of Purple Martins (*Progne subis*) during breeding season — estimates using D_2O^{18} and time budget methods. Ecology 54:597–604

Valdivieso D, Conde E, Tamsitt JR (1968) Lactate dehydrogenase studies in Puerto Rican bats. Comp Biochem Physiol 27:133–138

Vaughan TA (1959) Functional morphology of three bats: *Eumops, Myotis, Macrotus*. Pubis Mus Nat Hist Univ Kans 12:1–153

Vaughan TA (1966) Morphology and flight characteristics of molossid bats. J Mammal 47:249–260

Vaughan TA (1970a) The skeletal system. In: Wimsatt WA (ed) Biology of bats, vol 1. Academic Press, London New York, pp 97–138

Vaughan TA (1970b) The muscular system. In: Wimsatt WA (ed) Biology of bats, vol 1. Academic Press, London New York, pp 139–194

Vaughan TA (1970c) Flight patterns and aerodynamics. In: Wimsatt WA (ed) Biology of bats, vol 1. Academic Press, London New York, pp 195–216

Vaughan TA, Bateman MM (1980) The molossid wing: some adaptations for rapid flight. Proc 5th Int Bat Res Conf, Texas Tech Press, Lubboch, pp 69–78

Videler JJ, Weihs D, Daan S (1983) Intermittent gliding in the hunting flight of the kestrel, *Falco tinnunculus* L. J Exp Biol 102:1–12

Viohl G (1985) Geology of the Solnhofen lithographic limestones and the habitat of *Archaeopteryx*. In: Hecht MK, Ostrom JH, Viohl G, Wellnhofer P (eds) The beginning of birds. Proc Int *Archaeopteryx* Conf Eichstätt 1984, pp 31–44

Voitkevich AA (1966) The feathers and plumage of birds. October House, New York

Wagner H (1925) Entstehung des dynamisches Auftriebes von Tragflügeln. Z Angew Math Mech 5, 19 pp

Walcott C (1978) Anomalies in the Earth's magnetic field increase the scatter of pigeons' vanishing bearings. In: Schmidt-Koenig K, Keeton WT (eds) Animal migration, navigation, and homing. Springer, Berlin Heidelberg New York, pp 143–151

Walker GT (1925) The flapping flight of birds. J R Aeronaut Soc 29:590–594

Ward-Smith AJ (1984) Aerodynamic and energetic consideration relating to undulating and bounding flight in birds. J Theor Biol 111:407–417

Watson DR (1973) Aerodynamic experiments with wing leading-edge barbs. Ms thesis, School of Aerospace, Mechanical and Nuclear Engineering. Univ of Oklahoma, Norman, Oklahoma

Weis-Fogh T (1960) A rubber-like protein in insect cuticle. J Exp Biol 37:889–907

Weis-Fogh T (1965) Elasticity and wing movements in insects. Proc 12th Int Congr Ent, pp 186–188

Weis-Fogh T (1972) Energetics of hovering flight in hummingbirds and in *Drosophila*. J Exp Biol 56:79–104

Weis-Fogh T (1973) Quick estimates of flight fitness in hovering animals, including novel mechanisms for lift production. J Exp Biol 59:169–230

Weis-Fogh T (1977) Dimensional analysis of hovering flight. In: Pedley TJ (ed) Scale effects in animal locomotion. Academic Press, London New York, pp 405–420

Weis-Fogh T, Alexander R McN (1977) The sustained power output from striated muscle. In: Pedley TJ (ed) Scale effects in animal locomotion. Academic Press, London New York, pp 511–525

Welch A, Welch L, Irwing FG (1968) New soaring pilot. Murray, London

Wellnhofer P (1975a) Die Rhamphorhynchoidea (Pterosauria) der Oberjura-Plattenkalke Süddeutschlands. II. Systematische Beschreibung. Palaeontographica 148:132–186

Wellnhofer P (1975b) Die Rhamphorhynchoidea (Pterosauria) der Oberjura-Plattenkalke Süddeutschlands. III. Palökologie und Stammesgeschichte. Palaeontographica 149:1–30

Wellnhofer P (1978) Pterosauria. In: Kuhn O, Wellnhofer P (eds) Handbuch der Paläoherpetologie, Part 19:1–82. Fischer, Stuttgart

Wellnhofer P (1983) Solnhofer Plattenkalk: Urvögel und Flugsaurier. Solnhofer Aktien-Verein, Maxberg

Wellnhofer P (1987) Die Flughaut von *Pterodactylus* (Reptilia, Pterosauria) am Beispiel des Wiener Exemplares von *Pterodactylus kochi* (Wagner). Ann Naturhist Mus Wien 88:149–162

White DCS (1977) Muscle mechanics. In: Alexander R McN, Goldspink G (eds) Mechanics and energetics of animal locomotion. Chapman and Hall, London, pp 1–22

Wiltschko W (1983) Compasses used by birds. Comp Biochem Physiol A 76:709–717

Wiltschko W, Wiltschko R (1976) Die Bedeutung des Magnetkompasses für die Orientierung der Vögel. J Ornithol 117:362–387

Wiltschko W, Wiltschko R, Keeton WT, Madden R (1983) Growing up in an altered magnetic field affects the initial orientation of young homing pigeons. Behav Ecol Sociobiol 12:135–142

Wiltschko W, Wiltschko R, Walcott C (1987) Pigeon homing: Different effects of olfactory deprivation in different countries. Behav Ecol Sociobiol 21:333–342

Withers PC, Timko PL (1977) The significance of ground effect to the aerodynamic cost of flight and energetics of the black skimmer (*Rhyncops nigra*). J Exp Biol 70:13–26

Wolf LL, Hainsworth FR (1983) Economics of foraging strategies in sunbirds and hummingbirds. In: Aspey WP, Lustick SI (eds) Behavioral energetics: The cost of survival in vertebrates. Ohio State Univ Press, Columbus, pp 223–264

Wolf LL, Stiles FG, Hainsworth FR (1976) Ecological organization of a tropical, highland hummingbird community. J Anim Ecol 45:349–379

Yalden DW (1971) The flying ability of *Archaeopteryx*. Ibis 113:349–356

Yalden DW (1985) Forelimb function in *Archaeopteryx*. In: Hecht MK, Ostrom JH, Viohl G, Wellnhofer P (eds) The beginning of birds. Proc Int *Archaeopteryx* Conf Eichstätt. Freunde des Jura-Museums Eichstätt, Willibaldsburg, pp 91–97

Zimmer K (1943) Der Flug des Nektarvogels (*Cinnyris*). J Ornithol 91:371–387

Subject Index

Numbers in italics refer to figures on the given pages